A Collection of Papers on Chaos Theory and Its Applications

Edited by Paul Bracken and Dimo I. Uzunov

Published in London, United Kingdom

IntechOpen

Supporting open minds since 2005

A Collection of Papers on Chaos Theory and Its Applications
http://dx.doi.org/10.5772/intechopen.91599
Edited by Paul Bracken and Dimo I. Uzunov

Contributors
Vassilis Gaganis, Gbeminiyi M. Sobamowo, Getachew K. Befekadu, Jugal Mohapatra, Vladimir P. Pimenovich Dzyuba, Lal Mohan Saha, Paul Bracken, Jizhao Liu, Yide Ma, Jing Lian, Xinguo Zhang, Sumiyana Sumiyana, Sriwidharmanely Sriwidharmanely, Andrzej Gecow, Nahid Fatima, Neelam Gupta, Neel Kanth, Roman Romashko

First published in London, United Kingdom, 2021 by IntechOpen
IntechOpen is the global imprint of INTECHOPEN LIMITED, registered in England and Wales, registration number: 11086078, 5 Princes Gate Court, London, SW7 2QJ, United Kingdom
Printed in Croatia

British Library Cataloguing-in-Publication Data
A catalogue record for this book is available from the British Library

Additional hard and PDF copies can be obtained from orders@intechopen.com

A Collection of Papers on Chaos Theory and Its Applications
Edited by Paul Bracken and Dimo I. Uzunov
p. cm.
Print ISBN 978-1-83962-858-0
Online ISBN 978-1-83962-859-7
eBook (PDF) ISBN 978-1-83962-875-7

For EU product safety concerns:
IN TECH d.o.o., Prolaz Marije Krucifikse Kozulić 3, 51000 Rijeka, Croatia, info@intechopen.com or visit our website at intechopen.com.

Meet the editors

Professor Paul Bracken is currently a Professor in the Department of Mathematics, at the University of Texas RGV in Edinburg, TX. He obtained his BSc degree from the University of Toronto and holds a Ph.D. from the University of Waterloo in Canada. His research interests include mathematical problems from the area of quantum mechanics and quantum field theory, differential geometry, a study of partial differential equations as well as their overlap with other problems in physics. He has published more than 160 papers in journals and books and has given many talks at different levels over the years. This is the seventh volume he has worked on with IntechOpen publishers.

Dimo I. Uzunov was born in 1950, in Zlatograd, Bulgaria. He graduated from the Physics Faculty of St. Kliment Ohridski University of Sofia in 1974. He received his Ph.D. degree in Physics in 1981 and the Doctor of Physical Sciences degree in 1988. For many years he worked as a Research Fellow and Professor at the Bulgarian Academy of Sciences (Sofia). His research is in the broad area of condensed matter theory, statistical physics, superconductivity, magnetism, phase transitions, and critical phenomena. Professor Uzunov was a teacher and research adviser of a number of under- and post-graduate students, and young researchers. He has carried out a great number of remarkable international collaborations.

Contents

Preface

The subject of chaos theory has passed from the domain of scientific research to the general domain of interest among non-scientists. It is not unusual to hear of the subject of chaos discussed by professional scientists and laymen alike. It is well known that the subject touches on diverse areas such as population dynamics, the study of climates and weather, chaos in atoms and chemical reaction physics, and the subject of quantum chaos.

The current volume presents 12 very interesting contributions to this active area of research. One of the main contributions of the book is to illustrate the wide diversity of subjects that this area of research impacts. The papers themselves range over a wide collection of topics such as chaos and complex dynamics, several papers on nonlinear dynamics, another on quantum integrability, a paper that pertains to the Covid epidemic, chaotic behavior of ICT users, and as well as 3 papers related to perturbation theory and chaos.

The book has been assembled out of the hard work of an international group of invited authors. It is a pleasure to thank them for their efforts and scientific contributions. The editors are also grateful to acknowledge with much thanks to the continuous support and assistance of Mr. Josip Knapić, Author Services Manager, as well as the IntechOpen publishing group for the opportunity to participate in the assembly of this collection of papers.

Paul Bracken
Professor,
Department of Mathematics,
University of Texas RGV,
Edinburg, TX USA

Dimo I. Uzunov
Professor,
Bulgarian Academy of Sciences,
Bulgaria

Chapter 1

Classical and Quantum Integrability: A Formulation That Admits Quantum Chaos

Paul Bracken

Abstract

The concept of integrability of a quantum system is developed and studied. By formulating the concepts of quantum degree of freedom and quantum phase space, a realization of the dynamics is achieved. For a quantum system with a dynamical group G in one of its unitary irreducible representative carrier spaces, the quantum phase space is a finite topological space. It is isomorphic to a coset space G/R by means of the unitary exponential mapping, where R is the maximal stability subgroup of a fixed state in the carrier space. This approach has the distinct advantage of exhibiting consistency between classical and quantum integrability. The formalism will be illustrated by studying several quantum systems in detail after this development.

Keywords: classical, quantum, chaos, integrability, conservation law, algebra

1. Introduction

In classical mechanics, a Hamiltonian system with N degrees of freedom is defined to be integrable if a set of N constants of the motion U_i which are in involution exist, so their Poisson bracket satisfies $\{U_i, U_j\} = 0, i,j = 1, \dots, N$. For an integrable system, the motion is confined to an invariant two-dimensional torus in $2N$-dimensional phase space. If the system is perturbed by a small nonintegrable term, the KAM theorem states that its motion may still be confined to the N-torus but deformed in some way [1–3]. The first computer simulation of nonequilibrium dynamics for a finite classical system was carried out by Fermi and his group. They considered a one-dimensional classical chain of anharmonic oscillators and found it did not equilibrate.

Classically, chaotic motion is longtime local exponential divergence with global confinement, a form of instability. Confinement with any kind of divergence is produced by repeatedly folding, a type of mixing that can only be analyzed by using probability theory. The motion of a Hamiltonian system is usually neither completely regular nor properly described by statistical mechanics, but shows both regular and chaotic motion for different sets of initial conditions. There exists generally a transition between the two types of motion as initial conditions are changed which may exhibit complicated behavior. As entropy or the phase space area quantifies the amount of decoherence, the rate of change of the phase space area quantifies the decoherence rate. In other words, the decoherence rate is the rate at which the phase space area changes.

It is important to extend the study of chaos into the quantum domain to better understand concepts such as equilibration and decoherence. Both integrable as well as nonintegrable finite quantum systems can equilibrate [4, 5]. Integrability does not seem to play a crucial role in the structure of the quasi-stationary state. This is in spite of the fact that integrable and nonintegrable quantum systems display different level-spacing statistics and react differently to external perturbation. Although integrable systems can equilibrate, the main difference from nonintegrable systems may be longer equilibration times. This kind of behavior is contrary to integrable classical finite systems that do not equilibrate at all. Nonintegrable classical systems can equilibrate provided they are chaotic.

The properties of a quantum system are governed by its Hamiltonian spectrum. Its form should be important for equilibration of a quantum system. The equilibration of a classical system depends on whether the system is integrable or not. Integrable classical systems do not tend to equilibrate, they have to be nonintegrable. Quantum integrability in n dimensions may be defined in an analogous way requiring the existence of n mutually commuting operators, but there is no corresponding theorem like the Liouville theorem. An integrable system in quantum mechanics is one in which the spectral problem can be solved exactly, and such systems are few in number [6, 7].

In closed classical systems, equilibration is usually accompanied by the appearance of chaos. Defining quantum chaos is somewhat of an active area of study now. The correspondence principle might suggest we conjecture quantum chaos exists provided the corresponding classical system is chaotic and the latter requires the system to be nonintegrable. Classical chaos does not necessarily imply quantum chaos, which seems to be more related to the properties of the energy spectrum.

It was proposed that the spectrum of integrable and nonintegrable quantum systems ought to be qualitatively different. This would be seen in the qualitative difference of the density of states. At a deeper level, one may suspect that changes in the energy spectrum as a whole may be connected to the breaking of some symmetry or dynamical symmetry. This is the direction taken here [8–10].

It is the objective to see how algebraic and geometric approaches to quantization can be used to give a precise definition of quantum degrees of freedom and quantum phase space. Thus a criterion can be formulated that permits the integrability of a given system to be defined in a mathematical way. It will appear that if the quantum system possesses dynamical symmetry, it is integrable. This suggests that dynamical symmetry breaking should be linked to nonintegrability and chaotic dynamics at the quantum level [11–13].

Algebraic methods first appeared in the context of the new matrix mechanics in 1925. The importance of the concept of angular momentum in quantum mechanics was soon appreciated and worked out by Wigner, Weyl and Racah [14–16]. The close relationship of the angular momentum and the $SO(3)$ algebra goes back to the prequantum era. The realization that $SO(4)$ is the symmetry group of the Kepler problem was first demonstrated by Fock. A summary of the investigation is as follows. To familiarize those who are not familiar with algebraic methods in solving quantum problems, an introduction to the algebraic solution of the hydrogen atom is presented as opposed to the Schrödinger picture. This approach provides a platform for which a definition of quantum integrability of quantum systems can be established. Thus, at least one approach is possible in which a definition of concepts such as quantum phase space, degrees of freedom as well as how an idea of quantum integrability and so forth can be formulated [17–21]. After these issues are addressed, a number of quantum models will be discussed in detail to show how the formalism is to be used [22–24].

1.1 The hydrogen atom

The hydrogen atom is a unique system. In this system, almost every quantity of physical interest can be computed analytically as it is a completely degenerate system. The classical trajectories are closed and the quantum energy levels only depend on the principle quantum number. This is a direct consequence of the symmetry properties of the Coulomb interaction. Moreover, the properties of the hydrogen atom in an external field can be understood using these symmetry properties. They allow a parallel treatment in the classical and quantum formalisms.

The Hamiltonian of the hydrogen atom in atomic units is

$$H_0 = \frac{\mathbf{p}^2}{2} - \frac{1}{r}. \tag{1}$$

The corresponding quantum operator is found by replacement of $\mathbf{p}$ by $-i\nabla$. Due to the spherical symmetry of the system, the angular momentum components are constants of the motion,

$$\mathbf{L} = \mathbf{r} \times \mathbf{p}, \qquad [H_0, \mathbf{L}] = 0. \tag{2}$$

So $\{H_0, \mathbf{L}^2, L_z\}$ is a complete set of commuting operators classically, so three quantities in mutual involution, which implies integrability of the system.

The Coulomb interaction has another constant of the motion associated with the Runge-Lenz vector $\mathbf{R}$. This has the symmetrized quantum definition

$$\mathbf{R} = \frac{1}{2}(\mathbf{p} \times \mathbf{L} - \mathbf{L} \times \mathbf{p}) - \frac{\mathbf{r}}{r}, \qquad [H_0, \mathbf{R}] = 0. \tag{3}$$

If the $\mathbf{R}$ direction is chosen as the reference axis of a polar coordinate system in the plane perpendicular to $\mathbf{L}$, one deduces the equation of the trajectory as

$$r = \frac{\mathbf{L}^2}{1 + \|\mathbf{R}\| \cos\vartheta}. \tag{4}$$

The modulus determines whether the trajectory is an ellipse, a parabola or a hyperbola.

There are then 7 constants of the motion $(\mathbf{L}, \mathbf{R}, H_0)$ are not independent and satisfy

$$\mathbf{R} \cdot \mathbf{L} = 0, \qquad \mathbf{L}^2 - \frac{\mathbf{R}^2}{2H_0} = -\frac{1}{2H_0} - 1. \tag{5}$$

The minus one on the right in (5) is not present in classical mechanics. The mutual commutation relations are given in terms of ε_{ijk}, the fully antisymmetric tensor as follows,

$$[L_i, L_j] = i\varepsilon_{ijk} L_k, \qquad [L_i, R_j] = i\varepsilon_{ijk} R_k, \qquad [R_i, R_j] = i\varepsilon_{ijk}(-2H_0) L_k, \tag{6}$$

Let us look at the symmetry group of the hydrogen atom. The symmetry group is the set of phase space transformations which preserve the Hamiltonian and the equations of motion. It can be identified from the commutation relations between constants of motion. For hydrogen, for negative energies, the group of rotations in 4-dimensional space is called $SO(4)$.

The generators of the rotation group in an n-dimensional space are the $(n-1)n/2$ components of the n-dimensional angular momentum

$$\mathcal{L}_{ij} = x_i p_j - x_j p_i, \quad 1 \le i,j \le n. \tag{7}$$

In (7), $\mathcal{L}_{ij}$ is the generator of the rotations in the (i,j)-plane and has the following commutation relations

$$\left[\mathcal{L}_{ij}, \mathcal{L}_{kl}\right] = 0, \qquad \left[\mathcal{L}_{ij}, \mathcal{L}_{ik}\right] = i\mathcal{L}_{jk}. \tag{8}$$

The first bracket in (8) holds if all four indices are different.
Define the reduced Runge-Lenz vector to be

$$\mathbf{R}' = \frac{\mathbf{R}}{\sqrt{-2H_0}}. \tag{9}$$

The commutation relations (6) are those of a four-dimensional angular momentum with the identification

$$\begin{aligned} &\mathcal{L}_{12} = \mathcal{L}_z \quad \mathcal{L}_{23} = \mathcal{L}_z \quad \mathcal{L}_{31} = \mathcal{L}_y \\ &\mathcal{L}_{14} = R_{z'} \quad \mathcal{L}_{24} = R_{y'} \quad \mathcal{L}_{34} = R_{z'}, \end{aligned} \tag{10}$$

and Casimir operator

$$\mathcal{L}^2 = \sum_{i<j} \left(\mathcal{L}_{ij}\right)^2 = \mathbf{L}^2 + \mathbf{R}^2 = -\frac{1}{2H_0} - 1. \tag{11}$$

The classical trajectory is thus uniquely defined with the 6 components of $\mathcal{L}$ and $\mathbf{L} \cdot \mathbf{R}' = 0$. Any trajectory can be transformed into any other one having the same energy by a 4-dimensional rotation. An explicit realization of this four-dimensional invariance is to use a stereographic projection from the momentum space onto the 4-dimensional sphere with radius $p_0 = \sqrt{-2H_0}$. On this sphere, the solutions to Schrödinger's equation as well as the classical equations of motion are those of the free motion. Schrödinger's equation on the four-dimensional sphere can be separated into six different types of coordinates each associated with a set of commuting operators.

Spherical coordinates correspond to the most natural set, and choosing the quantization axis in the 4 direction and inside the (1,2,3) subspace, the z-axis or usual 3-axis as reference axis, the three operators can be simultaneously diagonalized,

$$\begin{aligned} \mathcal{L}^2 &= -\frac{1}{2H_0} - 1, \\ \mathbf{L}^2 &= \mathcal{L}_{12}^2 + \mathcal{L}_{31}^2 + \mathcal{L}_{23}^2 = L_x^2 + L_y^2 + L_z^2, \\ L_z &= \mathcal{L}_{12}, \end{aligned} \tag{12}$$

The respective eigenvalues of these operators are $n^2 - 1$, $l(l+1)$ and M such that $|M| \le l \le n-1$, so the total degeneracy is n^2. It corresponds to a particular subgroup chain given by

$$SO(4)_n \supset SO(3)_l \supset SO(2)_M. \tag{13}$$

Other choices are possible, such as other spherical coordinates obtained from the previous by interchanging the role of the 3 and 4 axes. This simultaneously diagonalizes the three operators

$$\begin{aligned} \mathcal{L}^2 &= -\frac{1}{2H_0} - 1, \\ \lambda^2 = \mathcal{L}_{12}^2 + \mathcal{L}_{14}^2 + \mathcal{L}_{24}^2 &= R_x'^2 + R_y'^2 + L_z'^2, \\ L_z &= \mathcal{L}_{12}. \end{aligned} \tag{14}$$

The respective eigenvalues of these operators are $n^2 - 1$, $\lambda(\lambda + 1)$ and M such that $|M| \leq \lambda \leq n - 1$. The subgroup chain for this situation is

$$SO(4)_n \supset SO(3)_\lambda \supset SO(2)_M. \tag{15}$$

Another relevant case is the adoption of cylindrical coordinates on the 4-dimensional sphere associated with the following set of commuting operators

$$\begin{aligned} \mathcal{L}^2 &= -\frac{1}{2H_0} - 1, \\ \mathcal{L}_{12} &= L_z, \\ \mathcal{L}_{34} &= R_z'. \end{aligned} \tag{16}$$

This set has the following associated subgroup chain,

$$SO(4) \supset SO(2) \otimes SO(2). \tag{17}$$

In configuration space, this is associated with separability in parabolic coordinates. This is a specific system but it exhibits many of the mathematical and physical properties that will appear here.

2. Quantum degrees of freedom

The time evolution of a system in classical mechanics in time is usually represented by a trajectory in phase space and the dynamical variables are functions defined on this space. The dimension of phase space is twice the number of degrees of freedom, and a point represents a physical state. The space is even-dimensional and it is endowed with a symplectic Poisson bracket structure. Dynamical properties of the system are described completely by Hamilton's equations within this space.

For a quantum system, on the other hand, the dynamical properties are discussed in the setting of a Hilbert space. Dynamical observables are self-adjoint operators acting on elements of this space. A physical state is represented by a ray of the space, so the Hilbert space plays a role similar to phase space for a classical system. The Hilbert space cannot play the role of a quantum phase space since its dimension does not in general relate directly to degrees of freedom. Nor can it be directly reduced to classical phase space in the classical limit. Let us define first the quantum degrees of freedom as well as giving a suitable meaning to quantum phase space.

Suppose $\mathcal{H}$ is a Hilbert space of a system characterized completely by a complete set of observables denoted $\mathcal{C}$. Set $\mathcal{C}$ is composed of the basic physical observables, such as coordinates, momenta, spin and so forth, but excludes the Hamiltonian. The basis vectors of the space can be completely specified by a set of quantum numbers which are related to the eigenvalues of what are usually referred to as the fully non-degenerate commuting observables $\mathcal{C}'$ of $\mathcal{C}$. A fully degenerate operator or observable $O \in C$ has for some constant λ the action

$$O\,|\psi_i\rangle = \lambda\,|\psi_i\rangle, \qquad |\psi_i\rangle \in \mathcal{H}. \tag{18}$$

Definition 1: (Quantum Dynamical Degrees of Freedom) Let $\mathcal{C}$: $\{O_j | [O_i, O_j] = 0; i, j = 1, \dots, N\}$ be a complete set of commuting observables of a quantum system. A basis set of its Hilbert space $\mathcal{H}$ can be labeled completely by M numbers $\{a_i : i = 1, \dots, M\}$ called quantum numbers which are related to the eigenvalues of the non-fully degenerate observables $\{O_i : i = 1, \dots, M(M \leq N)\}$, a subset of $\mathcal{C}$. Then the number M is defined to be the number of quantum dynamical degrees of freedom. □

Since the members of $\mathcal{C}$ are provided by the system, not including the Hamiltonian, it depends only on the structure of the system's dynamical group **G**. Thus the number of quantum dynamical degrees of freedom based on this definition is unique for a given system with a specific Hilbert space $\mathcal{H}$.

The physical and mathematical considerations for defining the dimension of the nonfully degenerate operator subset $\mathcal{C}'$ of $\mathcal{C}$, not the dimension of $\mathcal{C}$ itself, as the number of quantum dynamical degrees of freedom is as follows. In a given $\mathcal{H}$, all fully degenerate operators in $\mathcal{C}$ are equivalent to a constant multiple of the identity operator guaranteeing the irreducibility of $\mathcal{H}$. The expectation values of any fully degenerate operator is a constant and contains no dynamical information.

A given quantum system generally has associated with it a well-defined dynamical group structure due to the fact that the mathematical image of a quantum system is an operator algebra **g** in a linear Hilbert space. This was seen in the case of hydrogen. It comes about from the mathematical structure of quantum mechanics. The dynamical group G with algebra **g** is generated out of the basic physical variables, with the corresponding algebraic structure defined by the commutation relations.

The Hamiltonian $\mathcal{H}$ and all transition operators $\{O\}$ can be expressed as functions of a closed set of operators

$$H = H(T_i), \qquad O = O(T_i), \qquad [T_i, T_j] = \sum_k C_{ij}^k T_k. \tag{19}$$

The C_{ij}^k in (19) are called the structure constants of algebra **g**. The Hilbert space is decomposed into a direct sum of the carrier spaces of unitary irreducible (irrep) representations of the group. Consequently, the dynamical symmetry properties of the system can be restricted to an irreducible Hilbert space which acts as one of the irrep carrier spaces of G.

From group representation theory, it will be given that a total of σ subgroup chains exist for a given group

$$G^\alpha = \{G^\alpha_{s^\alpha} \supset G^\alpha_{s^\alpha - 1} \supset \cdots \supset G^\alpha_1\}, \qquad \alpha = 1, \dots, \sigma. \tag{20}$$

For each subgroup chain G^α of G, there is a complete set of commuting operators **C** which specifies a basis set of its irreducible basis carrier space $\mathcal{H}$, so the dimension of **C** for all subgroup chains of G is the same. A subgroup chain of dynamical group G serves to determine the M quantum dynamical degrees of freedom for a given quantum system with Hilbert space $\mathcal{H}$ an irrep carrier space of G.

Definition 2: For a quantum system with M independent quantum dynamical degrees of freedom the quantum phase space is defined to be a $2M$-dimensional topological space. The space is isomorphic to the coset space G/R with explicit symplectic structure. Here G is the dynamical group of the system and $R \subset G$ is the maximal stability subgroup of the Hilbert space. □

3. Quantum integrability and dynamical symmetry

Quantum phase space defined here can be compact or noncompact depending on the finite or infinite nature of the Hilbert space. A consequence of this development is that the classical definition of integrability can in general be directly transferred to the quantum case.

Definition 3: (Quantum Integrability) A quantum system with M independent dynamical degrees of freedom, hence a $2M$-dimensional quantum phase space, is integrable if and only if there are M quantum constraints of motion, or good quantum numbers, which are related to the eigenvalues of M non-fully degenerate observables: $O_1, O_2, \dots, O_n$. □

Any set of variables that commute may be put in the form of a complete set of commuting observables **C** by including certain additional observables with it. The definition then says that if the system is integrable, a complete set of commuting variables **C** can be found so that the Hamiltonian is always diagonal in the basis referred to by **C**. In the reverse sense, the definition implies that if the system is integrable, simultaneous accurate measurements of M non-fully degenerate observables in the energy eigenvalues can be carried out.

The link with the dynamical group structure can be developed. This specifies exactly the integrability of a quantum system. To this end the definition of dynamical symmetry is needed.

Definition 4: (Dynamical Symmetry) A quantum system with dynamical group G possesses a dynamical symmetry if and only if the Hamiltonian operator of the system can be written and presented in terms of the Casimir operators of any specific chain with α fixed

$$H = \mathcal{F}\left(C_{kj}^{\alpha}\right) \tag{21}$$

The index of a particular subgroup chain C_{kj}^{α} the i-th Casimir operator of subgroup G_k^{α}, $k = s^{\alpha}, \dots, 1$, $i = 1, \dots, l_k^{\alpha}$ and l_k^{α} denotes that the rank of subgroup G_k^{α} is l. It is now possible to state a theorem which gives a condition for integrability to apply.

Proposition 1: (Quantum Integrability) A quantum system with dynamical group G is said to be integrable if it possesses a dynamical symmetry of G.

To prove this, note that it can be broken down into two cases or subgroup classes for a given dynamical group G and are referred to as canonical and noncanonical.

First consider the case in which G^{α} is a canonical subgroup chain of G. The Casimir operators of G, $\{C_{Gi}\}$ and all Casimir operators $\left\{C_{ki}^{\alpha}\right\}$ corresponding to the subgroups in chain G^{α} form a complete set of commuting operators C^{α} of any carrier irrep space H of G^{α} so for fixed α,

$$C^{\alpha} : \{C_{Gi}\} \cup \left\{C_{ki}^{\alpha}\right\} \equiv \left\{Q_j, j = 1, \cdots, N\right\}. \tag{22}$$

When G^{α} is the dynamical symmetry of the system, all operators in C^{α} are constants of motion

$$\left[H, Q_j\right] = 0. \tag{23}$$

There are always M nonfully dynamical operators in C^α. By the third definition, the system is integrable.

For a non-canonical subgroup chain G^α the number of Casimir operators $\{C_{Gi}\}$ of G and all Casimir operators of $\{C_{ki}^\alpha\}$ of G^α is less than the number of the complete set of commuting operators C of any irrep carrier space of G. By definition, of any complete set of commuting operators, there must exist other commuting operators $\{O_j\}$ that commute with $\{C_{Gi}\}$ and $\{C_{ki}^\alpha\}$. These have to be included in the union as well when putting together C^α

$$C^\alpha : \{C_{Gi}\} \cup \{C_{ki}^\alpha\} \cup \{O_j\} \equiv \left\{Q_j : j = 1, \dots, N\right\}. \tag{24}$$

When the system is characterized by the dynamical symmetry of G^α, the operators in (24) satisfy relation (23) as well. In this case as well, there must exist M non-fully degenerate operators of constants of motion as in the previous case.

Based on this proposition, it can be stated that nonintegrability of a quantum system involves the breaking of the dynamical symmetry of the system. It may be concluded that dynamical symmetry breaking can be said to be a property which characterizes quantum nonintegrability. □

Let us summarize what has been found as to what quantum mechanics tells us. In a given quantum system with dynamical Lie group G which is of rank l and dimension n, the dimension of a complete set of commuting operators C of G with any particular subgroup chain is $d = l + (n - l)/2$ in which the l operators are Casimirs of G and are fully degenerate for any given irrep of G. The number M of the non-fully degenerate operators in C for a given irrep of G cannot exceed $M \leq (n - l)/2$. When dynamical symmetry is broken such that any of the M constants of the motion for the system is destroyed the system becomes nonintegrable.

4. Quantum phase space

It is of interest then to develop a model for phase space for quantum mechanics which may be regarded as an analogue to classical physics. By what has been said so far, the Hilbert space H of the system can be broken up into a direct sum of the unitary irreps carrier spaces of G,

$$H = \sum_\Lambda \oplus Y_\Lambda H_\Lambda. \tag{25}$$

In (25), the subscript Λ labels a particular irrep of Lie group G, Λ is the largest weight of the irrep and Y_Λ the degeneracy of Λ in H with no correlations existing between various H_Λ. The study of the dynamical properties of the system can be located on one particular irreducible subspace H_Λ of H. For a quantum system with M_Λ independent quantum dynamical degrees of freedom, the corresponding quantum phase space should be a $2M_\Lambda$-dimensional, topological phase space without additional constraints.

To construct the quantum phase space from the quantum dynamical degrees of freedom for an arbitrary quantum system, the elementary excitation operators can be obtained from the structures of G and H_Λ. Let $\{a_i^\dagger\}$ be a subset of generators of G such that any states $|\Psi\rangle$ of the system are generated for all $|\Psi\rangle \in H_\Lambda$ by means of

$$|\Psi\rangle = F\left(a_i^\dagger\right)|0\rangle. \tag{26}$$

Moreover $F\left(a_i^\dagger\right)$ is a polynomial in the operators $\left\{a_i^\dagger\right\}$ and $|0\rangle \in H_\Lambda$ is the reference state. The requirement placed on state $|0\rangle$ is that one can use a minimum subset of $\mathbf{g}$ to generate the entire subspace H_Λ from $|0\rangle$. In this event, the collection $\left\{a_i^\dagger\right\}$ is called the set of elementary excitation operators of the quantum dynamical degrees of freedom. If G is compact, $|0\rangle$ is the lowest $|\Lambda, -\Lambda\rangle$ or highest weight $|\Lambda, \Lambda\rangle$ state of H_Λ. If G is noncompact, it is merely the lowest state. The number of $\left\{a_i^\dagger\right\}$ is the same as the number of quantum dynamical degrees of freedom. Physically this has to be the case since that is how the operators are defined. Thus the set $\left\{a_i^\dagger\right\}$ and Hermitian conjugate $\{a_i\}$ in $\mathbf{g}_\Lambda$ form a dynamical variable subspace μ of g so we can write

$$\mu : \left\{a_i^\dagger, a_i; i = 1, \dots, M_\Lambda\right\}. \tag{27}$$

With respect to μ there exists a manifold whose dimension is twice that of the quantum dynamical degrees. It can be realized by means of a unitary exponential mapping of the dynamical variable operator subspace μ

$$\Omega = \exp\left(\sum_{i=1}^{M_\Lambda}\left(\eta_i a_i^\dagger - \eta_i^* a_i\right)\right) \in \coprod. \tag{28}$$

The η_i are complex parameters and $i = 1, \dots, M_\Lambda$. In fact, Ω is a unitary coset representation of G/R, where $R \subset G$ is generated by the subalgebra $\kappa = \mathbf{g} - \mu$. Thus (28) shows that q is isomorphic to the $2M_\Lambda$-dimensional coset space G/R, and will be denoted this way from now on. The discussion will apply just to semi-simple Lie groups whose $\mathbf{g}$ satisfies the usual Cartan decomposition $\mathbf{g} = \kappa + \mu$ and $[\kappa, \kappa] \subset \kappa$, $[\kappa, \mu] \subset \mu$ and $[\mu, \mu] \subset \kappa$. Thus G/R will be a complex homogeneous space with topology and a group transformation acting on G/R is a homomorphic mapping of G/R into itself.

The homogeneous space G/R has a Riemannian structure with metric

$$g_{ij} = \frac{\partial^2 \log \mathcal{K}(z, \bar{z})}{\partial z_i \partial \bar{z}_j} \tag{29}$$

The function $\mathcal{K}(z, \bar{z})$ is called the Bergmann kernel of G/R and can be represented as

$$\mathcal{K}(z, \bar{z}) = \sum_\lambda f_\lambda(z) f_\lambda^*(\bar{z}). \tag{30}$$

The functions $f_\lambda(z)$ in (30) constitute an orthogonal basis for a closed linear subspace $\mathcal{L}^2(G/R)$ of $L^2(G/R)$ such that

$$\int_{G/R} f_\lambda(z) f_{\lambda'}^*(\bar{z}) \mathcal{K}^{-1}(z, \bar{z}) d\nu(z, \bar{z}) = \delta_{\lambda\lambda'}, \tag{31}$$

and $d\nu(z, \bar{z})$ is the group invariant measure on the space G/R. It will be written

$$d\nu(z, \bar{z}) = \zeta\left[\det\left(g_{ij}\right)\right] \prod_{i=1}^{M_\lambda} \frac{dz_i d\bar{z}_i}{\pi}. \tag{32}$$

In (32) ζ is a normalization factor given by the condition that (32) integrated over the space G/R is equal to one. There is also a closed, nondegenerate two-form on G/R which is expressed as,

$$\omega = i\hbar \sum_{i,j} g_{ij} dz_i \wedge d\bar{z}_j. \tag{33}$$

Corresponding to this two form there is a Poisson bracket which is given by

$$\{f, h\} = \frac{1}{i\hbar} \sum_{i,j} g^{ij} \left[\frac{\partial f}{\partial z_i} \frac{\partial h}{\partial \bar{z}_j} - \frac{\partial f}{\partial \bar{z}_j} \frac{\partial h}{\partial z_i} \right]. \tag{34}$$

In (34) f and h are functions defined on G/R. By introducing canonical coordinates $(\mathbf{q}, \mathbf{p})$ these quantities can be rewritten in terms of these coordinates.

4.1 Phase space quantum dynamics

Based on what has been stated about G/R, it would be useful to describe the quantum phase space. This means for a given quantum system a phase space representation must exist. Such a representation can be found if there exists an explicit mapping such that

$$O(T_i) \to U(\mathbf{q}, \mathbf{p}), \quad |\Psi\rangle \to \rho(q + ip). \tag{35}$$

Here O is given by (19), and $\rho(\mathbf{q}, \mathbf{p}) \in L^2$. For a quantum system with a quantum phase space G/R, this mapping can be realized by coherent states. To construct coherent states of G and H_Λ defined on G/R, the fixed state $|0\rangle$ is chosen as the initial state

$$g|0\rangle = \Omega\mathbf{r}|0\rangle = |\Lambda, \Omega\rangle e^{i\varphi(\mathbf{r})}, \quad g \in G, \quad \mathbf{r} \in R, \quad \Omega \in G/R. \tag{36}$$

Then R is the maximal stability subgroup of $|0\rangle$ so any $\mathbf{r} \in R$ acting on $|0\rangle$ will leave $|0\rangle$ invariant up to a phase factor

$$\mathbf{r}|0\rangle = e^{i\varphi(\mathbf{r})}|0\rangle. \tag{37}$$

The $|\Lambda, \Omega\rangle$ are the coherent states which are isomorphic to G/R. Therefore,

$$\begin{aligned} |\Lambda, \Omega\rangle \equiv \Omega|0\rangle &= \exp\left(\sum_{i=1}^{M_\Lambda} (\eta_i a_i^\dagger - \eta_i^* a_i) \right) |0\rangle = \mathcal{K}^{1/2}(z, \bar{z}) \exp\left(\sum_{i=1}^{M_\Lambda} z_i a_i^\dagger \right) |0\rangle \\ &= \mathcal{K}^{-1/2}(z, \bar{z}) ||\Lambda, z\rangle. \end{aligned} \tag{38}$$

$$\begin{aligned} \mathcal{K}(z, \bar{z}) &= \left\langle 0| \exp\left(\sum_{i=1}^{M_\Lambda} \bar{z}_i a_i \right) \exp\left(\sum_{i=1}^{M_\Lambda} z_i a_i^\dagger \right) |0 \right\rangle = \langle \Lambda, z || \Lambda, z \rangle = |\rangle\, 0 | \Lambda, 0\rangle|^2 \\ &= \sum_\lambda f_{\Lambda\lambda}(z) f_{\Lambda\lambda}^*(z). \end{aligned}$$

The Bargmann kernel was introduced in (30), and for a semisimple Lie group, the parameters z_i are given by

$$z = \begin{cases} \eta \dfrac{\tan\left(\eta^\dagger \eta\right)^{1/2}}{\left(\eta^\dagger \eta\right)^{1/2}}, & G\,\text{compact}, \\ \eta \dfrac{\tanh\left(\eta^\dagger \eta\right)^{1/2}}{\left(\eta^\dagger \eta\right)^{1/2}}, & G\,\text{noncompact}. \end{cases} \tag{39}$$

Here η represents the nonzero $k \times p$ block matrix of the operator $\sum_{i=1}^{M_\Lambda} \left(\eta_i a_i - \eta_i^* a_i^\dagger\right)$. The state $\|\Lambda, z\rangle$ in (38) is an unnormalized form of $|\Lambda, \Omega\rangle$ and $f_{\Lambda,\lambda}(z)$ is the orthogonal basis of $\mathcal{L}^2(G/R)$ the function space

$$f_{\Lambda,\lambda}(z) = \langle \Lambda, \lambda \| \Lambda, \lambda \rangle, \tag{40}$$

where $|\Lambda, \lambda\rangle$ is a basis for H_Λ, a particular irreducible subspace of the Hilbert space. The coherent states of (38) are over-complete

$$\int_{G/R} |\Lambda, \Omega\rangle\langle \Lambda, \Omega| d\nu(z) = I. \tag{41}$$

A classical-like framework or analogy has been established in the form of a quantum phase space specified by G and H_Λ. Variables which reside in this classical analogy are denoted thus $\tilde{c}$. The $2M_\Lambda$-dimensional quantum phase space G/R has all the required structures of a classical mechanical system. It is always possible a classical dynamical theory can be established in G/R whose motion is confined to G/R and is determined by the following equations of motion

$$\frac{d\tilde{\mathcal{U}}}{dt} = \{\tilde{\mathcal{U}}(q,p), \tilde{H}(q,p)\}, \qquad q,p \in G/R. \tag{42}$$

This equation can be replaced by Hamilton's equations

$$\frac{dq_i}{dt} = \frac{\partial \tilde{H}(q,p)}{\partial p_i}, \qquad \frac{dp_i}{dt} = -\frac{\partial \tilde{H}(q,p)}{\partial q_i}. \tag{43}$$

In (42) and (43), $\tilde{H}(q,p)$ is the Hamiltonian of the system, and $\tilde{U}(q,p)$ is a physical observable. A correspondence principle is implied here and requires that suitable conditions can be found such that the quantum dynamical Heisenberg equations can be written this way.

Clearly, if suitable conditions hold the phase space representation of the commutator of any two operators is equal to the Poisson bracket of the phase space representation of these two operators so that

$$\frac{1}{i\hbar} \langle \Lambda, \Omega | [A_H, B_H] | \Lambda, \Omega \rangle = \left\{ \tilde{\mathcal{A}}, \tilde{\mathcal{B}} \right\}. \tag{44}$$

Then the phase space representation of the Heisenberg equation

$$\frac{dA_H}{dt} = \frac{1}{i\hbar}[A_H, H_H], \tag{45}$$

given by (42) is therefore equivalent to (43). In (45), A_H is the Heisenberg operator

$$A_H = UAU^{-1}, \qquad U = e^{iHt/\hbar}, \tag{46}$$

and A is time-independent in the Schrödinger picture. The coherent state on the left of (44) is time-independent. Observables on the right side are the expectation values of the Schrödinger operators in the time-dependent coherent state. The quantum phase space maintains many of the quantum properties which are important, such as internal degrees of freedom, the Pauli principle, statistical properties and dynamical symmetry. Formally the equation of motion is classical. The phase space representation is based on the whole quantum structure of the coset space G/R.

Let us discuss integrability and dynamical symmetry. A quantum system with M_Λ independent degrees of freedom is integrable if and only if the M_Λ non-fully degenerate observables can simultaneously be measured in the energy representation. There exist non-fully degenerate observables $\{C_i : i = 1, \dots, M_\Lambda - 1\}$ which commute with each other and H

$$[C_i, C_j] = 0, \qquad [C_i, H] = 0. \tag{47}$$

It follows that in the classical limit which has been formulated,

$$\{\tilde{C}_i, \tilde{C}_j\} = 0, \qquad \{\tilde{C}_i, \tilde{H}\} = 0. \tag{48}$$

Together with the Hamilton equations, (47) also formally defines classical integrability, so quantum integrability is completely consistent with the classical theory. In the classical analogy, the group structure of the system is defined by Poisson brackets. The concept of dynamical symmetry is naturally preserved in the classical analogy, so the theorem on dynamical symmetry and integrability is also meaningful for the classical analogy. If the Hamiltonian has the symmetry S, then its phase space picture representation has the same symmetry. To see this, if

$$SHS^{-1} = H, \tag{49}$$

in the phase space representation, it holds that

$$\langle \Lambda, \Omega | H | \Lambda, \Omega \rangle = \langle \Lambda, \Omega | SHS^{-1} | \Lambda, \Omega \rangle = \langle \Lambda, \Omega' | H | \Lambda, \Omega' \rangle. \tag{50}$$

To put this concisely, we write

$$\tilde{H}(q, p) = \tilde{H}(q', p'), \tag{51}$$

where $S^{-1}|\Lambda, \Omega\rangle = S^{-1}\Omega|0\rangle = |\Lambda, \Omega'\rangle e^{i\varphi(h)}$.

5. Applications to physical systems

5.1 Harmonic oscillator

The harmonic oscillator has dynamical group H_4 and is a single-degree of freedom system [13–15, 23]. To the dynamical group corresponds the algebra h_4 defined by the set $\{a^\dagger, a, a^\dagger a, I\}$ with Hilbert space the Fock space V^F : $\{|n\rangle, n = 1, 2, \dots\}$, so the fixed state is the ground state $|0\rangle$, and elementary excitation operator $a^\dagger$. The quantum phase space is constructed from the unitary exponential mapping of the subspace $\mu : \{a^\dagger, a\}$ of h_4,

$$\Omega(z) = \exp\left(z a^\dagger - \bar{z} a\right) \in H_4/U(1) \otimes U(1). \tag{52}$$

With generators $a^\dagger a$ and I, $U(1)\otimes U(1)$ in (52) is the maximal stability subgroup of $|0\rangle$. As $H_4/U(1)\otimes U(1)$ is isomorphic to the one-dimensional complex plane, the quantum phase space has metric $g_{ij}=\delta_{ij}$ and $d\nu(z)=dzd\overline{z}/\pi$. It is noncompact due to the infiniteness of the Fock space. There is a well-known symplectic structure on the complex plane with Poisson bracket of two functions $\tilde{F}_1,\tilde{F}_2$ defined by

$$\{\tilde{F}_1,\tilde{F}_2\}=\frac{1}{i\hbar}\left(\frac{\partial\tilde{F}_1}{\partial z}\frac{\partial\tilde{F}_2}{\partial\overline{z}}-\frac{\partial\tilde{F}_1}{\partial\overline{z}}\frac{\partial\tilde{F}_2}{\partial z}\right). \tag{53}$$

It is useful to introduce the standard canonical position and momentum coordinates

$$z=\frac{1}{\sqrt{2\hbar}}(q+ip),\qquad \overline{z}=\frac{1}{\sqrt{2\hbar}}(q-ip). \tag{54}$$

The Glauber coherent states can be realized by the states $|z\rangle$ with the set of these states isomorphic to $H_4/U(1)\otimes U(1)$ and given as

$$|z\rangle\equiv\Omega(z)|0\rangle=\exp\left(za^\dagger-\overline{z}a\right)|0\rangle=e^{-|z|^2/2}\exp\left(za^\dagger\right)|0\rangle. \tag{55}$$

The normalization constant in (55) is the Bargmann kernel

$$\mathcal{K}(z,\overline{z})=e^{|z|^2}. \tag{56}$$

The phase space representation of the wavefunction $|\Psi\rangle\in V^F$ is

$$f(z)=\langle\Psi||z\rangle=\sum_{n=0}^{\infty}f_n\frac{z^n}{\sqrt{n!}}. \tag{57}$$

By Wick's Theorem, it is always possible to write an operator A in normal product form

$$A=A\left(a^\dagger,a\right)=\sum_{k,l}A^n_{k,l}\left(a^\dagger\right)^k(a)^l. \tag{58}$$

The phase space representation of A is just

$$\tilde{U}(z,\overline{z})=\langle z|A|z\rangle=\sum_{k,l}A^n_{k,l}\overline{z}^k z^l. \tag{59}$$

In the case A is simply a generator of H_4, we can write (59) as

$$\begin{aligned}\tilde{a}^\dagger&=\langle z|a^\dagger|z\rangle, & \tilde{a}&=\langle z|a|z\rangle,\\ \tilde{a}^\dagger\tilde{a}&=\langle z|a^\dagger a|z\rangle=|z|^2, & \tilde{I}&=\langle z|I|z\rangle=I.\end{aligned} \tag{60}$$

The corresponding algebraic structure of H_4 in the phase-space representation is

$$i\hbar_c\{\tilde{a},\tilde{a}^\dagger\}=\tilde{I},\qquad i\hbar_c\{\tilde{a}^\dagger\tilde{a}\}=-\tilde{a},\qquad i\hbar_c\{\tilde{a}^\dagger\tilde{a},\tilde{a}^\dagger\}=\tilde{a}^\dagger. \tag{61}$$

Here $\hbar_c$ is used in the classical analogy. The algebraic structure of the H_4 generators is preserved when commutators are replaced by Poisson brackets in phase space. Using (54) the Dirac quantization condition and $\tilde{H}$ are given by

$$[q,p]=i\hbar_c\{q,p\}, \qquad \tilde{H}(q,p)=\langle z|H|z\rangle. \tag{62}$$

For the forced harmonic oscillator, the classical analogy of the Hamiltonian is given by

$$\begin{aligned}\tilde{H}(q,p)&=\frac{\omega}{2}(p^2+q^2)+i\sqrt{2}\mathfrak{R}(\lambda(t)q)-\sqrt{2}\mathfrak{I}(\lambda(t)p)\\&=\frac{\omega}{2}(p^2+q^2)+\frac{1}{\sqrt{2}}(\lambda(t)+\bar{\lambda}(t))q)+\frac{1}{\sqrt{2}i}((\lambda(t)-\bar{\lambda}(t))p).\end{aligned} \tag{63}$$

Hamilton's equations in (44) can be used to evaluate the t derivatives of q and p:

$$\frac{dq}{dt}=\omega p+\frac{1}{\sqrt{2}i}(\lambda(t)-\bar{\lambda}(t)), \qquad \frac{dp}{dt}=-\omega q-\frac{1}{\sqrt{2}}(\lambda(t)+\bar{\lambda}(t)). \tag{64}$$

Hence combining these two derivatives, we obtain

$$\frac{d}{dt}(q+ip)=-i\omega(q+ip)-\sqrt{2}i\lambda(t). \tag{65}$$

Multiplying both sides by the integrating factor $e^{i\omega t}$ and then integrating with respect to t, the solution is

$$q(t)+ip(t)=e^{-i\omega t}(q(0)+ip(0))-i\sqrt{2}e^{-i\omega t}\int_0^t\lambda(\tau)e^{i\omega\tau}\,d\tau=z(t)\sqrt{2\hbar_c}. \tag{66}$$

If the initial state is $|0\rangle$ or a coherent state $|z(0)\rangle$, then the exact quantum solution is

$$|\psi(t)\rangle=|z(t)\rangle e^{i\varphi(t)} \tag{67}$$

and $z(t)$ is given by (66). The phase φ is a quantum effect obtained from $z(t)$

$$\varphi(t)=-\frac{1}{2}\omega t-\int_0^t\mathfrak{R}[\lambda(\tau)z(\tau)]\,d\tau. \tag{68}$$

This seems to imply the classical analogy provides an exact quantum solution if the Hamiltonian is a linear function of the generators of G.

5.2 $SU(2)$ spin system

The phase space structure of a spin system will be constructed and as well the phase-space distribution and classical analogy.

Since the dynamical group of the spin system is $SU(2)$ and the Hilbert space is described by the states $V^{2j+1}=\{|j,m\rangle\}$ where $m=-j,\ -j+1,\ \dots,j$ and j is an integer or half-integer, the fixed state is $|j,\ -j\rangle$. This is the lowest weight state of V^{2j+1}. Thus the elementary excitation operator of the spin system is J_+ and the explicit form of $|j,m\rangle$ is

$$|j,m\rangle=\frac{1}{(j+m)!}\binom{2j}{j+m}^{-1/2}(J_+)^{j+m}|j,\ -j\rangle. \tag{69}$$

Any state $|\Psi\rangle = \sum_{m=-j}^{j} f_m |j,m\rangle \in V^{2j+1}$ can be generated by a polynomial of J_+ acting on $|j,-j\rangle$. This means the number of quantum dynamical degrees of freedom is equal to the number of elementary excitation operators. The quantum phase space can be found by mapping $\mu : \{J_+, J_-\}$ to the coset space $SU(2)/U(1)$ by means of $(\eta J_+ - \bar{\eta} J_-) \to \exp(\eta J_+ - \bar{\eta} J_-)$ where $\eta = (\vartheta/2)e^{-i\varphi}$, $0 \le \vartheta \le \pi$, $0 \le \varphi \le 2\pi$. The coset space $SU(2)/U(1)$ is isomorphic to a two-dimensional sphere. The coherent states of $SU(2)/U(1)$ are well known

$$|j\Omega\rangle = \exp(\eta J_+ - \bar{\eta} J_-)|j,-j\rangle = \left(1+|z|^2\right)^{-j} \exp(zJ_+)|j,-j\rangle = \left(1+|z|^2\right)^{-j} \| jz\rangle,$$
$$z = \tan\frac{\vartheta}{2} e^{-i\varphi}. \tag{70}$$

The generalized Bargmann kernel on S^2 is $\mathcal{K}(z,\bar{z}) = \left(1+|z|^2\right)^{2j}$. Then the metric g_{ij} and measure are given by

$$g_{ij} = \delta_{ij} \frac{2j}{\left(1+|z|^2\right)^2}, \qquad dv = \frac{1}{\pi}(2j+1)\frac{dz d\bar{z}}{1+|z|^2)^2}. \tag{71}$$

Given the canonical coordinates

$$\frac{1}{\sqrt{4j\hbar}}(q+ip) = \frac{z}{\sqrt{1+|z|^2}} = \sin\left(\frac{\vartheta}{2}\right)e^{-i\varphi}, \tag{72}$$

there obtains the bracket

$$\{\tilde{F}_1, \tilde{F}_2\} = \frac{\partial \tilde{F}_1}{\partial q}\frac{\partial \tilde{F}_2}{\partial p} - \frac{\partial \tilde{F}_1}{\partial p}\frac{\partial \tilde{F}_2}{\partial q}, \tag{73}$$

where $q^2 + p^2 \le 4j\hbar$, which implies the phase space of a spin system is compact. The phase space representation of the state $|\Psi\rangle \in V^{2j+1}$ is for $f \in L^2(S^2)$,

$$f(z) = \langle \Psi \| j\Omega\rangle = \sum_{n=0}^{\infty} f_n \binom{2j}{j+m}^{1/2} z^{j+m}, \tag{74}$$

The phase space representation of an operator $B = B(J_i)$ is

$$\tilde{B}(z,\bar{z}) = \langle j\Omega|B|j\Omega\rangle. \tag{75}$$

When the operator B in (75) is chosen to be one of the three operators J_+, J_- or J_0, the results are

$$\tilde{J}_+ = \langle j\Omega|J_+|j\Omega\rangle = \frac{2j\bar{z}}{1+|z|^2}, \quad \tilde{J}_- = \langle j\Omega|J_-|j\Omega\rangle = \frac{2jz}{\left(1+|z|^2\right)},$$
$$\tilde{J}_0 = \langle j\Omega|J_0|j\Omega\rangle = j\frac{|z|^2-1}{1+|z|^2}. \tag{76}$$

These can also be given in terms of q, p by using (72). The algebraic structure of $SU(2)$ in the phase space representation is given by the Poisson bracket

$$i\hbar_c\{\tilde{J}_-, \tilde{J}_+\} = -2\tilde{J}_0, \qquad i\hbar_c\{\tilde{J}_0, \tilde{J}_\pm\} = \pm\tilde{J}_\pm. \tag{77}$$

The classical analogy of an observable $B(J_i)$ is given by the following expression

$$\tilde{B}(q,p) = \langle j,\Omega|B(J_i)|j,\Omega\rangle. \tag{78}$$

The classical limit is found by taking $j \to \infty$ and the classical Hamiltonian function is

$$\tilde{H}_C(q,p) = H(\langle j,\Omega|J_i|j,\Omega\rangle) = H(\tilde{J}_+, \tilde{J}_-, \tilde{J}_0). \tag{79}$$

5.3 $SU(1,1)$ quantum systems and a two-level atom

A two-level atom is considered which interacts with two coupled quantum systems that can be represented in terms of a $su(1,1)$ Lie algebra. When for example mixed four-waves are injected into a cavity containing a single two level-atom an interaction occurs between the four waves and the atom that is electromagnetic radiation and matter. The Hamiltonian has the form

$$\frac{1}{\hbar}H = \sum_{i=1}^{2} \omega_i \left(a_i^\dagger a_i + \frac{1}{2} \right) + \frac{1}{2}\omega_0 \sigma_z + \lambda \left(a_1^2 a_2^2 \sigma_+ + a_1^{\dagger 2} a_2^{\dagger 2} \sigma_- \right). \tag{80}$$

It is similar to **5.1**, so we sketch the physical situation. The $\sigma_\pm$, σ_z are raising lowering and inversion operators which satisfy the commutation relations $[\sigma_z, \sigma_\pm] = 2\sigma_\pm$, $[\sigma_+, \sigma_-] = \sigma_z$, whereas the $a_i^\dagger$, a_i are basic creation and annihilation operators with $\left[a_j, a_j^\dagger\right] = \delta_{ij}$. The interaction term in (80) can be thought of as the interaction between two different second harmonic modes. This can be cast in terms of three $su(1,1)$ Lie algebra generators K_+, K_- and K_z which satisfy the commutation relations,

$$[K_z, K_\pm] = \pm K_\pm, \qquad [K_-, K_+] = 2K_z. \tag{81}$$

The corresponding Casimir K which has eigenvalue $k(k-1)$ given by

$$K^2 = K_z^2 - \frac{1}{2}(K_+K_- + K_-K_+). \tag{82}$$

Given that this is the Lie algebra, it can be said that the Fock space is spanned by the set of vectors $V^F : \{|m;k\rangle\}$ and the operators in (81) act on these states as follows,

$$\begin{aligned} K_z|m;k\rangle &= (m+k)|m;k\rangle, & K^2|m;k\rangle &= k(k-1)|m;k\rangle, \\ K_+|m;k\rangle &= \sqrt{(m+1)(m+2k)}|m+1;k\rangle, & K_-|m;k\rangle &= \sqrt{m(m+2k-1)}|m-1;k\rangle. \end{aligned} \tag{83}$$

It is the case that $K_-|0;m\rangle = 0$ so this is the lowest level state. The $su(1,1)$ Lie algebra can be realized in terms of boson annihilation and creation operators and it is isomorphic to the Lie algebra of the non-compact $SU(1,1)$ group. For the Hamiltonian (80) define operators $K_\pm^{(i)}$ and $K_z^{(i)}$ as

$$K_+^{(i)} = \frac{1}{2} a_i^{\dagger 2}, \quad K_-^{(i)} = \frac{1}{2} a_i^2, \quad K_z^{(i)} = \frac{1}{2} \left(a_i^\dagger a_i + \frac{1}{2} \right), \qquad i = 1, 2, \tag{84}$$

where the Bargmann index k is either $1/4$ for the even parity states while $3/4$ applies to the odd-parity states.

Using these operators, (80) is written in terms of $su(2)$ and $su(1,1)$ operators such that it has the form,

$$\frac{1}{\hbar} H = \sum_{i=1}^{2} \eta_i K_z^{(i)} + \frac{\omega}{2} \sigma_z + \lambda \left(K_+^{(1)} K_+^{(2)} \sigma_- + K_-^{(1)} K_-^{(2)} \sigma_+ \right). \tag{85}$$

The Heisenberg equations of motion obtained from (85) gives

$$i \frac{d}{dt} K_z^{(1)} = \lambda \left(K_+^{(1)} K_+^{(2)} \sigma_- - K_-^{(1)} K_-^{(2)} \sigma_+ \right), \quad i \frac{d}{dt} K_z^{(2)} = \lambda \left(K_+^{(1)} K_+^{(1)} \sigma_- - K_-^{(1)} K_-^{(2)} \sigma_+ \right),$$
$$i \frac{d}{dt} \sigma_z = \lambda \left(K_-^{(1)} K_-^{(2)} \sigma_+ - K_+^{(1)} K_+^{(1)} \sigma_- \right). \tag{86}$$

The following two operators N_1 and N_2 are constants of the motion

$$N_1 = K_z^{(1)} + \sigma_z, \qquad N_2 = K_z^{(2)} + \sigma_z. \tag{87}$$

Hamiltonian (80) can now be put in the equivalent form

$$\frac{1}{\hbar} H = N + C + I, \tag{88}$$

where I is the identity operator and N and C are the operators

$$N = \sum_{i=1}^{2} \eta_i N_i, \qquad C = \Delta \sigma_z + \lambda \left(K_+^{(1)} K_+^{(2)} \sigma_- + K_-^{(1)} K_-^{(2)} \sigma_+ \right). \tag{89}$$

The constant Δ is the detuning parameter defined as

$$\Delta = \frac{\omega}{2} - \eta_1 - \eta_2. \tag{90}$$

As N and C commute, each commutes with the Hamiltonian H so N and C are constants of the motion. The time evolution operator $U(t)$ is given by

$$U(t) = \exp\left(-i \frac{H}{\hbar} t \right) \cdot \exp\left(-iNt \right) \cdot \exp\left(iCt \right). \tag{91}$$

In the space of the two-level eigenstates

$$e^{-iNt} = \begin{pmatrix} e^{-iW_1 t} & 0 \\ 0 & e^{-iW_2 t} \end{pmatrix}. \tag{92}$$

The operators W_i, $i = 1, 2$ are defined by $W_1 = \eta_1 K_z^{(1)} + \eta_2 K_z^{(2)} + 1$ and $W_2 = \eta_1 K_z^{(1)} + \eta_2 K_z^{(2)} - 1$. The second exponential on the right of (91) takes the form,

$$\exp(-iCt) = \begin{pmatrix} \cos\tau_i t - \frac{i\Delta}{\tau}\sin\tau_1 t & -i\lambda\frac{\sin\tau_1 t}{\tau_1}K_-^{(1)}K_-^{(2)} \\ -i\lambda K_+^{(1)}K_+^{(2)}\frac{\sin\tau_1 t}{\tau_1} & \cos\tau_2 t - \frac{i\Delta}{\tau_2}\sin\tau_2 t \end{pmatrix} \tag{93}$$

where $\tau_j^2 = \Delta^2 + \nu_j, j = 1, 2$ and

$$\tau_1 = \lambda^2 K_-^{(1)}K_+^{(1)}K_-^{(2)}K_+^{(2)}, \qquad \tau_2 = \lambda^2 K_+^{(1)}K_-^{(1)}K_+^{(2)}K_-^{(2)} \tag{94}$$

The coherent atomic state $|\vartheta, \varphi\rangle$ is considered to be the initial state that contains both excited and ground states and has the structure,

$$|\vartheta, \varphi\rangle = \cos\left(\frac{\vartheta}{2}\right)|e\rangle + \sin\left(\frac{\vartheta}{2}\right)e^{-i\varphi}|g\rangle. \tag{95}$$

where ϑ is the coherence angle, φ the relative phase of the two atomic states. The excited state is attained by taking $\vartheta \to 0$, while the ground state of the atom is derived from the limit $\vartheta \to \pi$. The initial state of the system that describes the two $su(1,1)$ Lie algebras is assumed to be prepared in the pair correlated state $|\xi, q\rangle$ defined by

$$K_-^{(1)}K_-^{(2)}|\xi, q\rangle = \xi|\xi, q\rangle, \qquad \left(K_z^{(1)} - K_z^{(1)}\right)|\xi, q\rangle = q|\xi, q\rangle. \tag{96}$$

Since the operators $K_-^{(1)}K_-^{(2)}$ and $(K_z^{(1)} - K_z^{(2)})$ commute, $|\xi, q\rangle$ can be introduced which is simultaneously an eigenstate of both operators,

$$|\xi, q\rangle = \sum_{n=0}^{\infty} C_n |q + n + k_2 - k_1; k_1; n, k_2\rangle. \tag{97}$$

Then applying $K_-^{(2)}$ and then $K_-^{(1)}$ we obtain,

$$\begin{aligned} &K_-^{(1)}K_-^{(2)}|\xi, q\rangle \\ &= \sum_{n=0}^{\infty} C_n \sqrt{n(n + 2k - 1)(q + n + k_2 - k_1)(q + n + k_2 - k_1 + 2k_1 - 1}|q + n + k_2 - k_1 - 1, k_1; n - 1, k_2\rangle, \\ &= \sum_{n=0}^{\infty} C_{n+1} \sqrt{(n + 1)(n + 2k)(q + n + k_2 - k_1 + 1)((q + n + k_2 - k_1 + 2k_1)}|q + n + k_2 - k_1, k_1; n, k_2\rangle. \end{aligned} \tag{98}$$

This calculation implies that the normalization constant C_n can be obtained by solving

$$\sqrt{(n + 1)(n + 2k)(q + n + k_2 - k_1 + 1)(n + q + k_1 + k_2)C_{n+1}} = \xi C_0. \tag{99}$$

The new state is of the form,

$$|\xi, q\rangle = N_q \sum_{n=0}^{\infty} C_n |q + n + k_2 - k_1, k; n, k_2\rangle, \qquad N_q^{-2} = \sum_{n=0}^{\infty} |C_n|^2. \tag{100}$$

If it is assumed that at $t = 0$ the wave function of the system is $|\psi(0)\rangle = |\vartheta, \varphi\rangle \otimes |\xi, q\rangle$, using (91) on $|\psi(0)\rangle$, the state can be calculated for $t > 0$ can be determined

$$|\psi(t)\rangle = e^{-iW_1 t}\left[\left(\cos\tau_1 t\right) - i\frac{\Delta}{\tau_1}\sin\tau_1 t\right)\cos\left(\frac{\vartheta}{2}\right) - i\frac{\lambda}{\tau_1}\sin\left(\tau_1 t\right)K_-^{(1)}K_-^{(2)}e^{-i\varphi}\sin\frac{\vartheta}{2}\right]|e\rangle\otimes|\xi,q\rangle$$
$$+e^{iW_2 t}\left[\left(\cos\tau_2 t + i\frac{\Delta}{\tau_2}\sin\tau_2 t\right)e^{-i\varphi}\sin\frac{\vartheta}{2} - i\frac{\lambda}{\tau_2}\sin\tau_2 t K_+^{(1)}K_+^{(2)}\cos\frac{\vartheta}{2}\right]|g\rangle\otimes|\xi,q\rangle. \tag{101}$$

The reduced density matrix is constructed from this

$$\rho_f(t) = \mathrm{Tr}_{atom}\,|\psi(t)\rangle\langle\psi(t)|. \tag{102}$$

6. Summary and conclusions

Explicit structures for quantum phase space have been examined. Quantum phase space provides an inherent geometric structure for an arbitrary quantum system. It is naturally endowed with sympletic and quantum structures. The number of quantum dynamical degrees of freedom has a great effect on determining the quantum phase space. Inherent properties of quantum theory, the Pauli principle, quantum internal degrees of freedom and quantum statistical properties are included. A procedure can be stated for constructing this quantum phase space and canonical coordinates should be derivable for all semi-simple dynamical Lie groups with Cartan decomposition. The coset space G/R provides a way to define coherent states which link physical Hilbert space and quantum phase space. This motivates the study of the algebraic structure of the phase space representation of observables. The algebraic structure of operators is preserved in phase space if the operators are those of the dynamical group G. Through this approach, this property results in an explicit realization of the classical limit of quantum systems. A classical analogy was developed and seen in the examples as well for an arbitrary quantum system independently of the existence of the classical counterpart, so the classical limit of the quantum system can be obtained explicitly if it exists. The classical analogy will contain the first-order quantum correlation. A theorem which pertains to the relationship between dynamical symmetry and integrability has been proved, and is also valid in classical mechanics. It is then possible to construct a way to look for the quantum manifestation of chaos. Finally, it is then consistent with Berry's definition, the study of semi-classical but nonclassical, behavior characteristic of systems whose classical motion exhibits chaos.

Author details

Paul Bracken
Department of Mathematics, University of Texas, Edinburg, TX, USA

*Address all correspondence to: paul.bracken@utrgv.edu

References

[1] Baker GL Gollub JP. Chaotic Dynamics. Cambridge: Cambridge University Press 1990.

[2] Ott E. Chaos in Dynamical Systems. Cambridge: Cambridge University Prees 1993.

[3] Eckhart B. Quantum Mechanics of Classically Non-Integrable Systems. Phys Rep 1988; **163**: 205–297.

[4] Arnold VI. Mathematical Methods of Classical Mechanics. Springer, New York. 1978.

[5] Giannoni MJ, Voros A, Zinn-Justin J. eds. Chaos and Quantum Physics. Les Houches, Session LII. Amsterdam, North Holland. 1991.

[6] Berry MV. Semiclassical mechanics of regular and irregular motion, in: Chaotic Behavior in Deterministic Systems. eds Iooss G Helleman RHG Stora R. Amsterdam North Holland 171–271.

[7] Berry MV. Semiclassical theory of spectral rigidity. Proc R Soc London. 1995; **A 400**: 229–251.

[8] Berry MV Mount KE. Semiclassical wave mechanics. Rep Prog Physics. 1972; **35**: 315–397.

[9] Simon B. Nonclassical eigenvalue asymptotics. J. Fuct Anal. 1983; **53**: 84–98.

[10] Simon B. Holonomy, the quantum adiabatic theorem and Berry's phase. Phys. Rev. Lett. 1983; **51**: 2167–2170.

[11] Nelson P Alvarez-Gaumé L. Hamiltonian interpretation of anomalies. Commun Math Phys. 1985; **99**: 103–114,

[12] Teller E. Crossing of potential surfaces. J Chem Phys 1937; **41**: 109–116.

[13] Zhang W-M Feng PH Yuan J-M Wang S-J. Integrability and nonintegrability of quantum systems: Quantum integrability and dynamical symmetry. Phys Rev 1989; **A 40**: 438–447.

[14] Zhang W-M Feng DH Yuan J-M. Integrability and nonintegrability of quantum systems II. Dynamics in quantum phase space. Phys Rev 1990; **A 42**: 7125–7150.

[15] Zhang W-M Feng DH Pan Q Tjon J. Quantum fluctuations in quantum chaos. Phys Rev 1990; **42**: 3646–3649.

[16] Adams BG Cizek J Paldus J. Lie Algebraic Methods and Their Application to Single Quantum Systems. Adv Quantum Chem. 1988; **1988**: 1–85.

[17] Stone AD. Einstein's Unknown Insight and the Problem of Quantizing Chaos. Physics Tod. 2005; **8**: 37–43.

[18] Burić N. Hamiltonian quantum dynamics with separability constraints. Ann Phys. 2008; **323**: 17–33.

[19] Yazbashyan EA. Generalized microcanonical and Gibbs ensembles in classical and quantum integrable dynamics. Ann Phys. 2016; **367**: 288–296.

[20] Balain R Block C. Distribution of eigenfrequencies for the wave equation in a finite domain III. Ann Phys. 1972; **69**: 76–160.

[21] Dirac PAM. The adiabatic invariance of the quantum integrals. Proc Roy Soc. 1925; **107**: 725–734.

[22] Clemente-Gallardo J Marmo G. Towards a Definition of Quantum Integrability, Int. J. of Geometric Methods in Modern Physics. 2009; **6**: 120–172.

[23] Bohm *A. Quantum* Mechanics. Springer, New York, 1979.

[24] Abdalla MS Khalil EM Obada AS-F. Interaction between two $SU(1,1)$ quantum systems and a two-level atom. Physics A. 2016; **454**: 99–109.

Chapter 2

The Chaotic Behavior of ICT Users

Sumiyana Sumiyana and Sriwidharmanely Sriwidharmanely

Abstract

This paper describes how chaos theory was implemented to explain a behavioral aspect in an information system. The chaos theory was developed from the physical sciences and has been widely applied to many fields. However, this theory may also be applied to the social sciences. For certain types of human behavior, the chaos theory could comprehensively explain the phenomena of the use of information and communications technology (ICT). It means that this theory could clarify all the different kinds of human interactions with ICT. When the researchers used the chaos theory integratively, they could explain the distressed behavior of ICT users comprehensively. This theory argues that an individual acts randomly, even though the system is deterministic. When individuals use ICT, they could get technostress due to either the information systems or other users. This paper explains that ICT users could use information systems, with their complicated procedures and outputs. They were also probably disturbed by other users. The users, furthermore, experience chaotic pressures through their experiential values. This paper shows that users' behavior when facing chaotic pressure depends upon their personality dimensions. The authors finally propose a new paradigm that this chaos theory could explain the chaotic actions of ICT users.

Keywords: chaos theory, chaotic situation, technostress, coping strategy, creativity, controlling

1. Introduction

When individuals interact with information and communications technology (ICT) in either an information system or an application, they will relate to its complicated connections. They should try to have collaborative relationships with ICT. This relationship between users and ICT can lead them in either a circular motion or a non-linear direction that depends on the complexity of the problem. Meanwhile, the complexity of the problem is a result of the science used, and technology's progress, which sometimes makes surprising leaps forward. Thus, the problem requires not only the individual user's control and creativity but also his/her subtlety [1] to find alternative answers to the problems. Therefore, it is crucial to appreciate the potential for individuals to continue the interaction and influence the organizational direction and innovation. These individual are the people who can overcome an administration's dissolution and create workflow systems procedurally [2].

The chaos theory attempts to explain the complex and unexpected movements or system dynamics that depend on the initial condition. Wheatley [3] suggested that chaotic situations occurred when an organization left its ICT users to perceive the information system's devices themselves. The ICT users will usually follow inherent patterns and structures, based on their perceived procedures and rules. The users continue to stay within a particular gap, to define and shape their direction. Thus, chaos can become an ally when the information system requires to integrate its quality into the organizational workflows [4]. It means that the organization strives to find someone to innovate and develop this system's workflows [4].

Both an accounting information system (AIS) and other applications are dynamic workflows. Complex interactions and collaborations between the systems' elements can cause unexpected and dramatic changes that create chaos. In other words, complex interactions and collaborations between users and ICTs in an AIS cause chaos (among others, i.e. technostress). In this condition, if the users cannot adapt to this technological progress and complexity, they will feel frustrated and depressed, experiencing what is called technostress. Then, this technostress will have an impact on decreasing the users' satisfaction with this ICT [5–12], their performance [5, 13–15], productivity [16–18], innovation [12, 13], commitment to the organization [11, 12], and role conflicts [12, 16]. They could survive in these chaotic conditions if their organizations facilitate the users with flexibility and adaptability in the ICT systems [19]. Briggs and Peat [1] described that chaos would not reoccur in organizations when the ICT users have three techniques, which are: control, creativity and subtlety.

The chaos theory could be used to highlight the initial use of an information system and its complexity by organizations. These complexities could destroy the user experience because these information systems could produce some unexpected consequences for the ICT users in their organizational environment. This paper takes into account that a user will interpret the information he/she obtains in different ways to the other users, due to the dominant characteristics of their personality traits. It means that each user personality triggers various complex responses [20]. Thus, individuals with different personality traits will evaluate and assess the destructive events caused by ICT in different ways. Unequal evaluations and assessments are due to the various intrinsic and extrinsic needs of each user.

The authors argue that an ICT user could make either a positive or negative evaluation. We noted that ICT users when facing technostress creators, would be influenced by their extrinsic needs since those are the situational factors. ICT users will continuously choose available mitigating strategies. From another side, the ICT users are affected by their intrinsic conditions, which are the dimensions of their personalities [21]. Finally, the chaos theory suggests that an individual could act randomly, although the systems are deterministic. These random actions are profoundly possible because of an individual's creativity or innovative capability, personality traits, or how well he/she can control him/herself.

From another perspective of the mobile internet, the authors explain that an ICT user probably faces technostress creators that are from other users. We took into account that the other users could either deface the infrastructure of the ICT [22] or act in an iconoclastic manner, [23] that could hurt some individuals. However, the authors define the defacement and iconoclasm are in the context of ICT users' communication, either orally or written. We accentuate that the other users utilize linguistic communications that destroy an individual's cognition. In other words, different users employ sarcastic messages that destroy a person's cognitive flow. This means the victimized user will suffer from technostress because of what the other users did. Consequently, this user will, most probably, stop working with the other users and the information system or application. The authors, moreover, argue that whether or not the user continues using the mobile internet depends upon his/her personality's dimensions.

The latest discussion of this paper is that technostress causes variations in the ICT users' behavior through the state of their cognitive flow. In other words, technostress's creators influence ICT users' experiential values. The authors argue that chaotic pressure, due to the complicated information systems and applications, affects the users' enjoyment, entertainment, social affiliations, visual appeal, and escapism [24–26]. ICT users experiencing a technostress creator find it affects their motivation to achieve task-related performance levels and satisfaction. Ultimately, we re-emphasize that the chaotic experiential values of ICT users could be from a complicated interconnection and collaboration with the information systems, or the systems' defacement, or iconoclastic actions. The authors, in other words, propose that the technostress creators could support the users' continued use of, or abandonment of, information systems and applications. However, this continued use of ICT is related to each user's personality dimensions.

The remains of this paper will discuss the chaotic behavior of ICT users in four subsections. SubSection 2 presents a resume of chaos theory. The chaos theory, chaotic situations, and usefulness of this theory as an idea in explaining ICT users will be explored in subsections 3 and 4 consecutively. The last subsection has a conclusion.

2. The resume of chaos theory

In 1961, the meteorologist Edward Lorenz was the first to introduce the chaos theory. This theory tried to find a form of uniformity from seemingly random data [27]. Lorenz discovered this theory accidentally. He was looking for a reason why the weather was unpredictable. He used computer assistance and 12 formulation models. He created a program which could not predict the weather but can illustrate what the weather will be like if its starting point is known. One time Lorenz wanted to see the results of the weather model's sequence.

He started from the middle, not from the beginning. For simplicity, Lorenz entered a value consisting of three decimal numbers (0.506), while the number of the sequence was 0.506127. Because the rounding was correct, then the pattern formed by the two numbers should be similar, but it turned out that the design which appeared was more and more different from before. Based on this discovery, Lorenz re-experimented, this time using a simpler model with only three formulations. The result of the data, when displayed, again appeared to be random, but when the data were entered in graphical form, a phenomenon called the butterfly effect was created. A small difference at the starting point (only 0.000127 difference) changed the overall pattern.

The chaos theory refers to the tendency of dynamic, non-linear systems toward disorder or chaos, sometimes behaving unexpectedly, but always deterministically [27]. This theory also refers to the underlying linkages, which exist in random events, which are calculated from the initial conditions [1, 28, 29]. Chaos science focuses on hidden patterns, nuances, sensitivity to things, and rules about how something that cannot be predicted leads to human behaviors.

This theory is not only applied to exact sciences but also social ones, such as the social sciences, psychology, finance, decision making, management and behavioral or information systems. McBride [29] was the first researcher to use a framework based on the chaos theory in the field of information systems. This framework consisted of interaction domains, initial conditions, foreign attractors, events and choices, peak clutter, bifurcation, looping, and connectivity. The focus of this interpretive model was on the value of building descriptions of information systems' interactions in organizations.

Levy [19] applied the chaos theory when making theoretical frameworks to understand the dynamics of the industrial evolution and the complex interactions among industry players. An industry can be conceptualized and modeled as a complex and dynamic system, which shows both uncertainty and underlying order. Levy created a simulation model to illustrate the interactions between computer manufacturers, their suppliers, and their markets. The simulation's results showed how managers might underestimate the costs of international production. He concluded that, by understanding that any industry is a complex system, managers could improve their decision making and find innovative solutions.

Meanwhile, Ayers [28] mentioned that, in the field of psychology, the concept of chaos had been explored extensively. This concept is primarily in the area of psychoanalysis, on a symbolic level. Outside of psychoanalysis, the chaos theory has been applied to a variety of clinical subjects to varying degrees. Still, almost all of its applications appear metaphorical (although one cannot always make this statement explicitly). This theory was also used for psychological processes through the practical application of chaos methodology, e.g. [30, 31]. It showed that, even though the application of metaphors is useful in providing appropriate ways of looking at psychological disorders, the successful application of future psychopathological changes depends on whether it is validated by practical work demonstrating chaos in the associated psychological phenomena.

Moreover, Radu et al. [32] presented the application of chaos theory in management. Also, they explained the positive and negative sides of this theory in a company's current strategic management, in organizational change projects or the management of highly dynamic projects. Furthermore, Klioutchnikov et al. [33] explained that the chaos theory is very suitable for understanding financial perspectives because several circumstances determine the behavior of the financial markets, which are relative to the needs, and internal and external reasons can cause those circumstances to arise. They tried to clarify several points related to the possibility of using chaos theory in finance. Its mechanism of implementation in finance was in macro- and micro-processes. This mechanism also used specific methods and instruments, such as fractal and stochastic processes and predictions.

The latest work by Sauermann [34] involved chaos theorems drawn from the social choice theory and used to investigate the relationship between the indeterminacy of majority rule leads and voting cycles and to make democratic decisions. The study's results contradicted Riker's interpretation of the chaos theorems' implications. This core exhibited less attraction than generally assumed. Then, an empty core is not associated with majority rule's increased instability. Instead, conflicting preferences lead to more instability irrespective of the existence of an equilibrium.

3. Chaos theory and the chaotic situation

3.1 Technostress

Brod [35] introduced technostress as a disease caused by a person's inability to adapt to new computer technology. This paper argues that ICT users feel unhealthy and have little or no motivation to use information systems or applications anymore. This technostress could be manifested by ICT users getting either excessive fear or computer anxiety. Ragu-Nathan et al. [11] suggested that information technology created many problems which ICT users cannot overcome. In other words, ICT users feel that they cannot become familiar with the information systems or applications and what is required for them to follow the procedural tasks.

Srivastava et al. [36] suggested that technostress occurred when the requirements for using ICT exceeded a user's capabilities to cope with or mediate such stress. Moreover, Stich et al. [37] also concluded that technostress is impaired experiential cognition experienced by users because of the complicated ICT. Stich et al. [37] constructed two main concepts: stressors (the creators of technostress) and strains (the results of technostress). Finally, Tarafdar et al. [10, 16] conceptualized five main categories of techno-creators, which are:

1. Techno-overload describes situations where ICT users are forced to accomplish their work in the allotted time. Otherwise, there would be a massive workload placed on the information systems.

2. Techno-invasion refers to the information systems which could probably invade ICT users' privacy. This technostress also illustrates the effect of ICT's invasion in creating insufficient motivation, where ICT users have to continue or stop, using the information systems or applications.

3. Techno-complexity describes a situation where the complicated tasks associated with ICT users makes them feel inadequate. Thus, this technostress forces them to spend more time and effort. It could also be explained that these tasks need ICT users to learn and understand various aspects of the information technologies they use.

4. Techno-insecurity refers to conditions where ICT users feel threatened with losing their jobs due to the presence of new information systems and applications. This technostress is caused by the ability of ICT to replace human working processes. It also applies to ICT users that do not have a great deal of knowledge.

5. Techno-uncertainty describes situations where new information systems or applications disturb the users due to the needs of their additional capabilities. This technostress would probably occur during the implementation of a new ICT system, for which the users have to learn new things.

From another perspective, Tarafdar et al., [10, 16], and Ayyagari et al. [38] identified and then clustered five technostress triggers, which are:

1. Work-home conflict - ICT users perceived their intra-personal conflict to be between their work and family needs.

2. Invasion of privacy - information systems would probably not protect ICT users' privacy. The users perceive that internet systems must not compromise their privacy due to their data being saved by a third party.

3. Work overload - ICT users think that their capabilities and competencies do not match with the requirements of the information systems. In other words, ICT users feel that their abilities or skill levels are not skilled enough to operate this ICT.

4. Role ambiguity - many users say that they feel uncertainty when accomplishing their work using information systems. They do not know a procedurally work order or its consequences on their performance. This paper also explains that ICT users suffer from a lack of information when they want to expedite their roles and authority.

5. Job insecurity - the presence of ICT means the ICT users may lose their jobs because the information technology could replace them and do their job.

Ayyagari et al. [38] classified three technological characteristics that could influence techno-stressors, which are: usability, dynamic features, and intrusive features. Usability, complexity, and reliability are generally associated with the use of information technology. These three characteristics of information technology are part of its usability features. The rate and frequency of technological changes relate to the nature of ICT, which are dynamic features. The ICT feature refers to the extent to which a person feels a shift in the technological environment is happening quickly. In contrast, the presentism and anonymity of the invasion by ICT represent the intrusion feature. Presentation's characteristics describe the extent to which technology allows users to either reach it or not. In contrast, anonymity describes the times when ICT users feel that they could not identify or trace the work they produced using ICT.

As mention earlier, some previous studies showed that technostress had harmed ICT users' outcomes, including reducing their satisfaction and performance [5–12, 14, 15, 18, 39]. ICT users, moreover, could not survive in these technostress conditions, which organizations must have facilitated using all the aspects of their skill, flexibility and adaptability [19]. Hwang and Cha [40] showed that security-related technostress creators in organizations negatively affect employees' organizational commitments, both indirectly and directly. This technostress occurred through their complex role and then reduced their intentions to comply with the information system's security. From another perspective, employee-focused promotions could moderate the relationship between technostress creators and role stress. Employees with a focus on gaining promotion are more resistant to the adverse effects of technostress creators, because they experienced lower role stress. Nimrod [41] made a new scale to measure technostress levels between younger and older workers. Technostress, moreover, must be considered a particular threat to the future well-being of ICT users.

Qi (2019), developed a theoretical framework to investigate the double-edged effects of using mobile devices. It used the sampling design of mobile devices among college students. This framework argued that positive results (an improvement in their academic performance) were investigated from their use, while adverse effects triggered technostress. This paper takes into account that Qi's study was based on the person-technology fit model (P–T fit model). It explained that the educational use of mobile devices by students does not lead to technostress. This use, however, could improve academic performance due to their high usage of ICT. The paper argued that students' self-efficiency and their skill level in using cellular technology affected their high-low technostress.

Human-technology interactions, especially during the development of information systems, are complex. To portrait this complex phenomenon, McBride [29] adopted the chaos theory to make a framework for interpreting the success of information systems' implementations in organizations. McBride's paper suggested that the chaos theory could explain the complicated phenomenon and the non-linear and dynamic systems [19] such as the technostress creators in the implementation of an information system's development. The chaos theory means there is an underlying interconnectedness that exists in random events; hence the ICT users are concerned with the initial conditions [1, 28].

Through the chaos theory, the authors portray the phenomena of technostress holistically. We noted that developments to information systems are the domain of human-computer interactions, in what is probably a chaotic space between humans and information technology. The implementation of new information systems and complex ICT by organizations could be regarded as destructive events, resulting in some unexpected and unpredictable consequences for the users' environments [20]. When humans and information technology interact, individuals have to learn the new processes that are required. These processes will flow according to the respective ICT users' methods. However, when ICT users encounter a disturbance, it will cause various impacts depending on their motivation to respond to it. Likewise, what happens when ICT users are facing technostress is also a chaotic situation.

3.2 Defacement and inconoclasm

The authors state in this paper that the interaction between humans and information technology is complex. Individuals could not deny this complexity is all around them, as a result of the increasingly digitalized world. The authors show some pieces of evidence about the destructive nature of technology, such as is found in the global digital infrastructure, social media, the Internet of Things, robotic processes' automation, digital business platforms, algorithmic decision making, and other digitally-enabled networks and ecosystems; all of which also fuel the complexity people feel around them [42]. Building up hyper-connections and mutual dependencies among the human actors, technical artifacts, processes, organizations, and institutions caused this complexity; which affects human experiences within their cognitive state in all magnitudes. Both organizations and individuals turn to digitally enabled solutions to cope with the problems arising from computerized digitalization.

In the digital world, complexity and digital solutions present new opportunities and challenges for research into the information systems. Systems-wide changes in natural open systems reveal how unorganized entities in a given system, subjected to an externally imposed tension, could engage in far-from-equilibrium dynamic actions. The entities, therefore, could self-organize into distinct phase transitions leading to new higher-level orders [43]. Defacing the machinery and sending out iconoclastic messages, for instance, could drive and hamper these changes in a chaotic situation.

A defacement is a physical act of vandalism or the destruction of a material thing. In the IT field, defacement has been bastardized to mean website destruction. Romagna and Hout [44] defined defacement as a kind of electronic graffiti and, like other forms of vandalism, it has been used to spread messages by "cyber protesters" or politically motivated hackers. Davanzo et al. [22] defined defacement as destruction in the form of a general attack on a website. In this case, the site's content is partially or entirely replaced, by the attacker, with content that is embarrassing to the site owner, for example, disturbing images, political messages, the attacker's signature form, and so on [22]. Meanwhile, Bellman [45] defined defacement as enlightenment. In short, defacement implies causing damage to something which, in this paper, is the ICT users' communication.

In behavioral research, defacement means as an attack aimed at changing users' behavior. Thompson et al. [46] explained that defacers try to make some changes in users' behavior, by manipulating their perceptions of reality. Criminals cannot achieve the desired results from their attack unless the users change their behavior in some way [46]. It is this modification of the users' actions that is an essential link in the cognitive attack sequence. In the case of the multiplayer online battle arena (MOBA) game player, we defined defacement as a communication breakdown that causes someone to decide to deface or vandalize something. In other words, the vandalism of communications equipment aims to destroy the recipient. This paper argues that defacement behavior causes damage that results in behavioral or cognitive changes in MOBA game players. In a game, defacement behavior occurs when a player deliberately throws out bad words to lure other players in and interfere with the game. A user could create chaos among the players so that the other players do not focus on playing the game and do not intend to play it again.

Iconoclasm is the social belief in the importance of destroying icons, images and monuments [47–49]. Latour [23] defined iconoclasm as an act of destruction, where the intention to do damage is apparent. Besides, Clapperton et al. [50] defined iconoclasm as the use of a strategy that represents a logical and instrumental means for using violence to achieve political goals. Furthermore, Clay [49] stated that they used iconoclasm to show domination and control over a group. During research into the field of communication, Smith [51] used iconoclasm via internet memes as a tool to display fake news to damage or reduce the image of a public institution.

In the MOBA game, it described iconoclasm as the destruction of an icon. In this case, it was the "hero." In this game, icons which describe the identity of the game players represent heroes [52]. Iconoclasm tends to harm or destroy the players. It usually occurs when a player chooses a hero that iconoclasts do not like. Iconoclastic players will insult the person because they feel that the hero is not suitable for use in the game. This incident will result in the players' fighting each other, which may also be carried over into the game. This paper argues that when people insult someone else's favorite heroes, the players could lose their cognition. The player is less motivated to play, and he/she stops playing, or continues playing, but not in a serious manner. This chapter also posits that the destruction of communications, either through defacement or iconoclastic actions, is a form of destruction in the MOBA game's communication channel. Both defacement or iconoclasm could destroy the players' cognition and cause chaos in the game.

3.3 Experiential value

The authors recall that chaos theory is supposed to explain complex, non-linear dynamic systems. From a theoretical perspective, this theory is also equivalent to the postmodern paradigm. This paradigm questions deterministic positivism because it recognizes the complexity and diversity of experience. Boccaletti et al. [53] suggested that advocates of the chaos theory enthusiastically highlight signs everywhere. These signs are pointing to the complex dynamic systems which are ubiquitous in the social world, and the similarities between the patterns produced by simulating non-linear systems and sequences. For example, this paper presents how share prices in the stock market and commodity prices fluctuated abruptly because these reactions always change seconds per second.

The diversity of experiential values of ICT users could be characterized by their optimal behavior [24], such as is seen in their flow experience [25]. Experiential values could also be explained as a result of sophisticated learning. Moreover, Moneta and Csikszentmihalyi [26] demonstrated that experiential values require total concentration and a great deal of interest in the activities characterized by optimal experience. The attributes of the experiential values of ICT users are as follows:

a. Escapism - escapism is a behavioral view related to the personal activities undertaken to avoid the realities that are challenging, impossible or un attainable [54]. Running away occurs when a person finds his/her life is spent in unsatisfactory conditions, which cause him/her to become detached from reality, and is done to reduce his/her anxiety [55]. Thus, the impact of chaos is felt when the individual cannot optimally realize the value of his/his experiences. The individual then experiences confusion which can act on his/her cognitive processes and causes the formation of affective disorders in the user.

b. Enjoyment - enjoyment is the pleasure that an individual feels objective when doing certain activities [56]. Based on the flow theory, Csikszentmihalyi [24] stated that enjoyment occurs when a person not only fulfills the expectations that occur before or satisfies his/her desires but also achieves unexpected needs, which may have been previously unimaginable. Enjoyment occurs when a person feels involved in pleasure from within. This condition, therefore, causes people to tend to experience flow processes that form their cognitive and affective processes. It means that if an individual does not experience an optimal level of enjoyment, he/she will tend to have a chaotic pattern.

c. Social affiliation - it is through his/her social affiliations that a person feels interested in society, these are usually generated by his/her employer's company services, as an efficient approach to marketing [57]. Social collaboration occurs automatically and experiences a flow when the feelings of the individuals affect each other. The presence of an individual's flow in a social affiliation does not create an optimal experience. There will be a pattern of chaos in the individual's cognitive and affective flow so that it will harm the interaction socially.

d. Visual appeal - the visual appeal is a reactive source of esthetic value [58]. Visual appeal is a dominant matter to attract consumers' attention. From a marketing perspective, the attractiveness refers to the selection of data and information, and their transformation and presentation. Most companies usually facilitate customers'explorations and understanding [59, 60]. It means that a person's visual attractiveness shapes his/her experiences in condemning his/her affective and cognitive flow through data and information's selection, transformation, and presentation. Therefore, the experiential values are an essential source for the optimal experience to avoid cluttering the visual power.

e. Entertainment - entertainment involves observing the customers in a performance which leads to a relaxed reaction [61, 62]. This entertainment is an attribute of the ICT users' experiential values because their pleasured responses that make the results optimally. Thus, if it is not in the optimal joy, the chaotic patterns emerge in the ICT users' affective and cognitive flows.

The constructivist theory of learning [63] may be aligned with experiential values in which the outcomes of the learning process are varied and often unpredictable. This paper argues that an individual plays a critical role in assessing his/her learning outputs. An individual receives his/her experiential values from use or appreciation of a product or service [60] as like as information systems or an application. In this assessment process, everyone will respond differently depending on their self-control, activity and subtlety [1]. This process will always follow inherent patterns and structures, based on intrinsic values and rules, i.e. experiential values. In other words, this process always stays within certain boundaries to define and shape the direction of ICT users; otherwise, chaotic situations could occur [3].

4. Inducing the chaos theory to explain behavioral phenomena

Generally, many organizations use ICT to improve their competitive advantage so that this could transform their organizational efficiency, productivity, and effectiveness. From another point of view, they intend to use ICT to change their social and corporate environments [39]. However, if they cannot manage their ICT correctly, they are shadowed by the adverse effects due to their low use of it [40]. This paper recalls the implementation of a new ICT system that consisted of complex and collaborative relationships. This implementation led to stress for the users as they could not cope with their organization's demand that they use the new ICT system. Brod [35] introduced technostress as an illness resulting from a person's inability to adapt to new computer technology. It is typified by over-identification or computer anxiety. Ragu-Nathan et al. [11] described technostress as a problem because the users could not overcome the difficulties with the new ICT system, or they could not become familiar with the new system. Technostress can affect the individual's orientation regarding the time he/she spends doing something, his/her communication mode, and his/her interpersonal relationships as well as his/her job outcomes, i.e., performance or satisfaction.

To explain this phenomenon, researchers into information systems conduct studies in various disciplines, including psychology, sociology, philosophy, and organizational studies. These disciplines explain the stress phenomenon as a source of contextual paradigms, and previous researchers often used the person-environment fit model to describe technostress [5, 39, 40]. This theory stated that when the relationship between people and their environment is beyond the equilibrium condition, it will create stress [41], i.e., technostress. This theory also portraits technostress as a linear system, while the interaction between humans and technology (i.e. computers) is problematic for the development of information systems.

This paper argues that ICT users have specific conditions with which they can interpret and understand the environmental conditions through their capabilities. ICT users' power triggers them to find various and complex responses. Thus, chaos can be an ally or a desired quality when integrated into an organizational system, especially when the ICT users try to innovate and develop [4]. This theory showed that the users' chaotic cognition triggers the relationship of their stressed transactions. ICT users, furthermore, must have strategies to deal with the chaos. Coping is a thing that individuals do, which sometimes allows them to solve problems and adapt to changes.

The inducement of the chaos theory in explaining the ICT users' behavior is not deniable. The authors demonstrate the chaotic behavior from two sides, which are complex interactions and the collaboration of the ICT system's elements [5–18], and both defacement [22] and iconoclastic methods [23]. These two sides affect ICT users' behavior when they have to face the technostress's creators. By these means, these sides influence the ICT users' performances and satisfaction when they are in a chaotic situation. Although the ICT users could mitigate this chaos, they may choose to face it, depending on how mature their personalities are. In other words, the ICT users have to cope with the complicated uncertainty or technostress creators by relying upon their personalities and emotions to overcome the chaotic problems.

This paper supports the undeniable inducement of the chaos theory to explain the ICT users' mitigation of the harmful effects of technostress. It argues that the technostress's creators at first settled on the ICT users' cognitive states. In other words, the ICT users got their experiential values, which are enjoyment, escapism, visual appeal, social affiliation, and entertainment, when they faced situations with technostress. From the perspective of learning, the authors propose that chaos theory relates to the ICT users' learning processes [63]. We take into account that chaotic mitigation affects the ICT users and may prevent them from dealing with the technostress efficiently and effectively. We recommend that information systems or applications must be developed with consideration given to facilitating the ICT users' experiential values. It means that the information systems and applications make the ICT users increase their enjoyment, entertainment, social affiliation and visual appeal as well as decreasing their escapism. The authors argue that technostress for ICT users would otherwise have occurred.

4.1 Technostress and a proactive personality

Personality is a characteristic of an individual, and this determines the person's thinking and behavior. Every individual has a unique personality, which differs from that of other people. Bateman and Crant [64] defined a proactive personality as someone who is relatively unrestricted by the situational forces which influence environmental change. Someone with a proactive nature identifies opportunities and demonstrates initiative, takes action when appropriate, and persists until meaningful change occurs. Parker and Sprigg [65] explained that proactive personalities usually engage in activities that affect themselves and their environment.

From the perspective of the chaos theory, whenever individuals face technostress, they are either in a chaotic situation or not. It means that the users' performance and satisfaction would be explained when both the chaos and technostress theories work concurrently. To overcome this chaotic situation, the user has to be creative [1, 4, 66], because his/her behavior will vary based on experiences. Personal innovativeness means that individual traits have a role in technology's adoption. This innovativeness entails the implementation of creativity or the generation of novel and useful ideas for the development of new products and processes [67]. Thus, in the implementation of advanced ICT systems, a proactive personality can boost the creativity of the users. Therefore, we posit that a proactive personality can play a role in mitigating the harmful technostress to a user's satisfaction.

Based on the chaos theory, Sumiyana and Sriwidharmanely [68] demonstrated that individuals work randomly or differently because of their creativity or personal innovations [1, 69, 70]. They can mitigate the adverse effect of technostress on ICT users' performance by inducing their proactive personalities. This study shows that when users interact with new technologies, and the users feel there is a mismatch (cognitive impairment) between their abilities and the requirements of the latest technology, this condition creates discomfort during their interactions (a chaotic situation, known as technostress). However, this sense of discomfort will be minimized if they have the creativity to use technology to help them complete their tasks. So, in the end, they can maintain their performance levels. In other words, they can turn a threat into an opportunity.

Specifically, this study's result shows that proactive-transform personalities maintained their performance better than proactive-conform personalities did when the ICT users experienced high technostress. It meant that the creativity of the users was more active when they faced high levels of technostress than low levels, which offered significantly more benefits for the proactive-transform personalities. The ICT users can take advantage of the work overload and deadline times in the system, so they can still maintain their performance. Even for the same proactive-transform personalities, the user faced with high levels of technostress performed better than the user who experienced the lower levels.

4.2 Technostress and positive emotions

ICT users probably feel that their capabilities are not compatible with the requirements of the new ICT and that they have limited control over them. They then feel uncomfortable because this creates technostress. So they will implement strategies to overcome these painful experiences (mitigation), whether they are related to the users' psychological expectations, rejection or wishful thinking (inward), or related to realizing and seeking support that affects their emotions directly (outward), or not. This strategy is called emotion-focused coping [71].

This strategy mainly focuses on the effort to restore emotional stability and reduce the tension caused by the implementation of a new ICT system. This paper highlights that cognitive dynamic instability results in the ICT users' adverse impacts. For instance, we infer that the users' coping strategies are based on the control theory [1, 72], which was mentioned earlier, and these can cope with a chaotic situation or technostress. We argue that users' self-control (inward) and feedback on the assigned task's performance (outward) are the types of strategies which have a direct impact.

Self-control gives ICT users the belief that they could implement the system successfully. It takes into account the users' self-control because the system's development process is complex, and needs intensive involvement and the interaction of various agents [73]. Meanwhile, feedback is a communication process that involves a source (sender) and destination (receiver) [74]. Concerning the performance aspects or understanding the system, the ICT itself could provide feedback to the users who search for answers and solutions, so that they can evaluate whether they have the correct response or not [75].

By applying a contrast analysis, we confirmed that the broaden-and-build theory [76] explains that positive emotions can improve ICT users' capabilities to cope with their technostress. Positive emotions are affective components which ICT users typically find pleasurable to experience. Positive emotions could help ICT users to broaden their horizons, and then widen the scope of their focus [77]. Positive emotions could also increase the users' performance of a cognitive task by lifting their spirits without distracting them [78].

Expressly, we undertook a study which indicated that positive task performance feedback could boost the positive feelings of ICT users. It documented that the users who have low self-control also perform their tasks poorly. If they receive some form of therapy and positive feedback, their understanding is better than that of the ICT users who receive negative feedback. Our study, furthermore, showed that positive emotions play an essential role when ICT users face the harmful effects of technostress on their performance [76, 79]. Moreover, this study found that positive emotions affect both those with low and high self-control. It found that ICT users' task performances, for those with both low and high levels of self-control, were not different. It means that positive emotions have a more profound effect on mitigating the adverse impacts of technostress. The authors, therefore, argue that positive feedback could enhance the users' self-efficacy and individual innovativeness.

5. The chaos theory in behavior research as a new paradigm

The chaos theory suggests that an individual could act randomly although the systems are deterministic. The individual acts randomly because of his/her level of self-control, creativity or personal innovativeness and subtlely [1, 69, 70]. If the individual is in a state of technostress, or a chaotic situation, his/her capabilities are shown by the coping strategies that he/she uses to accomplish a complicated task. The authors argue that coping behavior is a transaction carried out by an individual to overcome the various demands (internal and external) of the thing that burdens and interferes with his/her survival. Coping is a cognitive and behavioral effort to manage (reduce, minimize, or tolerate) the internal and external demands of the person-environment transactions that an individual judge to exceed his/her resources [80]. Each individual will have a unique coping strategy for overcoming or hinting at a way to solve his/her problem. It means that when ICT users experience technostress, they should adjust themselves to the system or organizational environment.

When dealing with stress triggers, individuals overcome these disorders by using two main processes that are continuous, and which influence each other [80, 81].

These are also known as cognitive appraisals and coping strategies. First, individuals evaluate the potential consequences of events by making a judgment. The central assessment is one's judgment regarding the significance of an event that is stressful, positive, controlled, challenging, or irrelevant. Subsequent inspections are assessments of the resources and choices of individual mitigation strategies. This second assessment addresses what individuals can do to control the situation. Individuals take different actions to deal with chaotic conditions. It means that their mitigation strategy is to face the harmful effects of technostress. Thus, a mitigation strategy is an adaptive action that individuals do in response to disturbing events that occur in their environment.

More broadly, the interactions between the socio-technical entities produce a lot of the results that appear in the information system. This paper presents an example, which includes the creation of collaborative online orders and technology's capabilities [82]. It demonstrates that the organizations need the information systems to be in alignment [83] and that new configurations between organizational, platform and participant dimensions exist [84]. The emergence perspective offers a lens to understand the many unpredictable socio-technical phenomena that reach the individual, group, organizational and community levels, in the context of expanding digitalization.

In practice, the chaos theory can help accountants, auditors, and educators understand their environment holistically so that they can control or behave creatively to adapt and continue to survive in their environment. Levy [19] suggested the need for innovativeness to be examined. The advantages of the chaos theory are that it can portray industrial phenomena holistically. In a complex system, managers must be creative to improve the quality of their decision making and to help them find innovative solutions. Not all accountants, auditors, or educators have the resources to keep pace with the development of new information systems or applications. The implementation of new information systems enables ICT users to experience technostress. Facing this condition, each individual will have a different coping response or behavior. Holistically, ICT users can utilize their creativity or innovation to mitigate the negative impact of information technology. The ICT users, therefore, would not allow a new ICT system to continue to interfere with them achieving the required performance. Managers can make policies related to their staff's dysfunctional behavior due to complicated information technology. Managers must consider who gets stressed and how it impacts on them and others. Furthermore, managers can accommodate ICT users' innovations for facing technostress. In other words, managers can recommend ICT media that can be used to improve the users' learning of coping strategies.

6. Conclusions

The chaos theory implies that an individual could act randomly although the systems are deterministic. The individual acts randomly because of his/her self-control, creativity or personal innovativeness and subtlely. We can recommend the chaos theory needs further research because this theory could be used to explain the phenomena of technostress. We propose that the chaos theory and its conceptual framework could overcome the weaknesses of some previous approaches that only investigated technostress phenomena from a single side. This paper argues for the proper way to apply the chaos theory so that future researchers could portray the technostress phenomena comprehensively.

Not all ICT users can meet the needs or requirements of new information technology in an organization. It means that coping behavior could occur in the unit analysis, either for individual or group users. This phenomenon still provides opportunities for further research. On the other hand, some research has also shown that the effects of information technology are not only harmful, just like other stressors, but they also have positive impacts. These positive consequences, due to technostress, also provide an opportunity to conduct further investigations because this impact could be not only linear but also non-linear.

From a different perspective, this chapter proposes the anti-thesis of the ICT users who had been hurt by technostress. It argues the use of the build and broadens theory for mitigating the harmful effects of technostress. When ICT users feel confused, due to the technostress's creators, the developers of information systems and applications could use this theory to facilitate them in coping with chaotic problems. This theory recommends that ICT users could be encouraged by information systems that improve the state of their cognitive flow. It then opens opportunities for future research to investigate the influence of this theory in reducing ICT users' emotional situations. Another future research possibility is the development of materials, tools or knowledge based on the build and broadens approach that could mitigate the negative experiential values of ICT users.

Acknowledgements

First of all, the authors' gratitude goes to the IntechOpen in allowing us to be involved in this book chapter. Secondly, our gratitude is also directed at the Faculty of Economics and Business, University Gadjah Mada, which has provided support for all the peripherals needed for the development process of this chapter. We thank all our assistants: NurHalimahSiahaan, Yanto Yanto, Rikhana Rikhana, Zhafirah Salsabil, and others who helped to correct this material and to provide some references for finishing and completing this chapter.

Conflict of interest

The authors declare no conflict of interest.

Author details

Sumiyana Sumiyana[1] and Sriwidharmanely Sriwidharmanely[2]*

1 Economics and Business Faculty of Gadjah Mada University, Yogyakarta, Indonesia

2 Economics and Business Faculty of Bengkulu University, Bengkulu, Indonesia

*Address all correspondence to: widharmanely@gmail.com

References

[1] J. Briggs and F. D. Peat, *Seven life lessons of chaos : spiritual wisdom from the science of change*, First Edit. HarperCollins Publishers, Inc, 1999.

[2] I. Nonaka, "Creating-Organizational-order-out-of-chaos," *California Management Review*. pp. 57-73, 1988.

[3] M. J. Wheatley, *Leadership and the new science : learning about organization from an orderly universe*. Berrett-Koehler Publishers, 1992.

[4] S. Smith and S. Paquette, "Creativity, chaos and knowledge management," *Bus. Inf. Rev.*, vol. 27, no. 2, pp. 118-123, Jun. 2010, doi: 10.1177/0266382110366956.

[5] R. K. Jena, "Technostress in ICT enabled collaborative learning environment: An empirical study among Indian academician," *Comput. Human Behav.*, vol. 51, no. Part B, pp. 1116-1123, Oct. 2015, doi: 10.1016/J.CHB.2015.03.020.

[6] C. Maier, S. Laumer, A. Eckhardt, and T. Weitzel, "Analyzing the impact of HRIS implementations on HR personnel's job satisfaction and turnover intention," *J. Strateg. Inf. Syst.*, vol. 22, no. 3, pp. 193-207, Sep. 2013, doi: 10.1016/J.JSIS.2012.09.001.

[7] C. Fieseler, S. Grubenmann, M. Meckel, and S. Muller, "The Leadership Dimension of Coping with Technostress," in *2014 47th Hawaii International Conference on System Sciences*, Jan. 2014, pp. 530-539, doi: 10.1109/HICSS.2014.73.

[8] A. Khan, H. Rehman, and Shafiq-ur-Rehman, "An Empirical Analysis of Correlation Between Technostress and Job Satisfaction: A Case of KPK, Pakistan," *Pakistan J. Libr. Inf. Sci.*, no. 14, 2013.

[9] M. U. Saganuwan, W. K. W. Ismail, and U. N. U. Ahmad, "Technostress: Mediating Accounting Information System Performance," *Inf. Manag. Bus. Rev.*, vol. 5, no. 6, pp. 270-277, 2013.

[10] M. Tarafdar, Q. Tu, and T. S. Ragu-Nathan, "Impact of Technostress on End-User Satisfaction and Performance," *J. Manag. Inf. Syst.*, vol. 27, no. 3, pp. 303-334, Dec. 2010, doi: 10.2753/MIS0742-1222270311.

[11] T. S. Ragu-Nathan, M. Tarafdar, B. S. Ragu-Nathan, and Q. Tu, "The Consequences of Technostress for End Users in Organizations," vol. 19, no. 4, pp. 417-433, 2008, doi: 10.1287/isre.1070.0165.

[12] M. Tarafdar, Q. Tu, T. S. Ragu-Nathan, and B. S. Ragu-Nathan, "Crossing to the dark side: creators, outcomes, examining and inhibitors of technostress," *Commun. ACM*, vol. 54, no. 9, pp. 113-120, Sep. 2011, doi: 10.1145/1995376.1995403.

[13] M. Tarafdar, E. Bolman Pullins, and T. S. Ragu-Nathan, "Examining impacts of technostress on the professional salesperson's behavioural performance," *J. Pers. Sell. Sales Manag.*, vol. 34, no. 1, pp. 51-69, Jan. 2014, doi: 10.1080/08853134.2013.870184.

[14] M. . Saganuwan, W. K. . Ismail, and U. N. . Ahmad, "Technostress of Accounting Information System and its Effect on Task Performance," *Aust. J. Basic Appl. Sci.*, vol. 8, no. 16, pp. 30-37, 2014, Accessed: Oct. 18, 2018. [Online]. Available: https://www.academia.edu/11196175/Technostress_of_Accounting_Information_System_and_its_Effect_on_Task_Performance?auto=download.

[15] M. U. Saganuwan, W. K. W. Ismail, and U. N. U. Ahmad, "Conceptual Framework: AIS Technostress and Its

Effect on Professionals' Job Outcomes," *Asian Soc. Sci.*, vol. 11, no. 5, 2015, doi: 10.5539/ass.v11n5p97.

[16] M. Tarafdar, Q. Tu, B. S. Ragu-Nathan, and T. S. Ragu-Nathan, "The Impact of Technostress on Role Stress and Productivity," *J. Manag. Inf. Syst.*, vol. 24, no. 1, pp. 301-328, Jul. 2007, doi: 10.2753/MIS0742-1222240109.

[17] M. Tarafdar, E. Pullins, and T. S. Ragu-Nathan, "Examining Impacts of Technostress on the Professional Salesperson' s Performance," in *AMCIS 2011 Proceedings- All Submissions*, 2011, p. paper 107, Accessed: Nov. 18, 2018. [Online]. Available: http://aisel.aisnet.org/amcis2011_submissions.

[18] W.-H. Hung, K. Chen, and C.-P. Lin, "Does the proactive personality mitigate the adverse effect of technostress on productivity in the mobile environment?" *Telemat. Informatics*, vol. 32, no. 1, pp. 143-157, Feb. 2015, doi: 10.1016/J.TELE.2014.06.002.

[19] D. Levy, "Chaos theory and strategy: Theory, application, and managerial implications," *Strateg. Manag. Journal; Chicago*, vol. 15, pp. 167-178, 1994, Accessed: Oct. 18, 2018. [Online]. Available: https://search.proquest.com/docview/224966458?pq-origsite=summon.

[20] T. L. Griffith, "Technology Features as Triggers for Sensemaking," *Acad. Manag. Rev.*, vol. 24, no. 3, pp. 472-488, Jul. 1999, doi: 10.2307/259137.

[21] R. S. Lazarus and S. Folkman, *Stress, Appraisal and Coping*. New York: Springer Publishing Company, Inc., 1984.

[22] G. Davanzo, E. Medvet, and A. Bartoli, "Anomaly detection techniques for a web defacement monitoring service," *Expert Syst. Appl.*, vol. 38, no. 10, pp. 12521-12530, 2011, doi: 10.1016/j.eswa.2011.04.038.

[23] B. Latour, "What is iconoclast? or is there a world beyond the Image Wars?" in *What is Iconoclash*, 2002, pp. 16-40.

[24] Mihaly Csikszentmihalyi, *Flow_ The Psychology of Optimal Experience*. 2008.

[25] G. B. Moneta, "The Flow Experience Across Cultures," *J. Happiness Stud.*, vol. 5, no. 2, pp. 115-121, 2004, doi: 10.1023/b:johs.0000035913.65762.b5.

[26] G. B. Moneta and M. Csikszentmihalyi, "The Effect of Perceived Challenges and Skills on the Quality of Subjective Experience," *J. Pers.*, vol. 64, no. 2, 1996, doi: 10.1111/j.1467-6494.1996.tb00512.x.

[27] E. N. Lorenz, "Deterministic Nonperiodic Flow," *J. Atmos. Sci.*, vol. 20, no. 2, pp. 130-141, 1963, Accessed: Oct. 26, 2018. [Online]. Available: https://journals.ametsoc.org/doi/pdf/10.1175/1520-0469%281963%29020%3C0130%3ADNF%3E2.0.CO%3B2.

[28] S. Ayers, "The Application of Chaos Theory to Psychology," *Theory Psychol.*, vol. 7, no. 3, pp. 373-398, Jun. 1997, doi: 10.1177/0959354397073005.

[29] N. McBride, "Chaos theory as a model for interpreting information systems in organizations," *Inf. Syst. J.*, vol. 15, pp. 233-254, 2005, Accessed: Oct. 18, 2018. [Online]. Available: http://commonweb.unifr.ch/artsdean/pub/gestens/f/as/files/4660/35107_094822.pdf.

[30] A. Combs, M. Winkler, and C. Daley, "A Chaotic Systems Analysis of Rhythms in Feeling States," *Psychol. Rec.*, vol. 44, no. 3, pp. 359-368, 1994, doi: 10.1007/bf03395920.

[31] S. P. Reidbord and D. J. Redington, "Psychophysiological Processes During Insight-Oriented Therapy," *The Journal of Nervous and Mental Disease*,

vol. 180, no. 10. pp. 649-657, 1992, doi: 10.1097/00005053-199210000-00007.

[32] B. Ştefan Radu, M. Liviu, and G. Cristian, "Aspects Regarding the Positive and Negative Sides of Chaos Applied to the Management Science in Projects of Organizational Change," *Procedia Econ. Financ.*, vol. 15, pp. 1543-1548, 2014, doi: 10.1016/S2212-5671(14)00623-6.

[33] I. Klioutchnikov, M. Sigova, and N. Beizerov, "Chaos Theory in Finance," *Procedia Comput. Sci.*, vol. 119, pp. 368-375, 2017, doi: 10.1016/j.procs.2017.11.196.

[34] J. Sauermann, "On the instability of majority decision-making: testing the implications of the 'chaos theorems' in a laboratory experiment," *Theory Decis.*, vol. 88, no., pp. 505-526, 2020, doi: 10.1007/s11238-019-09741-4.

[35] C. Brod, *Technostress: The Human Cost of the Computer Revolution*, First Prin. Addison-Wesley Publishing Company, 1984.

[36] S. C. Srivastava, S. Chandra, and A. Shirish, "Technostress creators and job outcomes: theorizing the moderating influence of personality traits," *Inf. Syst. J.*, vol. 25, no. 4, pp. 355-401, Jul. 2015, doi: 10.1111/isj.12067.

[37] J.-F. Stich, M. Tarafdar, C. L. Cooper, and P. Stacey, "Workplace stress from actual and desired computer-mediated communication use: a multi-method study," *New Technol. Work Employ.*, vol. 32, no. 1, pp. 84-100, Mar. 2017, doi: 10.1111/ntwe.12079.

[38] R. Ayyagari, V. Grover, and R. Purvis, "Technostress: Technological Antecedents and Implications," *MIS Q.*, vol. 35, no. 4, pp. 831-858, 2011, doi: 10.2307/41409963.

[39] J. D'Arcy, A. Gupta, M. Tarafdar, and O. Turel, "Reflecting on the 'Dark Side' of Information Technology Use," *Commun. Assoc. Inf. Syst.*, vol. 35, 2014.

[40] I. Hwang and O. Cha, "Examining technostress creators and role stress as potential threats to employees' information security compliance," *Comput. Human Behav.*, vol. 81, pp. 282-293, Apr. 2018, doi: 10.1016/j.chb.2017.12.022.

[41] G. Nimrod, "Technostress: measuring a new threat to well-being in later life," *Aging Ment. Heal.*, vol. 22, no. 8, pp. 1080-1087, 2018, doi: 10.1080/13607863.2017.1334037.

[42] H. Benbya, N. Nan, H. Tanriverdi, and Y. Yoo, "Complexity and information systems research in the emerging digital world," *MIS Q.*, vol. 44, no. 1, pp. 1-17, 2020, doi: 10.25300/MISQ/2020/13304.

[43] I. Prigogine and I. Stengers, *Order out of chaos: man's new dialogue with nature*. Bantam, 1984.

[44] M. Romagna and N. J. van den Hout, "Hacktivism and Website Defacement: Motivations, Capabilities and Potential Threats," in *27th Virus Bulletin International Conference*, 2017, no. October, p. 10, [Online]. Available: https://www.researchgate.net/publication/320330579_Hacktivism_and_Website_Defacement_Motivations_Capabilities_and_Potential_Threats.

[45] B. Bellman, "Defacement: Public Secrecy and the Labor of the Negative," *Am. Anthropol.*, vol. 103, no. 3, pp. 878-879, 2008.

[46] P. Thompson, G. Cybenko, and A. Giani, "Cognitive hacking," in *Economics of Information Security*, 2004, pp. 255-287.

[47] W. J. T. Mitchell, *What do pictures want?* Vol. 1. The University of Chicago Press, 2005.

[48] J. A. González Zarandona, C. Albarrán-Torres, and B. Isakhan, "Digitally Mediated Iconoclasm: the Islamic State and the war on cultural heritage," *Int. J. Herit. Stud.*, vol. 24, no. 6, pp. 649-671, 2018, doi: 10.1080/13527258.2017.1413675.

[49] R. Clay, *Iconoclasm in Revolutionary Paris: The Transformation of Signs*. Voltaire Foundation in association with Liverpool University Press, 2012.

[50] M. Clapperton, D. M. Jones, and M. L. R. Smith, "Iconoclasm and strategic thought: Islamic State and cultural heritage in Iraq and Syria," *Int. Aff.*, vol. 93, no. 5, pp. 1205-1231, 2017, doi: 10.1093/ia/iix168.

[51] C. A. Smith, "Weaponized iconoclasm in Internet memes featuring the expression 'Fake News,'" *Discourse Commun.*, vol. 13, no. 3, pp. 303-319, 2019, doi: 10.1177/1750481319835639.

[52] C. Kim, S. G. Lee, and M. Kang, "I became an attractive person in the virtual world: Users' identification with virtual communities and avatars," *Comput. Human Behav.*, vol. 28, no. 5, pp. 1663-1669, 2012, doi: 10.1016/j.chb.2012.04.004.

[53] S. Boccaletti, C. Grebogi, Y.-C. Lai, H. Mancini, and D. Maza, "The control of chaos: theory and applications," *Phys. Rep.*, vol. 329, no. 3, pp. 103-197, May 2000, doi: 10.1016/S0370-1573(99)00096-4.

[54] B. Henning and P. Vorderer, "Psychological escapism: Predicting the amount of television viewing by the need for cognition," *J. Commun.*, vol. 51, no. 1, pp. 100-120, 2001, doi: 10.1093/joc/51.1.100.

[55] E. C. Hirschman, "Aesthetics, Ideologies and the Limits of the Marketing Concept," *J. Mark.*, vol. 47, no. 3, p. 45, 1983, doi: 10.2307/1251196.

[56] J. W. Moon and Y. G. Kim, "Extending the TAM for a World-Wide-Web context," *Inf. Manag.*, vol. 38, no. 4, pp. 217-230, 2001, doi: 10.1016/S0378-7206(00)00061-6.

[57] A. Chaudhuri and M. B. Holbrook, "The chain of effects from brand trust and brand affect to brand performance: The role of brand loyalty," *J. Mark.*, vol. 65, no. 2, pp. 81-93, 2001, doi: 10.1509/jmkg.65.2.81.18255.

[58] E. Greussing and H. G. Boomgaarden, "Simply Bells and Whistles?: Cognitive Effects of Visual Aesthetics in Digital Longforms," *Digit. Journal.*, vol. 7, no. 2, pp. 273-293, 2019, doi: 10.1080/21670811.2018.1488598.

[59] N. H. Lurie and C. H. Mason, "Visual representation: Implications for decision making," *J. Mark.*, vol. 71, no. 1, pp. 160-177, 2007, doi: 10.1509/jmkg.71.1.160.

[60] C. Mathwick, N. Malhotra, and E. Rigdon, "Experiential value: Conceptualization, measurement and application in the catalogue and Internet shopping environment," *J. Retail.*, vol. 77, no. 1, pp. 39-56, 2001, doi: 10.1016/S0022-4359(00)00045-2.

[61] W. J. Ladeira, W. M. Nique, D. C. Pinto, and A. Borges, "Running for pleasure or performance? How store attributes and hedonic product value influence consumer satisfaction," *Int. Rev. Retail. Distrib. Consum. Res.*, vol. 26, no. 5, pp. 502-520, 2016, doi: 10.1080/09593969.2016.1182934.

[62] B. J. Pine II, and J. H. Gilmore, *The Experience Economy*. 2011.

[63] D. V and Y. A, "Constructivism: A Paradigm for Teaching and Learning," *Arts Soc. Sci. J.*, vol. 7, no. 4, pp. 1-5, 2016, doi: 10.4172/2151-6200.1000200.

[64] T. S. Bateman and J. M. Crant, "The proactive component of organizational

behavior: A measure and correlates," *J. Organ. Behav.*, vol. 14, no. 2, pp. 103-118, Mar. 1993, doi: 10.1002/job.4030140202.

[65] S. K. Parker and C. A. Sprigg, "Minimizing strain and maximizing learning: the role of job demands, job control, and proactive personality.," *J. Appl. Psychol.*, vol. 84, no. 6, pp. 925-39, Dec. 1999, Accessed: Oct. 26, 2018. [Online]. Available: http://www.ncbi.nlm.nih.gov/pubmed/10639910.

[66] F. Maimone and M. Sinclair, "Dancing in the dark: creativity, knowledge creation and (emergent) organizational change," *J. Organ. Chang. Manag.*, vol. 27, no. 2, pp. 344-361, Apr. 2014, doi: 10.1108/JOCM-12-2012-0197.

[67] H. Sarooghi, D. Libaers, and A. Burkemper, "Examining the relationship between creativity and innovation: A meta-analysis of organizational, cultural, and environmental factors," *J. Bus. Ventur.*, vol. 30, no. 5, pp. 714-731, Sep. 2015, doi: 10.1016/j.jbusvent.2014.12.003.

[68] S. Sumiyana and S. Sriwidharmanely, "Mitigating the harmful effects of technostress: inducing chaos theory in an experimental setting," *Behav. Inf. Technol.*, pp. 1-15, Jul. 2019, doi: 10.1080/0144929X.2019.1641229.

[69] R. Agarwal and J. Prasad, "A Conceptual and Operational Definition of Personal Innovativeness in the Domain of Information Technology," *Inf. Syst. Res.*, vol. 9, no. 2, pp. 204-215, 1998, Accessed: Oct. 26, 2018. [Online]. Available: http://web.a.ebscohost.com/ehost/pdfviewer/pdfviewer?vid=1&sid=fbc78923-cd30-44a7-9a96-63a0c5d49817%40sdc-v-sessmgr06.

[70] R. Agarwal and J. Prasad, "A Field Study of the Adoption of Software Process Innovations by Information Systems Professionals," *IEEE Trans. Eng. Manag.*, vol. 47, no. 3, pp. 295-308, 2000, doi: 10.1109/17.865899.

[71] A. Beaudry and A. Pinsonneault, "Understanding user responses to information technology," *MIS Q.*, vol. 29, no. 3, pp. 493-524, 2005, [Online]. Available: http://aisel.aisnet.org/misq/vol29/iss3/7/.

[72] J. R. Edward, "The Determinants and Consequences of Coping with Stress," in *Causes, Coping and Consequences of Stress at Work*, 1988, pp. 322-263.

[73] L. J. Kirsch and L. L. Cummings, "Contextual influences on self-control of is professionals engaged in systems development," *Accounting, Manag. Inf. Technol.*, vol. 6, no. 3, pp. 191-219, Jan. 1996, doi: 10.1016/0959-8022(96)00018-5.

[74] D. R. Ilgen, *C. fisher*, and M. S. Taylor, "Consequences of Individual Feedback on behaviour," *J. Appl. Psychol.*, vol. 64, no. 4, pp. 349-371, 1979, [Online]. Available: https://www.researchgate.net/profile/Cynthia_Fisher2/publication/232557703_Consequences_of_individual_feedback_on_behavior_in_organizations/links/0deec51dca0195bc4d000000.pdf.

[75] J. Hattie and H. Timperley, "The Power of Feedback - ProQuest," *Rev. Educ. Res. Washingt.*, vol. 77, no. 1, pp. 81-112, 2007, Accessed: Oct. 18, 2018. [Online]. Available: https://search.proquest.com/docview/214113991?pq-origsite=summon.

[76] B. L. Fredrickson, "Positive emotions broaden and build," in *Advances in Experimental Social Psychology*, vol. 47, North Carolina, USA, 2013, pp. 1-54.

[77] B. L. Fredrickson, "NIH public access author manuscript: The role of positive emotions in positive psychology: The broaden-and-build

theory of positive emotions," *Am. Psychol.*, vol. 56, no. 3, pp. 218-226, 2001.

[78] C. M. Tyng, H. U. Amin, M. N. M. Saad, and A. S. Malik, "The influences of emotion on learning and memory," *Front. Psychol.*, vol. 8, no. AUG, 2017, doi: 10.3389/fpsyg.2017.01454.

[79] H. Liang, Y. Xue, A. Pinsonneault, and Y. "Andy" Wu, "What Users Do Besides Problem-Focused Coping When Facing IT Security Threats: An Emotion-Focused Coping Perspective," *MIS Q.*, vol. 43, no. 2, pp. 373-394, 2019, Accessed: May 20, 2019. [Online]. Available: https://misq.org/what-users-do-besides-problem-focused-coping-when-facing-it-security-threats-an-emotion-focused-coping-perspective.html.

[80] S. Folkman, R. S. Lazarus, R. J. Gruen, and A. DeLongis, "Appraisal, Coping, Health Status, and Psychological Symptoms," *J. Pers. Soc. Psychol.*, vol. 50, no. 3, pp. 571-579, 1986, doi: 10.1037/0022-3514.50.3.571.

[81] R. S. Lazarus, "Coping theory and research: past, present, and future.," *Psychosom. Med.*, vol. 55, no. 3, pp. 234-247, May 1993, doi: 10.1097/00006842-199305000-00002.

[82] N. Nan and Y. Lu, "Harnessing the Power of Self-organization in an Online Community during Organizational Crisi," *MIS Q.*, vol. 38, no. 4, pp. 1135-1158, 2014.

[83] H. Benbya, D. E. Leidner, and D. Preston, "MIS Quarterly Research Curation on Information System Alignment," *MIS Q.*, pp. 141-157, 2019.

[84] H. Benbya and D. Leidner, "How Allianz UK used an idea management platform to harness employee innovation," *MIS Q. Exec.*, vol. 17, no. 2, pp. 139-155, 2018.

Chapter 3

Perturbation Theory and Phase Behavior Calculations Using Equation of State Models

Vassilis Gaganis

Abstract

Equations of State (EoS) live at the heart of all thermodynamic calculations in chemical engineering applications as they allow for the determination of all related fluid properties such as vapor pressure, density, enthalpy, specific heat, and speed of sound, in an accurate and consistent way. Both macroscopic EoS models such as the classic cubic EoS models as well as models based on statistical mechanics and developed by means of perturbation theory are available. Under suitable pressure and temperature conditions, fluids of known composition may split in more than one phases, usually vapor and liquid while solids may also be present, each one exhibiting its own composition. Therefore, computational methods are utilized to calculate the number and the composition of the equilibrium phases at which a feed composition will potentially split so as to estimate their thermodynamic properties by means of the EoS. This chapter focuses on two of the most pronounced EoS models, the cubic ones and those based on statistical mechanics incorporating perturbation analysis. Subsequently, it describes the existing algorithms to solve phase behavior problems that rely on the classic rigorous thermodynamics context as well as modern trends that aim at accelerating computations.

Keywords: statistical mechanics, perturbation theory, equation of state, phase behavior, phase stability, phase split, algorithms

1. Introduction

Equations of State (EoS) have been widely used in the chemical engineering industry for the calculation of process fluids phase properties. EoS models are algebraic expressions of the form $f(p, T, v_m) = 0$ which relate molar volume v_m to pressure and temperature. Since the derivation of the ideal gas law and following the pioneering work of Van der Waals, dozens of EoS models of various complexity and thermodynamic considerations have been presented to accurately estimate thermophysical properties. Among them, basic and extended cubic equations of state, virial forms, EoS models with association terms and models based on statistical physics. Of them, the ones most widely used in the chemical engineering industry are the cubic ones [1] due to their simplicity and speed of calculations, thus minimizing the computing time required for flow simulations in processes, porous media and pipelines. Less simple but more accurate models incorporating associating theory are often used in midstream and downstream applications [2].

IntechOpen

EoS models based on the application of statistical mechanics in conjunction to perturbation theory to describe the thermodynamic behavior of substances at a microscopic level are commonly used to estimate properties of liquids [3]. This approach is based on studying the microscopic behavior of a set of molecules by considering ensembles comprising of many instances of the set. Subsequently, the system energy and eventually all thermodynamic properties of interest are obtained by treating statistically the ensemble properties. However, as the derivation of a closed form of the energy function is usually intractable, perturbation theory greatly simplifies that task. A known closed form solution for a simple reference system is firstly adopted, and the additional energy terms required to improve the simple reference system to the complex one are considered as a perturbation of the original reference system. Perturbation theory utilizes linearization to lead to approximate closed form solutions of the combined complex system.

To obtain estimates of the thermophysical properties using EoS models, it is necessary that the composition of the fluid is known and that a reliable characterization of the mixture components, by means of specific components properties values, is available. Cubic EoS models though simple they are not predictive, and the reliability of their predictions can only be ensured by "tuning" the model, i.e. varying the components properties so that the model predictions match accurately the available experimental measurements.

Once a tuned EoS model is available, properties such as density, fugacity coefficient, enthalpy, heat capacity, Joule-Thomson coefficient and speed of sound can be easily computed by simple expressions. Calculations become more complex when the phase state of a mixture is not known a priori. As an example consider a control volume, i.e. a grid block, in a flow simulation model where the pressure and saturation change of each coexisting phase at current timestep need to be determined in order to get a description of the fluid state. The pressure change in the control volume is related to mass influx and outflux through fluids density and compressibility. When the control volume content is a single-phase fluid both properties can be easily computed by means of the EoS model. However, when the content is saturated, it will split into two or more phases, each one exhibiting its own properties, thus introducing the need to identify the number and composition of the equilibrating phases, hence their density and compressibility.

In such cases, a test to determine if the fluid appears in a single or two phases needs to be run, known as stability test [4]. If the test indicates the presence of two or more phases in equilibrium, the phase split problem further needs to be solved to compute the composition and the amount of the two coexisting properties [5]. By knowing their composition, all properties of the equilibrium phases can then be computed regularly.

In this chapter, the utilization of EoS models of the cubic form and those based on perturbation theory is discussed and their application to compute fluids thermophysical properties is presented. Algorithms to run phase stability and phase split in the classic context as well as in the reduced variables one are also discussed. Additionally, the chapter discusses the recent developments in the use of soft computing techniques to accelerate the solution of the stability and phase split problems in flow simulations.

2. The PR and SRK cubic EoS models

2.1 Development of the cubic EoS models

The ideal gas law $pv_m = RT$, where the gas constant $R = k_B N_A$ is defined as the product the Boltzmann constant and the Avogadro number, only considers the elastic collision of molecules thus considering the thermodynamic behavior of the fluid as a purely kinetic process. As a result, it exhibits accurate predictions of the molar volume only when gases at pressures and temperatures close to the atmospheric ones are considered. On the other hand, the real gas law $pv_m = ZRT$ can be used to describe accurately the properties of any fluid and at any conditions provided that the appropriate value of the compressibility factor Z (also known as deviation factor in the sense that it considers the deviation of the real gas law from the ideal gas one) can be computed. Clearly, the real gas law simplifies to the ideal one by simply setting $Z = 1$.

Van der Waals was first to recognize the need to separately consider attractive and repulsive forces between the fluid molecules thus leading to the first cubic equation

$$p = RT/(v_m - b) - a/v_m^2. \tag{1}$$

Indeed, the a term in Eq. (1) can be thought of as a term accounting for the attractive forces between molecules as it reduces pressure. Parameter b accounts for the molecules volume which becomes significant at high pressures (i.e. liquid state) as $\lim_{p\to\infty} v_m = b$. Both parameters are functions of the properties of the component or mixture under consideration. Clearly, by setting both parameters to zero we revert back to the ideal gas law.

Ever since, various new cubic EoS models have been proposed with the Soave-Redlich-Kwong (SRK) and the Peng-Robinson (PR) ones [6] being by far the most commonly used ones in the chemical engineering industry. Both are pressure explicit and are defined by the following expression

$$p = \frac{RT}{v_m - b} - \frac{a}{(v_m + \delta_1 b)(v_m + \delta_2 b)}, \tag{2}$$

where the parameters values are given in **Table 1**. The temperature dependent term in that Table is given by

$$\alpha(T) = \left(1 + m\left(1 - \sqrt{T/T_c}\right)\right)^2, \tag{3}$$

where m is a function of the component acentric factor ω defined by

$$m = \begin{cases} 0.48 + 1.574\omega - 0.176\omega^2 & \text{SRK} \\ 0.37464 + 1.54226\omega - 0.26992\omega^2 & \text{PR}, \omega \le 0.49\,. \\ 0.3796 + 1.485\omega - 0.1644\omega^2 + 0.01667\omega^3 & \text{PR}, \omega > 0.49 \end{cases} \tag{4}$$

The required properties of pure components can be found in any standard petroleum thermodynamics textbook [7]. When pseudo-components are used to describe the fluid composition, such as such as pseudo-C_8 and pseudo-C_{11} in petroleum mixtures, average values can also be obtained from the literature. Custom

EoS	δ_1	δ_2	a	b
SRK	0	1	$0.42747\alpha R^2T_c^2/p_c$	$0.08664RT_c/p_c$
PR	$1+\sqrt{2}$	$1-\sqrt{2}$	$0.45724\alpha R^2T_c^2/p_c$	$0.07780RT_c/p_c$

Table 1.
Cubic EoS models constants.

pseudo-components such as petroleum mixtures heavy end need to be treated by means of suitable correlations which utilize molar mass and density to provide estimates of the critical properties and the accentric factor or other required properties [8]. When it comes to mixtures, parameters mixing rules need to be utilized to estimate a and b. For a mixture of known composition z_i, they are given by

$$\begin{aligned} a_{mix} &= \sum_{i=1}^{n}\sum_{j=1}^{n} z_i z_j \sqrt{a_i a_j}(1-k_{ij}) \\ b_{mix} &= \sum_{i=1}^{n} z_i b_i. \end{aligned} \tag{5}$$

The Binary Interaction Parameters (BIP) k_{ij} account for the interaction between different constituents and are usually initialized either to zero or by the Prausnitz [9] rule

$$k_{ij} = 1 - \left(\frac{2 v_{c_i}^{1/6} v_{c_j}^{1/6}}{v_{c_i}^{1/3} + v_{c_j}^{1/3}} \right)^{\theta}, \tag{6}$$

where the critical molar volume is obtained by solving the EoS at critical conditions

$$v_c = Z_c R T_c / p_c, \tag{7}$$

and the critical value Z_c of the compressibility factor for the PR EoS equals to 0.3074. Parameter θ is user dependent and is usually set to 1.2. Note Eq. (6) is only used to determine BIPs between hydrocarbon components. BIPs between nonhydrocarbons or between hydrocarbon and nonhydrocarbon components are taken from Tables [6].

Once all parameters have been estimated for a mixture of known composition at fixed pressure and temperature, the EoS can be solved for volume. Usually a dimensionless form that can be solved for $Z = pv_m/RT$ rather than for v_m is preferred

$$\begin{aligned} & Z^3 + ((\delta_1 + \delta_2 - 1)B - 1)Z^2 + (A + \delta_1\delta_2 B - (\delta_1 + \delta_2)B(B+1))Z \\ & \quad - \left(AB + \delta_1\delta_2 B^2(B+1)\right) \\ & \quad = 0, \end{aligned} \tag{8}$$

where the dimensionless EoS constants are given by

$$A = a_{mix} p/(RT)^2, \quad B = b_{mix} p/(RT). \tag{9}$$

2.2 Use of the cubic EoS models

As soon as the EoS constants have been defined, the compressibility factor Z can be obtained by solving the cubic polynomial Eq. (8) [10]. When more than one real positive roots are obtained, the smallest one is selected when the fluid is a liquid whereas the largest one is used for a gas. Molar volume and density can be easily computed by

$$v_m = ZRT/p, \quad \rho = pM/ZRT, \tag{10}$$

where M denotes the fluid molar mass. Components fugacity coefficients φ_i, hence fugacity $f_i = \varphi_i z_i p$, can be computed by the following expressions

$$\ln\phi_i = B_i/B(Z-1) - \ln(Z-B) + A/((\delta_1-\delta_2)B)\left(1/A\{\partial A/\partial \mathbf{z}\}_i - B_i/B\right)\times$$
$$\times \ln(Z+\delta_1 B)/(Z+\delta_2 B), \tag{11}$$

where $A_i = a_i p/(RT)^2$, $B_i = b_i p/(RT)$ and $\{\partial A/\partial \mathbf{z}\}_i = \sum_{j=1}^{n} z_j\left(1-k_{ij}\right)\sqrt{A_i A_j}$. Derivative properties such as the Joule-Thomson coefficient μ_{JT} can be computed by differentiating the EoS and incorporating the derivatives in the rigorous thermodynamic definitions of the properties. For example,

$$\mu_{JT} = \frac{v_m}{c_p}\left(\frac{T}{v_m}\frac{\partial v_m}{\partial T}\bigg|_p - 1\right). \tag{12}$$

2.3 Volume translation

Cubic EoS models are notoriously known for their deficiency in estimating liquid density. A simple modification, known as volume shifting or volume translation, originally proposed by Peneloux [11], can greatly improve the capabilities of cubic EoS. The idea lies in "shifting" the predicted phase molar volumes v_m^{EoS} by some amount that depends on the fluid composition and its components properties. More specifically, the shifted volume is given by

$$v_m = v_m^{EoS} - \sum_{i=1}^{n} z_i c_i. \tag{13}$$

Parameters c_i are component specific and they are usually given as functions of the covolume parameters b_i, that is

$$s_i = c_i b_i, \tag{14}$$

where values of s_i for common pure components are available in Tables [6].

It should be noted that "shifting" (or "translating") the volume also affects the Z factor which needs to be updated to ensure calculations consistency

$$Z = Z^{EoS} - p/RT\sum_{i=1}^{n} z_i c_i. \tag{15}$$

It can be shown that when applying volume translation to two phases that equilibrate, the fugacities of the components do change but they do in the same amount so that they remain equal, thus not disturbing the equilibrium. As a result, volume translation does not affect phase compositions in flash calculations or saturation conditions but only phase density.

3. EoS models in the thermodynamic perturbation theory context

Unlike macroscopic EoS models such as those described in the previous section, major efforts have been oriented toward the development of microscopic approaches based on statistical mechanics where the individual behavior of each particle in a fluid substance is considered. The repulsive and attractive forces are handled separately and combined to provide a description of the thermodynamic properties of fluids through methods based on statistical physics.

The basic idea is to study the microscopic behavior of a set of molecules by considering many instances of the set, each one corresponding to one possible state. This ensemble is described through the statistical properties averaged over all possible states. The basic components for this task is the pair potential function $u(r)$ and the pair correlation function $g(r)$ respectively, both functions of the distance r away from the center of some molecule. By defining them one can generate expressions to compute the system free energy and eventually all thermodynamic properties of interest [3].

Arriving to the energy expression while starting from $u(r)$ and $g(r)$ is a very complex task from the mathematical treatment point of view. Complex expressions of the two functions might correspond to more accurate description of the molecules dynamics but they also lead to intractable mathematical expressions. For this task perturbation theory has greatly enhanced the derivation of EoS models by firstly utilizing known closed form solutions for simple reference functions. Subsequently, the small changes between the accurate $u(r)$ and $g(r)$ functions and the reference ones are treated in a very elegant way by means of perturbation theory thus leading to approximate closed form solutions for complex pair functions [12].

3.1 The correlation function formalism to derive EoS models

Consider a thermodynamically large system comprising of a fixed number of molecules, at fixed temperature and volume, which is allowed to exchange heat with the environment. Subsequently, consider a collection of many such probable systems forming what is known as a *canonical ensemble*. The aggregate thermodynamic properties of such systems can be described as functions of the statistic properties of the ensemble. For this task the *canonical partition function* is defined by

$$Q = \sum \exp\left(-\beta E_i\right), \tag{16}$$

where E_i corresponds to the energy of each possible microstate, $\beta = 1/k_B T$ is the thermodynamic beta and k_B is the Boltzmann constant. Note that Q is dimensionless and as it will be shown later it relates macroscopic thermodynamic properties of the system to the energy of the microscopic systems forming the ensemble.

For a system comprising of N identical molecules the partition function $Q_N(V,T)$ at given volume and temperature is given by [3].

$$Q_N(V,T) = \frac{Z_N(V,T)}{N!\Lambda^3}, \tag{17}$$

where

$$Z_N(V,T) = \int_N \exp\left(-\beta U_N\left(\mathbf{r}^N\right)\right) d\mathbf{r}^N, \quad \Lambda = \sqrt{\frac{h^2}{2\pi m k_B T}}, \tag{18}$$

and $Z_N(V,T)$ is known as the configuration integral. It is easy to show that if the system potential energy U_N is assumed to be zero then the configuration integral Z_N simplifies to the system volume and the application of the related partition function leads simply to the ideal gas law. On the other hand, when $U_N \neq 0$, it is often represented by a sum of pair-wise potentials, i.e.

$$U\left(\mathbf{r}^N\right) = \sum_i \sum_{j>i} u\left(r_{ij}\right). \tag{19}$$

Various pair potential models $u(r)$ have been presented with the hard-sphere, the square well and the Lennard-Jones being the most pronounced ones [12]. The hard sphere model assumes that the particles are perfect spheres of diameter σ, the potential at distances less than the sphere diameter is equal to infinite (hence the "hard" sphere) and zero beyond that. Therefore, $u(r)$ and the corresponding Boltzmann factor are given by

$$u(r) = \begin{cases} \infty & r < \sigma \\ 0 & r > \sigma \end{cases}, \quad \exp\left(-\beta u(r)\right) = \begin{cases} 0 & r < \sigma \\ 1 & r > \sigma \end{cases}. \tag{20}$$

The square-well model [13] further allows for a negative value at some distance beyond the hard sphere diameter:

$$u(r) = \begin{cases} \infty & r < \sigma \\ -\varepsilon & \sigma < r < \gamma\sigma \\ 0 & r > \gamma\sigma \end{cases}. \tag{21}$$

The Lennard-Jones model [14] offers the advantage of being defined by a continuous function of the distance r:

$$u(r) = 4\varepsilon\left[(l/r)^{12} - (l/r)^{6}\right]. \tag{22}$$

In the equations above σ is the sphere diameter, parameter γ is used to scale the well width, l is the length parameter and ε is the energy parameter. A detailed description on how to use the hard-sphere model pair potential function to develop an EoS model is given in Section 3.3.

3.2 Derivation of fluid properties for specific pair functions

By selecting the pair potential model $u(r)$ and incorporating it the configuration integral Z_N and eventually to the canonical partition function expression, internal energy can be obtained by noting that

$$E = k_B T^2 \frac{\partial}{\partial T} Q_N(V, N). \tag{23}$$

By utilizing the pair-wise potential energy model of Eq. (19) it can be shown that

$$E = \frac{3}{2} N k_B T + 2\pi\rho N \int_0^\infty u(r) g(r) r^2 dr. \tag{24}$$

Therefore, internal energy can be obtained as a function of the particle properties $u(r)$ and $g(r)$. Clearly, the first term corresponds to the kinetic energy of the particles, that is the ideal gas contribution of the system.

Using similar arguments, pressure can be obtained by as the volume derivative of the configuration integral Z_N, that is

$$p = k_B T \frac{\partial}{\partial V} Z_N(V, N) = k_B T \frac{N}{V} - \frac{2\pi}{3}\rho^2 \int_0^\infty r^3 \frac{\partial u(r)}{\partial r} g(r) dr. \tag{25}$$

Again, the first term corresponds to the ideal gas pressure term. The system Helmholtz energy is defined by

$$H = E - TS = -k_B T \ln Q_N(V, T). \tag{26}$$

The chemical potential corresponds to the energy required to add one more particle in the collection it is given by

$$\mu = H(V, T, N) - H(V, T, N-1) = \left.\frac{\partial H}{\partial N}\right|_{V,T}, \tag{27}$$

and eventually

$$\mu = k_B T \ln \rho \Lambda^3 + 4\pi\rho \int_0^1 \int_0^\infty r^2 u(r) g(r, \lambda) dr d\lambda. \tag{28}$$

Given the expression above, entropy can be obtained by

$$S = \frac{E - H}{T}. \tag{29}$$

3.3 The hard-sphere model

The generic fluid properties expressions derived in the previous section are now applied to the hard sphere model for the pair potential energy. By noting that the derivative of the Boltzmann factor of the hard-sphere model is simply the Dirac delta function [15] and replacing it to the generic properties' expressions of the previous paragraph, it follows for pressure that

$$p = p^{IG} + p^{EX} = \rho k_B T + \rho k_B T \frac{4\eta - 2\eta^2}{(1-\eta)^3} = \rho k_B T \frac{1 + \eta + \eta^2 + \eta^3}{(1-\eta)^3}, \tag{30}$$

where the pressure is now split into the ideal gas and the excess part and the packing function η which corresponds to the ratio of the particles volume over the total one is given by

$$\eta = \frac{1}{V} N \frac{4\pi}{3} \left(\frac{\sigma}{2}\right)^3 = \frac{\pi}{6} \rho \sigma^3. \tag{31}$$

The expressions for the other properties of interest are obtained similarly and they are given by

$$\begin{aligned} H &= H^{IG} + H^{EX} = N k_B T \left(\ln \rho \Lambda^3 - 1\right) + N k_B T \frac{4\eta - 3\eta^2}{(1-\eta)^2} \\ S &= S^{IG} + S^{EX} = -N k_B \left(\ln \rho \Lambda^3 - \frac{5}{2}\right) - N k_B \frac{4\eta - 3\eta^2}{(1-\eta)^2} \\ \mu &= \mu^{IG} + \mu^{EX} = k_B T \left(\ln \rho \Lambda^3 - 1\right) + k_B T \frac{1 + 5\eta - 6\eta^2 + 2\eta^3}{(1-\eta)^3}. \end{aligned} \tag{32}$$

3.4 Thermodynamic perturbation theory

Although the hard-sphere pair potential model allows for an explicit calculation of thermodynamic properties of interest, its results are not that accurate mostly due to the inherent simplicity of the hard-sphere model itself. Nevertheless, many researchers have pointed out that the comparison between the experimental structure factor and the one obtained computationally from the hard-sphere model indicates that the two curves are quite close to each other. To get a better match a more complex pair potential model could be sought which, however, would inevitably lead to mathematically intractable expressions of the properties. Alternatively, perturbation methods can be applied to the original simple hard-sphere model to add thermodynamic complexity under controlled extra computational burden.

The idea, firstly presented by Zwanzig [16], is to divide the total potential energy into two terms, U_0 and U_p respectively, where the first term corresponds to a reference system and the second one corresponds to the perturbation, which needs to be significantly smaller than the reference one for the perturbation method to be applied successfully. The total energy is then given by

$$U = U_0 + \lambda U_p. \tag{33}$$

The perturbation parameter λ allows for various mixtures of U_0 and U_p whereas the original fluid energy is obtained for $\lambda = 1$. By replacing that expression to the configuration integral we obtain

$$Z_N(V,T) = Z_N^{(0)}(V,T)\langle \exp\left(-\beta\lambda U_p\right)\rangle_0, \tag{34}$$

where the $\langle . \rangle$ operator denotes the statistical average of the reference system. Replacing Eq. (34) to the expression for the Helmholtz energy we obtain

$$-\beta H = \ln \frac{Z_N(V,T)}{N!\Lambda^{3N}} + \ln\langle \exp\left(-\beta\lambda U_1\right)\rangle_0 = -\beta H_0 - \beta H_p, \tag{35}$$

where the first term corresponds to a multiple of the Helmholtz energy of the reference system and the second term accounts for the energy of the perturbation. By combining the Taylor expansion forms of the exponential term and of the logarithmic term we obtain

$$-\beta H_p = \ln\left\langle \exp\left(-\beta\lambda U_p\right)\right\rangle_0 = -(\lambda\beta)c_1 + (\lambda\beta)^2 c_2 - (\lambda\beta)^3 c_3 + \dots, \tag{36}$$

where

$$\begin{aligned} c_1 &= \langle U_p \rangle_0 \\ c_2 &= \frac{1}{2!}\left(\left\langle U_p^2\right\rangle_0 - \langle U_p\rangle_0^2\right) \\ c_3 &= \frac{1}{3!}\left(\left\langle U_p^3\right\rangle_0 - 3\langle U_p\rangle_0\left\langle U_p^2\right\rangle_0 + 2\langle U_p\rangle_0^3\right). \end{aligned} \tag{37}$$

The treatment above has allowed the energy to be described by simpler expressions of the perturbation energy term based on the c_1, c_2, c_3 parameters. Therefore, to get the full energy expression one needs to choose the U_p model, compute the values of the c_1, c_2, c_3 parameters and replace then in Eq. (36) while setting $\lambda = 1$. All thermodynamic properties of interest can then be computed as functions of the energy as shown in Section 3.2.

The beauty of the perturbation theory is that although the calculation of the c_1, c_2, c_3 parameters, which have been introduced by the application of perturbation theory and the assumption of a simple reference system, is not an easy task it still is significantly easier than replacing a complex pair potential and pair correlation function and running mathematical operations in Eq. (18).

As an example application of perturbation theory in statistical mechanics based thermodynamics, consider the use of the model obtained by perturbation theory to generate the Van der Waals EoS, this time from a statistical mechanics point of view instead from the classic macroscopic one. Firstly, let us state the following assumptions:

1. The potential energy consists of the sum of all pair potentials: $U_p = \sum_i \sum_{j>i} u_p(r_{ij})$.

2. The pair potentials are equal between any pair of molecules: $U_p = \sum_i \sum_{j>i} u_p(r_{ij}) = \frac{N(N-1)}{2} u_p(r_{12})$.

By introducing those assumptions to Eq. (37) the calculation of coefficient c_1 simplifies to

$$c_1 = \frac{\rho^2}{2} \int d\mathbf{r} \int u_p(\mathbf{r}) g_o(\mathbf{r}) d\mathbf{r}. \tag{38}$$

To proceed we further need to introduce the following assumptions:

1. The reference system to describe U_0 is the hard-sphere one

2. The particles are uniformly distributed which implies that the pair correlation function $g_0(r)$ is equal to one at any distance beyond the limits of the particle and equal to zero inside that

3. The free fluid volume is $V - Nb$ where b is the particle volume that equals to $b = 2/3\pi\sigma^3$

and we end up with

$$c_1 = -a\rho N, \quad a = -2\pi \int_\sigma^\infty u_p(r) r^2 dr. \tag{39}$$

Finally, by utilizing first order approximation only (up to c_1), replacing c_1 in the free energy Eq. (36) and differentiating over volume to obtain pressure (i.e. $p = -\partial H/\partial V|_T$) the well known Van der Waals equation is obtained

$$p = \frac{N k_B T}{V - Nb} - a \frac{N^2}{V^2}. \tag{40}$$

Interestingly, the perturbed energy term u_p has not been defined explicitly but it has been incorporated into the EoS a parameter. From a perturbation theory point of view, the accuracy of the Van der Waals equation of state can be improved by further considering the c_2 term, the calculation of which, however, is quite more complicated.

4. Conventional phase behavior calculations

4.1 The stability test

The question answered by a stability test is whether a mixture of given composition, at given pressure and temperature, will appear as a single phase or as a multi-phase one. Clearly, the question can be answered if the bubble point/upper dew point pressure and the lower dew point pressure (if any) of the mixture at operating temperature is known. Any fluid above its bubble point pressure will appear as single-phase liquid whereas when above its upper dew point or below its lower dew point pressure it will appear as single-phase gas.

As the saturation pressure calculation is very costly, a brilliant approach by Michelsen [4] is most preferably used. The idea lies in the fact that if a mixture is unstable, i.e. if it splits into two or more phases when in equilibrium, there exists at least one composition which when forms a second phase in an infinitesimal quantity leads to a reduction of the system's Gibbs energy. Therefore, one should try a bubble/drop of any possible composition, consider that as a second phase that coexists with the original fluid and examine whether the system Gibbs energy is reduced compared to that of the original single phase fluid. To avoid looking over all possible compositions, Michelsen suggested that one should only look for compositions that minimize the mixture's Gibbs energy rather than simply reduce it. If all minima lie above the single-phase fluid Gibbs energy, then there is no composition that allows for an energy reduction, hence the fluid is single phase, and otherwise it lies in two-phase equilibrium. The Gibbs energy difference, the sign of the minimum of which is used to determine the fluid phase state, is referred to as the Tangent Plane Distance (TPD).

Locating the minima of the TPD is not an easy task as any optimization algorithm may be trapped in a local positive rather than the global negative minimum, thus leading to wrong conclusion about the number of phases present. Additionally, the stability problem has a natural "trivial solution", the one corresponding to a second phase composition same to that of the original fluid. This solution leads to a zero TPD value and it may attract any optimization algorithm, thus misleading the stability algorithm away from the true TPD minimum.

To overcome those issues two approaches can be envisaged. Firstly, one might use global minimization algorithms which ensure that the minimum found is the global one [17]. Such algorithms take significant time to run hence they can only be applied to single calculations rather than batch ones, as is the case in fluid flow simulation. The second approach considers the repeated run of simple optimization algorithms, each time with appropriate initial values so that the global minimum will be located by at least one of those tries.

Based on the above observations, Michelsen [4] presented an algorithm which constitutes the standard approach to treat phase stability. To simplify calculations, it is recommended to optimize TPD by varying the equilibrium coefficients $k_i = y_i/z_i$, also known as distribution coefficients, rather than the bubble composition y_i itself. The algorithm is as follows

1. Compute fugacity of each component of the feed $f_i^{(z)}$ using the EoS model

2. Initialize k_i using Wilson's correlation [18]

3. Assume feed is a liquid and look for a bubble, i.e. compute $Y_i = k_i z_i$

4. Compute trial bubble composition sum $S_V = \sum Y_i$

5. Normalize composition $y_i = Y_i/S_V$ and compute its fugacity $f_i^{(y)}$

6. Compute correction factor $R_i = 1/S_V f_i^{(z)}/f_i^{(y)}$

7. Check for convergence by evaluating $\sum (R_i - 1)^2 < \varepsilon$

8. If convergence has not been achieved, update the equilibrium coefficients by applying $k_i \leftarrow k_i R_i$

9. After convergence has been achieved check if the algorithm has arrived at a trivial solution by evaluating $\sum (\ln k_i)^2 < \delta$

The algorithm needs to be repeated, this time by assuming that the feed is a gas and the second phase is a drop. In that case, the algorithm is as follows

1. Assume feed is a liquid and look for a drop, i.e. compute $X_i = z_i / k_i$

2. Compute drop composition sum $S_L = \sum X_i$

3. Normalize composition $x_i = X_i / S_L$ and compute its fugacity $f_i^{(x)}$

4. Compute correction factor $R_i = 1/S_L f_i^{(x)} / f_i^{(z)}$

5. Check for convergence by evaluating $\sum (R_i - 1)^2 < \varepsilon$

6. If convergence has not been achieved, update the equilibrium coefficients by applying $k_i \leftarrow k_i R_i$

7. After convergence has been achieved check if the algorithm has converged to a trivial solution by evaluating $\sum (\ln k_i)^2 < \delta$

As soon as both calculations have been completed, **Table 2** can be used to reckon on the phase state.

The algorithm described above is known as the "two-sided" stability test as the trial phase is tested both from the bubble as well as from the drop side. The bubble test converges to nontrivial negative solutions only when the test pressure and temperature conditions lie within the phase envelope in the range where the feed is predominantly liquid. Similarly, the drop test converges to nontrivial negative solutions only when the feed is predominantly gas. The two ranges overlap in a region known as "the spinodal" [19] where both tests converge to solutions with a negative TPD value indicating instability of the feed (**Figure 1**).

Vapor phase test	Liquid phase test	Result
Trivial solution	Trivial solution	Stable
$S_V \leq 1$	Trivial solution	Stable
Trivial solution	$S_L \leq 1$	Stable
$S_V \leq 1$	$S_L \leq 1$	Stable
$S_V > 1$	Trivial solution	Unstable
Trivial solution	$S_L > 1$	Unstable
$S_V > 1$	$S_L > 1$	Unstable
$S_V > 1$	$S_L \leq 1$	Unstable
$S_V \leq 1$	$S_L > 1$	Unstable

Table 2.
Stability test result selection.

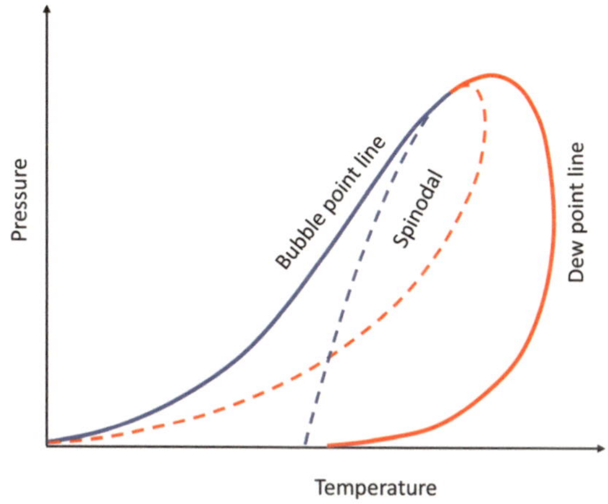

Figure 1.
The spinodal.

4.2 The phase split

Once the stability test has indicated that the feed is split into two or more phases, a phase split algorithm, also known as flash, needs to be run to determine the composition and relative amount of each phase present in equilibrium. For the simple case of vapor-liquid equilibrium (VLE) which is most commonly encountered in petroleum engineering applications, the phase split algorithm will provide the compositions of the gas and liquid phase, y_i and x_i respectively, as well as the vapor phase molar fraction β. At equilibrium, the two phases should satisfy two conditions, namely the mass balance and the minimization of the system Gibbs energy. The first condition simply requires that the mass of each component in the feed should equal to sum of their mass in the resulting two phases in equilibrium, i.e.

$$z_i = (1-\beta)x_i + \beta y_i \quad i = 1, \dots, n. \tag{41}$$

The second condition additionally requires the phase compositions to be so that the two-phase system's Gibbs energy, defined by

$$G = (1-\beta)\sum_{i=1}^{n} x_i \ln f_i^{(x)} + \beta \sum_{i=1}^{n} y_i \ln f_i^{(y)}, \tag{42}$$

is at its minimum. It is easy to show that setting the Gibbs energy gradient equal to zero, an equivalent condition is obtained which requires that the fugacity of each component in the vapor phase is equal to its fugacity in the liquid phase, i.e.

$$f_i^{(x)} - f_i^{(y)} = 0 \Rightarrow \phi_i^{(y)} y_i p - \phi_i^{(x)} x_i p = 0 \Rightarrow \frac{\phi_i^{(x)}}{\phi_i^{(y)}} = \frac{y_i}{x_i} = k_i, \quad i = 1, \dots, n. \tag{43}$$

Finally, we need to ensure that the composition of each equilibrium phase is consistent by summing up to unity. Equivalently, we may require that

$$\sum_{i=1}^{n} (x_i - y_i) = 0. \tag{44}$$

Summarizing, the solution of the flash problem can be seen as the solution of a system of $2n + 1$ equations, that is Eq. (41), (43) and (44), in $2n + 1$ unknowns, i.e. y_i, x_i and β.

Note that the mass balance equations are linear in the phase compositions, hence they can be solved and replaced in Eq. (44) thus allowing the flash problem to be reformulated in terms of the k-values and the molar fraction β. Indeed, by incorporating Eq. (41) and the k-values definition in Eq. (43) to Eq. (44), the famous Rachford-Rice equation is obtained

$$r(\beta) = \sum_{i=1}^{n} \frac{z_i}{\beta - \overline{\beta}_i} = 0, \tag{45}$$

where $\overline{\beta}_i = 1/(1 - k_i)$, which can be solved for the molar fraction β. Given β and the k-values, the equilibrium phase compositions can then be obtained by

$$x_i = \frac{1}{k_i - 1} \frac{z_i}{\beta - \overline{\beta}_i}, \quad y_i = k_i x_i, \quad i = 1, \ldots, n \tag{46}$$

Therefore, the phase split problem can be treated as the solution of a system of $n + 1$ equations, that is Eq. (43) and (45), in $n + 1$ unknowns, i.e. k_i and β. Of course, if the k-values are known by any means, the problem simplifies to the solution of the Rachford-Rice Eq. (45) to compute β and phase compositions are obtained from Eq. (46).

From Eq. (45) it can be seen that the Rachford-Rice equation is a monotonically decreasing one, as its derivative is always negative, and that it is nonlinear in the molar fraction. In fact, as shown in example **Figure 2**, it is a sum of many decreasing hyperbolas each one defined by its own asymptote $\overline{\beta}_i$, hence it comprises of $n + 1$ branches and exhibits $n - 1$ distinct roots. The only physically sound one is bounded in the [0, 1] range and it can be proved that the asymptotes which enclose that range are the ones corresponding to the maximum and to the minimum k-values, i.e. $[\overline{\beta}_{min} = 1/(1 - k_{max}), \overline{\beta}_{max} = 1/(1 - k_{min})]$. Beyond the obvious option of using the Newton-Raphson method to find the root, various alternative methods have been presented taking advantage of its special form to ensure safe and rapid convergence to the desired root [20, 21].

Alternatively, the phase split problem can be treated as a constrained optimization problem where the system Gibbs energy in Eq. (42) needs to be minimized by varying k_i under the mass balance constraint in Eq. (45).

4.2.1 Using k-values from correlations and charts

Equilibrium coefficients are functions of pressure, temperature and composition. However, at low pressures and temperatures, such as those prevailing at

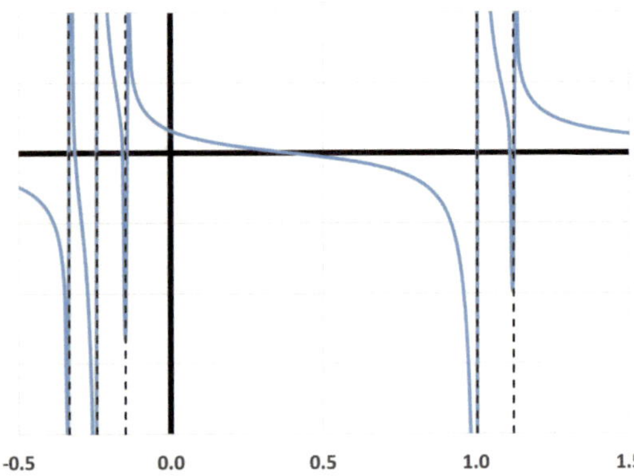

Figure 2.
The Rachford-Rice equation and its asymptotes.

surface separators, the dependency on composition is very loose, thus allowing for the derivation of k-values correlations which only utilize pressure and temperature such as the one by Wilson [18].

$$k_i(p, T) = \frac{\exp\left(5.37(1+\omega_i)(1-T_{c_i}/T)\right)}{p/p_{c_i}}. \tag{47}$$

Similar correlations by Standing [22] and Whitson and Torp [23] have also been presented. An alternative approach is based on the utilization of charts which provide k-values at various pressures and temperatures. The generation of those charts is based on the observation that at high pressures k-values approach unity. In fact, there exists a composition dependent pressure value, known as the convergence pressure p_k, at which all k-values become equal to unity. Charts for various convergence pressure values and system temperatures provide plots of the k-values as functions of pressure [24]. To utilize them in flash calculations, the user needs to determine the convergence pressure by means of any of the available methods [23, 25, 26] and select the appropriate chart where from the prevailing k-values can be obtained.

The solution algorithm is as follows

1. Estimate convergence pressure p_k

2. Get k_i from convergence pressure-based correlations or Tables

3. Solve the Rachford-Rice equation (Eq. (45)) for the vapor phase molar fraction.

4. Compute phase compositions using Eq. (46).

The Rachford-Rice equation needs to be solved by means of any iterative function-solving method such as the Newton-Raphson one and the molar fraction update is given by

$$\beta \leftarrow \beta - \left.\frac{dr}{d\beta}\right|_{\beta} r(\beta), \tag{48}$$

where

$$\frac{dr}{d\beta} = -\sum_{i=1}^{n} \frac{z_i}{\left(\beta - \overline{\beta}_i\right)^2}. \tag{49}$$

4.2.2 Using composition dependent k-values from an EoS model

When an EoS model is available, components fugacity f_i, hence fugacity coefficients φ_i and k-values $k_i = \varphi_i^{(x)}/\varphi_i^{(y)}$ can be accurately computed rather than been read from charts. Apart from the nonlinearity of the Rachford-Rice equation (Eq. (45)), the complex formulae (Eq. (11)) relating phase composition to fugacity through the EoS model introduces additional nonlinearity to the calculation of the k-values thus imposing the need for iterative solution methods.

Computations may involve any one of the three methods available, i.e. Successive Substitution (SS), numerical solution of the systems of equations in Eq. (43) and (45) by means of the Newton Raphson method or direct minimization of the system Gibbs energy in Eq. (42) by means of optimization algorithms.

The SS method starts with an estimation of the k-values, solves the Rachford-Rice equation for β to ensure mass balance and computes the phase composition and components fugacity. If phase fugacities are not equal, k-values are updated by the inverse fugacity coefficient ratio in Eq. (43). The algorithm is as follows

1. Initialize k_i

2. Solve the Rachford-Rice equation (Eq. (45)) for the molar fraction β

3. Compute phase compositions using Eq. (46)

4. Solve the cubic polynomial of each phase (Eq. (8)) and compute components fugacity (Eq. (11))

5. Check for convergence by evaluating $\sum \left(ln\, f_i^{(y)} / f_i^{(x)} \right)^2 < \varepsilon$

6. If convergence has not been achieved, update the equilibrium coefficients by applying Eq. (43), i.e. $k_i \leftarrow k_i \varphi_i^{(x)} / \varphi_i^{(y)}$, and return to step 2

As mentioned above, the flash problem is governed by $n + 1$ equations in $n + 1$ unknowns. The problem can be further split to the solution of a system of n nonlinear equations (Eq. (43)) subject to one more nonlinear one (Eq. (45)). This way one needs to apply the Newton-Raphson method to solve the n nonlinear thermodynamic equilibrium equations and at each iteration compute β to ensure mass balance and composition consistency. To describe this algorithm, we define

$$g_i = f_i^{(y)} - f_i^{(x)}, \tag{50}$$

or equivalently, in a vector format:

$$\mathbf{g}(\mathbf{z}, \mathbf{k}) = \mathbf{f}^{(\mathbf{y})} - \mathbf{f}^{(\mathbf{y})}, \tag{51}$$

which needs to be driven to zero, i.e. $\mathbf{g}(\mathbf{z}, \mathbf{k}) = \mathbf{0}$, by varying k_i. The algorithm is identical to the SS one except step 6 which now reads.

6. If convergence has not been achieved, update the equilibrium coefficients by the Newton-Raphson method $\mathbf{k} \leftarrow \mathbf{k} - \mathbf{J}^{-1}\mathbf{g}(\mathbf{z}, \mathbf{k})$ and return to step 2. The nxn Jacobian matrix is defined by

$$\mathbf{J} = \frac{\partial \mathbf{g}(\mathbf{z}, \mathbf{k})}{\partial \mathbf{k}} = \left\{ \frac{\partial g_i}{\partial k_j} \right\} = \left\{ \frac{\partial f_i^{(x)}}{\partial k_j} - \frac{\partial f_i^{(y)}}{\partial k_j} \right\}. \tag{52}$$

The optimization approach uses any optimization method to minimize Gibbs energy subject to mass balance. Quasi-Newton methods such as the BFGS [27] only require computation of the Gibbs energy gradient with respect to the k-values, whereas a Newton method also requires the Hessian [27]. Hence, step 6 now reads.

6a. If convergence has not been achieved, compute the Gibbs energy gradient, update k-values by means of the BFGS method and return to step 2 or.

6b. If convergence has not been achieved, compute the Gibbs energy gradient and Hessian, update k-values by means of the Newton method and go to step 2.

The gradient and Hessian are defined by

$$\frac{\partial G(\mathbf{z},\mathbf{k})}{\partial \mathbf{k}} = \left\{\frac{\partial G}{\partial k_i}\right\} \tag{53}$$

$$\frac{\partial^2 G(\mathbf{z},\mathbf{k})}{\partial \mathbf{k}\partial \mathbf{k}^T} = \left\{\frac{\partial^2 G}{\partial k_i \partial k_j}\right\} = \left\{\frac{\partial^2}{\partial k_i \partial k_j}\left((1-\beta)\sum_{i=1}^{n} x_i \ln f_i^{(x)} + \beta \sum_{i=1}^{n} y_i \ln f_i^{(y)}\right)\right\}. \tag{54}$$

Although the Jacobian, gradient and Hessian formulae are rather complex to compute they allow for the very quick convergence of the optimization algorithm to its solution.

4.2.3 k-value initialization

Flash equations are always satisfied by a "trivial" solution which simply implies that $x_i = y_i = z_i$, hence $k_i = 1$. Clearly, that solution satisfies mass balance and equilibrium conditions (Eq. (41) and (43)) and it also satisfies composition consistency (Eq. (44)) for any vapor phase molar fraction value. Converging to the physically sound rather than the trivial solution can only be ensured by utilizing appropriate initial estimates of the equilibrium coefficients. SS has proved to be more robust, yet slow, when initialized away from the true solution, as opposed to the Newton-Raphson, BFGS and Newton methods which perform rapidly only provided that they are initialized close to the solution.

To benefit from the advantages of each method most flash algorithms run a few SS iterations until the convergence criterion $\sum \left(ln\, f_i^{(y)}/f_i^{(x)} \right)^2$ becomes sufficiently small. Then the algorithm switches to any other method that converges rapidly to the solution. To initialize SS, Wilson's correlation (Eq. (47)) might be used. If a stability test has been run before the phase split, the k-values obtained can be used as a very good estimate of the final solution.

In flow applications where physical properties are obtained by EoS models, k-values are often initialized to the values they exhibited at the same point in the previous timestep, thus taking advantage of the fact that flow in petroleum engineering applications is a slow varying process with time. Even more accurate estimations can be obtained by extrapolating the converged k-values obtained in the previous 2 or 3 timesteps using linear or quadratic interpolation respectively [28].

4.3 Saturation condition calculations

The estimation of saturation pressure or temperature can be considered as a special case of a flash calculation where the molar ratio is known, i.e. $\beta = 0$ for a bubble point or $\beta = 1$ for a dew one, whereas pressure or temperature needs to estimated. The bubble or drop composition, known as incipient phase, needs to be estimated as well. At saturation conditions, the Rachford-Rice equation reads

$$\sum_{i=1}^{n} z_i k_i - 1 = 0 \quad \text{for}\, p_b \tag{55}$$

$$\sum_{i=1}^{n} z_i / k_i - 1 = 0 \quad \text{for}\, p_d\,. \tag{56}$$

Equilibrium, i.e. equality of fugacity between the feed and the incipient phase, needs to be respected. Therefore, the $n+1$ equations that need to be solved for the case of a bubble point calculation are Eqs. (55) and (43) where $x_i = z_i$. The $n+1$ unknowns are the bubble composition $y_i = z_i k_i$, or equivalently the prevailing k-values, and the saturation pressure or temperature. For the case of a dew point calculation, the equations are (56) and (43) where $y_i = z_i$ and the drop composition is $x_i = z_i / k_i$.

An alternative, more elegant approach is based on the fact that the TPD at a saturation point needs to be equal to zero. In other words, forming a bubble with the incipient phase composition, different than the feed one, retains the system Gibbs energy. A zero TPD value implies $f_i^{(y)} = f_i^{(z)}$, hence $Y_i = z_i k_i = z_i \varphi_i^{(z)} / \varphi_i^{(y)} = y_i f_i^{(z)} / f_i^{(y)}$ which in turn implies $\sum Y_i = \sum y_i = 1$. When dealing with a dew point, i.e. Eq. (56), a similar result is obtained, $\sum X_i = \sum x_i = 1$. Michelsen's algorithm [29] varies pressure until the following condition is met

$$Q(p, k_i) = 1 - \sum Y_i = 0 \,\text{or}\, Q(p, k_i) = 1 - \sum X_i = 0. \tag{57}$$

In detail, the algorithm is as follows

1. Initialize p_{sat} to a pressure guaranteed to be in the two-phase region. This can be done by running a stability test at various pressures
2. Initialize k_i
3. Compute $Y_i = z_i k_i$ for a bubble point or $X_i = z_i / k_i$ for a dew point
4. Compute $S_V = \sum Y_i$ or $S_L = \sum X_i$
5. Normalize incipient phase composition using $y_i = Y_i / S_V$ or $x_i = X_i / S_L$
6. Compute incipient phase fugacity $f_i^{(y)}$ or $f_i^{(x)}$
7. Update incipient phase composition using $Y_i = y_i f_i^{(z)} / f_i^{(y)}$ or $X_i = x_i f_i^{(z)} / f_i^{(x)}$
8. Update pressure by running a Newton-Raphson iteration $p \leftarrow p - \frac{Q}{\partial Q / \partial p}$
9. Check for convergence by evaluating $\sum \left(ln\, f_i^{(y)} / f_i^{(z)} \right)^2 < \varepsilon$ or $\sum \left(ln\, f_i^{(x)} / f_i^{(z)} \right)^2 < \varepsilon$
10. Check trivial solution by evaluating $\sum \left(ln\, y_i / z_i \right)^2 < \delta$ or $\sum \left(ln\, \chi_i / z_i \right)^2 < \delta$

The Newton-Raphson derivative is given by

$$\begin{aligned} \frac{\partial Q}{\partial p} &= \sum_{i=1}^{n} y_i \frac{f_i^{(z)}}{f_i^{(y)}} \left(\frac{\partial f_i^{(y)}}{\partial p} \frac{1}{f_i^{(y)}} - \frac{\partial_i^{(z)}}{\partial p} \frac{1}{f_i^{(z)}} \right) \quad \text{for } p_b \\ \frac{\partial Q}{\partial p} &= \sum_{i=1}^{n} x_i \frac{f_i^{(z)}}{f_i^{(x)}} \left(\frac{\partial f_i^{(x)}}{\partial p} \frac{1}{f_i^{(x)}} - \frac{\partial f_i^{(z)}}{\partial p} \frac{1}{f_i^{(z)}} \right) \quad \text{for } p_d \end{aligned}. \tag{58}$$

4.4 Negative flash calculations

Whitson and Michelsen [30] extended the regular phase split algorithm beyond the limits of the phase envelope to allow flash calculations at conditions where the fluid is physically single phase. They showed that the phase split equations can still be satisfied, this time with negative β values at pressures above the bubble point or with β values above unity at pressures above the upper or below the lower dew point. The more is the distance from the phase boundary the more is the absolute value of the molar fraction, eventually approaching $-\infty$ and $+\infty$ at the convergence pressure p_k. At convergence pressure, the equilibrium coefficients become equal to unity whereas beyond p_k the flash equations have only one solution, the trivial one. Algorithms to compute the locus of the convergence pressure over a temperature range, known as "convergence locus" (CL), have been developed [31]. The negative flash area between the regular phase envelope and the CL is often referred to as the "shadow region" [32].

They also showed that stability tests can also be interpreted outside the phase envelope. Each of the two trial phases converges to a nontrivial solution (i.e. the TPD distance is positive) up to a locus in the shadow region, known as "stability test limit locus", STLL) which is enclosed by the CL. Such stability test results can be used to initialize negative flash calculations. Beyond STLL, the stability test only converges to the trivial solution. The regions discussed are shown in **Figure 3** for a black oil, where the phase envelope interior is shown in red, cyan and yellow color and the latter corresponds to the spinodal. The shadow regions above the bubble point and the dew point lines are shown in pink and blue color respectively. Green color indicates the area outside the CL where the trivial solution is the only one to the phase split problem.

To interpret physically the results of a negative flash we firstly need to note that a molar fraction value of $0 < \beta < 1$ in a regular flash calculation implies that β moles of gas of composition y_i need to be added to $1 - \beta$ moles of liquid of composition x_i to reconstruct the original feed composition z_i. In a negative flash with $\beta < 0$, $|\beta| = -\beta$ moles of gas need to be removed from $1 - \beta = 1 + |\beta|$ moles of liquid to reconstruct one mole of the original feed composition. Similarly, when $\beta > 1$, $\beta - 1$ moles of liquid need to be removed from β moles of gas.

Clearly, negative flash solutions are not of any direct use in fluid flow calculations. However, they can significantly improve the convergence properties of the regular flash calculations close to the phase boundary by allowing the solution at some iteration to escape temporarily outside the phase envelope while trying to arrive to the exact solution.

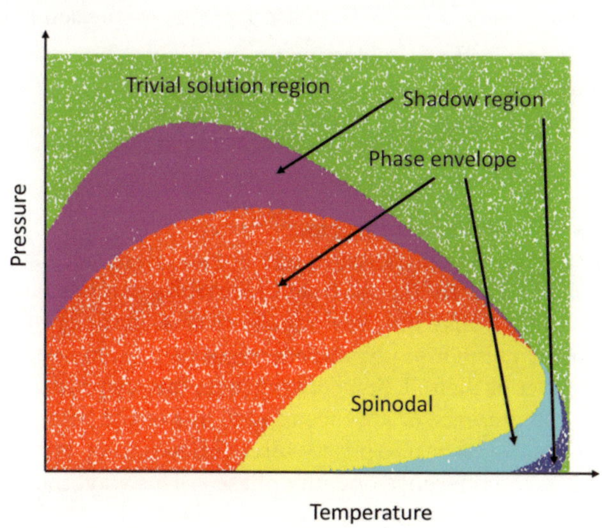

Figure 3.
Regular phase envelope, shadow region and trivial solution region.

4.5 Multiphase calculations

The need for multiphase calculations varies depending on the chemical engineering field. For example, in the upstream petroleum industry it is not that intense as multiphase equilibrium very rarely occurs in the reservoir and only when special studies in the wellbore and pipeline flow are considered. A case that is possible to happen in the reservoir is the presence of oil with high CO_2 content where two liquid phases (a CO_2-rich and a CO_2-poor one) and a vapor one could be formed. Things become more complicated when solids are considered as is the case with asphaltenes, waxes or hydrates. In the latter case, the phases that need to considered as possible to form are the solid one which may correspond to more than one hydrate structures (i.e. sI, sII and sH [33]), the aqueous phase which can be in liquid of solid form (ice) and the hydrocarbons phase (liquid, vapor or both). Nevertheless, multiphase equilibrium appears very often in chemical engineering processes taking place in process plants.

To identify such situations the standard approach is to repeatedly use the conventional two-phase Michelsen's stability test. Firstly, the test is run and if instability is detected then the vapor-liquid flash problem is solved. Subsequently, the equilibrium phase compositions are used as feeds (i.e. x_i and/or y_i instead of z_i) with suitable initial k-values to further detect if indeed they are stable or if one of them (e.g. the liquid one) will further split to two liquids.

Although many multiphase flash algorithms have been presented, the one developed by Michelsen is still considered as the most elegant one. By directly extending the two-phase flash requirements to a total of F phases, the mass balance, equilibrium and composition consistency expressions generalize to

$$\begin{aligned} &\sum_{j=1}^{F}\beta_j y_i^j = z_i \quad i = 1, \ldots, n \\ &f_i^{(y_1)} = f_i^{(y_2)} = \ldots = f_i^{(y_F)} \Leftrightarrow y_i^1\phi_i^{(y_1)} = y_i^2\phi_i^{(y_2)} = \ldots = y_i^F\phi_i^{(y_F)} \quad i = 1, \ldots, n \\ &\sum_{i=1}^{n} y_i^j = 1 \quad j = 1, \ldots, F, \end{aligned} \tag{59}$$

where β_j denotes the molar fraction of phase $1 \le j \le F$ and y_i^j denotes the concentration of component $1 \le i \le n$ in phase $1 \le j \le F$. Michelsen [34] proposed varying y_i^j and β_j to minimize the objective function given by

$$Q = \sum_{j=1}^{F}\beta_j - \sum_{i=1}^{n} z_i \sum_{k=1}^{F} \frac{\beta_k}{\phi_i^{(y_k)}}, \tag{60}$$

which satisfies Eq. (57) at its minimum.

An alternative approach that combines stability and flash calculations in a single algorithm [35] at the cost of an increased set of variables that need to be determined, has also been presented. Unlike the previous algorithms, in the one presented here F denotes the maximum number of phases that might be present in equilibrium rather than the actual number of them. Upon convergence, this algorithm will also provide information about the presence or absence of each one of the potential phases.

The algorithm requires that one phase, surely known to be present in the mixture, is considered as the reference one, say phase r. This way, the equilibrium coefficients of any other potential phase can be defined with respect to the reference one, i.e. $k_i^j = y_i^j / y_i^r$, where $k_i^r = 1$. Let θ_j be the stability variable of a phase, defined so that it is equal to zero when the phase is present (hence $\beta_j > 0$) or exhibits a positive value when the phase does not exist (i.e. when $\beta_j = 0$). Therefore, $\beta_j > 0$ and $\theta_j = 0$ for an existing phase whereas $\beta_j = 0$ and $\theta_j > 0$ for a nonexisting one.

To solve the phase split problem we need to determine all k-values k_i^j, the molar fractions β_j and the stability variables θ_j for all phases but the reference one. Indeed, once those variables have been determined, the composition of any equilibrium phase can be computed by

$$y_i^j = \frac{z_i}{1 + \sum_{\substack{j=1 \\ j \neq r}}^{F} \beta_j \left(k_i^j e^{\theta_j} - 1 \right)}, \quad i = 1, \dots, n, \ j = 1, \dots, F. \tag{61}$$

At the solution the mass balance and thermodynamic equilibrium conditions need to be simultaneously satisfied. For the first condition, the two-phase Rachford-Rice equation is extended to multiphase calculations as follows

$$r_k(\boldsymbol{\beta}, \boldsymbol{\theta}) = \sum_{i=1}^{n} \frac{z_i \left(k_i^k e^{\theta_k} - 1 \right)}{1 + \sum_{\substack{j=1 \\ j \neq r}}^{F} \beta_j \left(k_j^k e^{\theta_j} - 1 \right)}, \quad j = 1, \dots, F. \tag{62}$$

Note that the above equation needs to be satisfied for all $k = 1, \cdots, F$ and $k \neq r$. To satisfy the second condition, a minimum of the Gibbs energy is achieved when

$$\beta_j \theta_j = 0, \tag{63}$$

subject to $\beta_j \geq 0, \theta_j \geq 0$ for all phases. Note that Eq. (63) is satisfied by definition for the reference phase, i.e. $j = r$, as that phase is known to exist, hence $\theta_r = 0$.

To solve the numerical problem it is initially assumed that all phases are present, hence all θ_j are set to zero, the k-values are initialized using appropriate correlations or expected equilibrium phase compositions and molar fractions are equally spaced. Firstly, the mass balance and equilibrium equations are solved for the molar fractions and the stability variables using the currently estimates of the k-values. Subsequently, phase compositions and fugacities are computed using Eq. (61). Finally, k-values are updated in an inner loop by

$$k_i^j = \phi_i^{(r)} / \phi_i^{(j)}, \tag{64}$$

and calculations are repeated until convergence.

It is interesting to note that for the case of VLE phase split calculations, by defining the liquid phase to be the reference one, Eq. (61) simplifies to Eq. (46). Furthermore, the extended Rachford-Rice equation reduces to

$$r(\beta, \theta) = \frac{z_i (k_i e^{\theta} - 1)}{1 + \beta(k_i e^{\theta} - 1)}. \tag{65}$$

When both phases are present, $\theta = 0$ and Eq. (65) simplifies to Eq. (45).

5. Accelerated phase behavior calculations

When flow simulations are considered, reliability undoubtedly comes first as lack of convergence or obtaining unrealistic results during the calculations at any grid block would lead to a general failure of the reservoir simulation run. However, some tolerance can be shown to the accuracy of the EoS model produced results due to the latter's inherent simplicity, to the nonexhaustive fluid's compositional analysis available and to questionable tuning procedures. In fact, small inaccuracies in the fluid behavior calculations that might be introduced can be partially remediated by the history matching procedure of the field model.

On the other hand, the ever increasing demand for complex flow domain models in terms of both grid and fluid models complexity has rendered nowadays the speed of phase behavior calculations as one of the most critical issues of flow simulation, especially for cases of complex thermodynamic phenomena such as near critical phase behavior and multiphase equilibrium in the presence of solids. As a result, speeding up phase behavior calculations is considered as a major issue, even if this involves some sacrifice in the calculations accuracy.

5.1 Rigorous methods

Reducing the number of components used to describe the fluid composition through a splitting and lumping procedure is the standard way to obtain simpler, hence faster EoS models. Firstly, the heavy end, usually corresponding to a limited carbon number, needs to be replaced with a large number of pseudo-components defined by means of computational methods. This way the flexibility during the EoS model tuning increases. The most pronounced method is the one developed by Whitson that utilizes the Gamma distribution [6]. Subsequently, the extended number of components is reduced (lumped) to a small number of pseudo-components, usually 3 to 5, by means of algorithms which aim at preserving the EoS model's performance [7]. Finally, pure components are grouped together to minimize the composition vector size. Typical selections are N_2 with C_1, CO_2 with C_2, nC_4 with iC_4 and nC_5 with iC_5. When two or more components are lumped together, the new group's properties need to be rematched against the available PVT measurements. A very illustrative example is given my Ahmed [7] where a full C_{7+} composition that includes N_2 and CO_2, thus summing up to 11 components, reduces gradually the number of components to only 7 according to the lumping procedure shown in **Table 3**.

Other accelerating methods include different treatments of the mathematical form of the problem or of its variables [36, 37] and utilizing solution acceleration techniques such as the GDEM update one [5]. Rasmussen et al. [32] provided criteria to completely skip phase behavior calculations during a simulation run when the prevailing equilibrium conditions fall within specific regions of the fluid's phase diagram. Simply speaking, if the fluid is a single phase one, most probably it will keep so if its distance to the phase boundary is large enough. So is the case with fluids lying well inside the two-phase region. In both cases, the stability test can be skipped whereas in the former one the phase split can be skipped as well.

Finally, efforts have been concentrated on utilizing advanced code optimization [38] and High Performance Computing (HPC) techniques which take advantage of

Original components set										
CO_2	N_2	C_1	C_2	C_3	iC_4	nC_4	iC_5	nC_5	C_6	C_{7+}
Lumped components set										
$N_2 + C_1$	$CO_2 + C_2$	$C_3 + iC_4 + nC_4$	$iC_5 + nC_5 + C_6$	F_1	F_2	F_3				

Table 3.
Components' number reduction by splitting and lumping.

the parallel computing capabilities of modern computer architectures [39]. Despite the difficulties in distributing the work load and in optimizing memory transfer between clusters, impressive acceleration factors have been reported [40].

5.2 The reduced variable framework

Reduced variables methods are based on the fact that the intrinsic dimensionality of the stability and phase split calculations, hence the number of equations to be solved, is related to the rank of the complementary BIP matrix $\mathbf{\Gamma}$, defined by $\gamma_{ij} = 1 - k_{ij}$, rather than the number n of components used. Michelsen [41] derived the first reduced variables algorithm for cubic EoS models with zero BIPs ($k_{ij} = 0$) by showing that the equations to be solved could be reduced to only 3. Simply speaking, although the phase composition, e.g. y_i, is a vector with n components, it is incorporated in the mixing rules only through its scalar projections to the components' a_i and b_i constants vectors, thus forming only two variables, i.e. $a_{mix} = \left(\sum \sqrt{a_i} y_i\right)^2$ and $b_{mix} = \sum b_i y_i$. By further considering the molar fraction β, the number of variables to be determined reduces to only 3.

In general, the $n + 1$ original variables (i.e. the k-values and the molar fraction) are replaced by a set of $m + 2$ reduced ones, with $m \ll n$, thus significantly reducing the phase behavior problem dimensionality. Several authors extended Michelsen's idea to calculations with nonzero BIP [42–44] by applying Singular Value Decomposition to the BIP matrix so as to split it in a sum of rank-1 matrices. The less is the number of rank-1 matrices required to reconstruct accurately the original BIP matrix, the less is the number of reduced variables that need to be utilized, hence the less is m. Nichita and Graciaa [45] presented an alternative reduced variables set which allows for an easier Hessian matrix computation procedure and faster convergence while Gaganis and Varotsis [46] proposed a new procedure for generating improved reduced variables.

More specifically, let the complementary BIP matrix $\mathbf{\Gamma} = \{1 - k_{ij}\}$ be decomposed to a set of eigenvalues λ_i and eigenvectors $\mathbf{t}_i$ by use of the Singular Value Decomposition method [28], so that $\mathbf{\Gamma} = \sum_{i=1}^{m} \lambda_i \mathbf{t}_i \mathbf{t}_i^T$, where m denotes the rank of $\mathbf{\Gamma}$. For the vapor phase, we define the projection vectors $\mathbf{q}_i = \mathbf{t}_i \circ \sqrt{\mathbf{a}}$ and the reduced variables $\mathbf{h}_V = \mathbf{Q}^T \mathbf{y}$, where $\mathbf{a} = \{a_i\}$ is the vector containing the components energy parameters, $\mathbf{Q} = \begin{bmatrix} \mathbf{q}_1 & \cdots & \mathbf{q}_m \end{bmatrix}$ and operator $\circ$ denotes the Hadamard vector product (by element multiplication). The phase energy parameter a_V and its derivative (required for the computation of phase fugacity) can be computed as functions of the reduced variables, that is $a_V = \mathbf{h}_V^T \mathbf{\Lambda} \mathbf{h}_V$ and $\partial a_V / \partial \mathbf{y} = 2\mathbf{Q}\mathbf{\Lambda}\mathbf{h}_V$, where $\mathbf{\Lambda} = diag\{ \lambda_1 \quad \cdots \quad \lambda_m \}$. By further considering the vapor phase volume parameter b_V as an unknown variable all required quantities (i.e. compressibility factor Z_V from Eq. (7) and fugacity coefficients from Eq. (10)) can now be completed as functions of $\mathbf{h}_V$ and b_V .

The corresponding variables of the liquid phase can be easily computed by considering the vapor phase molar fraction β as an unknown variable and applying mass balance, i.e. $\mathbf{h}_L = \left(\mathbf{Q}^T \mathbf{z} - \beta \mathbf{h}_V\right)/(1 - \beta)$ and $b_L = \left(\mathbf{b}^T \mathbf{z} - \beta b_V\right)/(1 - \beta)$, thus allowing for the computation of the liquid phase properties as well.

To summarize, $\mathbf{h}_V$, b_V and β form an alternative set of variables in terms of which the phase split problem can be cast. The constraining equations that need to be satisfied are

$$\mathbf{h}_V - \mathbf{Q}^T \mathbf{y} = \mathbf{0} \tag{66}$$

$$b_V - \mathbf{b}^T\mathbf{y} = 0 \quad (67)$$

$$\sum \frac{z_i(k_i - 1)}{1 + \beta(k_i - 1)} = 0. \quad (68)$$

The solution algorithm is as follows

1. Initialize y_i and β
2. Compute $\mathbf{h}_V = \mathbf{Q}^T\mathbf{y}$
3. Compute $a_V = \mathbf{h}_V^T\mathbf{\Lambda}\mathbf{h}_V$, $\partial a_V/\partial \mathbf{y} = 2\mathbf{Q}\mathbf{\Lambda}\mathbf{h}_V$ and $b_V = \mathbf{b}^T\mathbf{y}$
4. Compute $\mathbf{h}_L = \left(\mathbf{Q}^T\mathbf{z} - \beta\mathbf{h}_V\right)/(1-\beta)$ and $b_L = \left(\mathbf{b}^T\mathbf{z} - \beta b_V\right)/(1-\beta)$
5. Compute $a_L = \mathbf{h}_L^T\mathbf{\Lambda}\mathbf{h}_L$, $\partial a_L/\partial \mathbf{x} = 2\mathbf{Q}\mathbf{\Lambda}\mathbf{h}_L$ and $b_L = \mathbf{b}^T\mathbf{x}$
6. Solve the cubic polynomial for both phases (Eq. (8))
7. Compute fugacity coefficients for both phases using (Eq. (11))
8. Compute $k_i = \varphi_i^{(x)}/\varphi_i^{(y)}$ and phase compositions using (Eq. (46))
9. Check convergence by evaluating if Eq. (66) are satisfied
10. If convergence has not been achieved, update $\mathbf{h}_V$, b_V and β by means of a Newton-Raphson step and return to step 3.

Eqs. (66) and (67) guarantee thermodynamic equilibrium whereas Eq. (68) ensures mass balance. Clearly, the equations are nonlinear and their solution still requires the utilization of iterative function solving methods. Nevertheless, the benefit of the reduced variables approach lies in the cardinality of the variables set which is usually smaller than that of the conventional approach as it equals to $m + 2$. When the BIP matrix contains many small or even zero values, as it is commonly the case with the EoS modeling of multicomponent fluids, the rank of matrix $\mathbf{\Gamma}$ is much smaller than its size ($m \ll n$) which implies that the number of equations that need to be solved is significantly reduced. Moreover, reduced variables h_i corresponding to very low eigenvalues λ_i can also be neglected at the cost of the truncation error of matrix $\mathbf{\Gamma}$. For the extreme case where all BIPs are equal to zero, $m = 1$, Eq. (66) simplifies to a scalar one and only three nonlinear equations need to be solved regardless of the number of the mixture components [41]. Nevertheless, there has been some questioning about the real benefit of reduction methods as modern computers architecture has significantly reduced their computing time gain against the conventional ones [47, 48].

5.3 Soft computing methods

Soft computing methods aim at solving phase equilibrium problems by utilizing data points rather than solving the thermodynamically rigorous equations discussed in the previous sections. Simply speaking, data related to the stability and phase split problems are generated and subsequently used to build correlations which provide directly the variables of interest such as the TPD value and the prevailing k-values for the stability and phase split problems respectively. Such flow-specific and fluid-specific soft computing models are case dependent as they are generated using data obtained either prior to the specific simulation of interest or during that.

The benefit lies in that the generated correlations consist of simple, noniterative calculations which are by orders of magnitude faster than the conventional iterative ones. Although the numerical treatment of the datapoints involves purely numerical techniques such as regression, classification and clustering [49], thermodynamics are still incorporated indirectly in the soft computing based models as the data points used to build the models have been generated in advance by conventional rigorous methods.

Composition independent correlations to estimate the equilibrium coefficients (k-values), such as those of Standing and Whitson as well as the convergence pressure method, all discussed in 3.2.1, can be thought of as the simplest soft computing method to treat the phase split problem as they provide k-values estimates without being based on a rigorous EoS model, hence avoiding the iterative solution of the fugacity equations or the minimization of the Gibbs energy.

Voskov and Tchelepi [50] proposed the generation and storage of the encountered tie-lines in Tables "on the fly". Initially, for each feed composition encountered during the simulation, the phase split problem is solved conventionally and the equilibrium compositions (i.e. the tie line endpoints) are stored. For each subsequent feed the algorithm searches quickly the Tables to identify the closest stored tie-lines and interpolate them linearly to get the equilibrium compositions. If no close enough tie lines can be found, the phase split problem is solved conventionally, and the table is enriched. Stability is determined by using the negative flash approach [30]. To reduce the computing time cost for accessing and further building-up the tie line Table, Belkadi et al. [51] proposed the Tie-line Distance Based Approximation which further accelerates the search procedure.

Gaganis and Varotsis [52, 53] presented the methodology to develop proxy models for treating both the phase stability and phase split problems using machine learning tools. Their approach aims at solving conventionally the phase behavior problem for a set of sampled operating points and using the obtained data to generate explicit proxy models using multivariate regression models such as neural networks to directly predict the prevailing equilibrium coefficients values given feed composition, pressure and temperature (for nonisothermal runs). For the

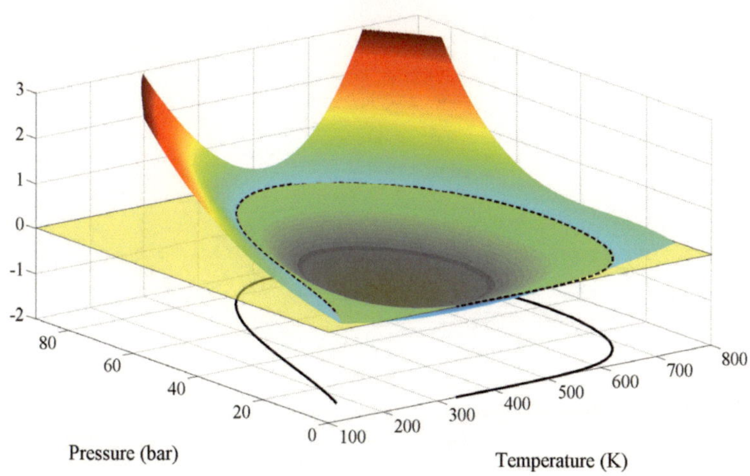

Figure 4.
The SVM output equals to zero at the phase boundary.

phase stability problem, their model outputs a positive nonlinear transformation of the conventional TPD value that exhibits the same sign as the former (**Figure 4**). Their model utilizes Support Vector Machines, SVM [54] to provide the same binary stable/unstable answers anywhere in the operating space even outside the stability test limit locus [31]. An improved stability test method has been presented by Gaganis [55] which reliefs the need to model accurately the phase boundary thus allowing for even simpler and faster to evaluate stability models. His approach develops two classifiers which only identify whether the point under question lies "far enough" from the phase boundary or not. If it lies far enough outside of the phase envelope, then the fluid is surely single phase whereas it is certainly at two-phase when lying well inside the phase envelope. If a certain answer cannot be obtained, a regular stability algorithm is invoked.

6. Conclusions

Equations of State of varying complexity and accuracy are nowadays available to describe the thermodynamic behavior of almost all types of fluids. Beyond the classic and easy-to-implement cubic EoS models, recent advances in perturbation theory have allowed its application to the derivation of models that describe accurately in a microscopic level the behavior of fluids.

Phase behavior calculations by means of EoS models are massively required during all types of flow simulations, thus rendering the availability of robust, thermodynamic rigorous algorithms as of major importance. However, as the required computational load can be very heavy, various accelerating methods have been developed, and they have been proved to perform very well.

Author details

Vassilis Gaganis[1,2]

1 National Technical University of Athens, Hellas, Athens, Greece

2 Foundation for Research and Technology, Hellas, Athens, Greece

*Address all correspondence to: vgaganis@metal.ntua.gr

References

[1] Michelsen M, Mollerup J. Thermodynamic Models: Fundamental and Computational Aspects. Denmark: Tie-Line Publications; 2007. p. 73. DOI: 10.1016/j.fluid.2005.11.032

[2] Novak N, Louli V, Skouras S, Voutsas E. Prediction of dew points and liquid dropouts of gas condensate mixtures. Fluid Phase Equilibria. 2018; **457**:62-73. DOI: 10.1016/j.fluid.2017.10.024

[3] Bretonnet JL. Thermodynamic Perturbation Theory of Simple Liquids in Thermodynamics: Interaction Studies - Solids, Liquids and Gases. Intech Open; 2011. p. 839. DOI: 10.5772/23477

[4] Michelsen M. The isothermal flash problem. Part I. Stability. Fluid Phase Equilibria. 1982;**9**:1-19. DOI: 10.1016/0378-3812(82)85001-2

[5] Michelsen M. The isothermal flash problem. Part II. Phase split calculation. Fluid Phase Equilibria. 1982; **9**:21-40. DOI: 10.1016/0378-3812(82) 85002-4

[6] Whitson C, Brule M. Phase Behavior. Richardson, TX: SPE Monograph; 2000. p. 47

[7] Ahmed T. Equations of State and PVT Analysis. Cambridge, MA: Gulf Publishing; 2016. p. 5. DOI: 10.1016/C2013-0-15511-0

[8] Twu C. An internally consistent correlation for predicting the critical properties and molecular weights of petroleum and coal-tar liquids. Fluid Phase Equilibria. 1984;**16**:137-150. DOI: 10.1016/0378-3812(84)85027-X

[9] Oellrich L, Plocker U, Prausnitz M, Knapp H. Equations of state methods for computing phase equilibria and enthalpies. International Chemical Engineering. 1981;**21**:1-16

[10] Neumark S. Solution of Cubic and Quartic Equations. UK: Elsevier; 1965. p. 5. DOI: 10.1016/C2013-0-05408-4

[11] Peneloux A, Rauzy E, Freze R. A consistent correlation for Redlich-Kwong-Soave volumes. Fluid Phase Equilibria. 1982;**8**:7-23. DOI: 10.1016/0378-3812(82) 80002-2

[12] Boublic T. Perturbation Theory in. In: Sengers JV, Kayser RF, Peters CJ, White HJ Jr editors. Equations of State for Fluids and Fluid Mixtures. NY, USA: Elsevier; 2000. pp. 127-168

[13] Barker JA, Henderson D. Perturbation theory and equation of state for fluids: The square-well potential. The Journal of Chemical Physics. 1967;**47**(8):2856-2861. DOI: 10.1063/1.1712308

[14] Barker JA, Henderson D. Perturbation theory and equation of state for fluids II: A successful theory of liquids. The Journal of Chemical Physics. 1967;**47**(11):4714-4721. DOI: 10.1063/1.1701689

[15] Mansoori GA, Canahan NF, Starling KE, Leland TW Jr. Equilibrium thermodynamic properties of the mixture of hard spheres. The Journal of Chemical Physics. 1971;**54**:4714-4721. DOI: 10.1063/1.1675048

[16] Zwanzig R. High-temperature equation of state by a perturbation method. I. Nonpolar gases. The Journal of Chemical Physics. 1954;**22**:1420-1428. DOI: 10.1063/1.1740409

[17] Nichita D, Gomez S, Luna E. Phase stability analysis with cubic equations of state by using a global optimization method. Fluid Phase Equilibria. 2002; **194–197**:411-437. DOI: 10.1016/S0378-3812(01)00779-8

[18] Wilson G. A modified Redlich-Kwong EOS, application to general

physical data calculations. In: Paper 15C, Presented at the Annual AIChE National Meeting. Cleveland, US; 4-7 May 1968

[19] Aursanda P, Gjennestad M, Aursand E, Hammera M, Wilhelmsen Ø. The spinodal of single- and multi-component fluids and its role in the development of modern equations of state. Fluid Phase Equilibria. 2017;**436**: 98-112. DOI: 10.1016/j.fluid.2016.12.018

[20] Gaganis V, Marinakis D, Varotsis N. A general framework of model functions for rapid and robust solution of Rachford-Rice type of equations. Fluid Phase Equilibria. 2012;**322–323**: 9-18. DOI: 10.1016/j.fluid.2012.03.001

[21] Nichita D, Liebovici C. A rapid and robust method for solving the Rachford–Rice equation using convex transformations. Fluid Phase Equilibria. 2013;**353**:38-49. DOI: 10.1016/j.fluid.2013.05.030

[22] Standing M. A set of equations for computing equilibrium ratios of a crude oil/natural gas system at pressures below 1,000 psia. Journal of Petroleum Technology. 1979;**31**(9):1193-1195. DOI: 10.2118/7903-PA

[23] Whitson C, Torp S. Evaluating constant volume depletion data. In: Paper SPE 10067, Presented at the SPE 56th Annual Fall Technical Conference. San Antonio, TX, US; 5-7 October 1981. DOI: 10.2118/10067-PA

[24] Standing M. Volumetric and phase behavior of oil field hydrocarbon systems. TX, USA: SPE of AIME; 1977. p. 56. DOI: 10.1126/science.117.3042.432

[25] Lohrenz J, Clark G. A compositional material balance for combination drive reservoirs with gas and water injection. Journal of Petroleum Technology. 1963; **15**(11):1233-1238. DOI: 10.2118/558-PA

[26] Rowe A. The critical composition method – A new convergence pressure method. SPE Journal. 1967;7:54-60. DOI: 10.2118/1631-PA

[27] Nocedal J, Wright S. Numerical Optimization. New York: Springer; 2006. p. 30. DOI: 10.1007/978-0-387-40065-5

[28] Press W, Flannery B, Teukolsky S, Vetterling W. Numerical Recipes in C: The Art of Scientific Computing. UK: Cambridge University Press; 1988. p. 108. DOI: 10.1002/9780470974704

[29] Michelsen M. Saturation point calculations. Fluid Phase Equilibria. 1985;**21**:181-192. DOI: 10.1016/0378-3812(85)90005-6

[30] Whitson C, Michelsen M. The negative flash. Fluid Phase Equilibria. 1989;**53**:51-71. DOI: 10.1016/0378-3812(89)80072-X

[31] Nichita D, Broseta D, Montel F. Calculation of convergence pressure/temperature and stability limit loci of mixtures with cubic equations of state. Fluid Phase Equilibria. 2007;**261**: 176-184. DOI: 10.1016/j.fluid.2007.07.041

[32] Rasmussen C, Krejberg K, Michelsen M, Bjurstrom K. Increasing the computational speed of flash calculations with applications for compositional transient simulation. SPE Reservoir Evaluation and Engineering. 2006;**2**:32-38. DOI: 10.2118/84181-MS

[33] Sloan E, Koh C. Chathrate Hydrates of Natural Gases. FL, USA: CRC Press; 2007. p. 206. DOI: 10.1201/9781420008494

[34] Michelsen M. Calculation of multiphase equilibrium. Fluid Phase Equilibria. 1994;**18**:545-550. DOI: 10.1016/0098-1354(93)E0017-4

[35] Gupta A, Bishnoi P, Kalogerakis N. A method for the simultaneous phase equilibria and stability calcs for

multiphase reacting and non-reacting systems. Fluid Phase Equilibria. 1991;**63**: 65-89. DOI: 10.1016/0378-3812(91) 80021-M

[36] Michelsen M. Phase equilibrium calculations. What is easy and what is difficult. Computers and Chemical Engineering. 1993;**17**:431-439. DOI: 10.1016/S0098-1354(09) 80006-9

[37] Michelsen M. Speeding up two-phase PT-flash, with applications for calculation of miscible displacement. Fluid Phase Equilibria. 1998;**143**:1-12. DOI: 10.1016/S0378-3812(97)00313-0

[38] Haugen K, Beckner B. Highly optimized phase equilibrium calculations. In: Paper SPE 163583, Presented at the SPE Reservoir Simulation Symposium. Woodlands, TX, US; 18-20 February 2013. DOI: 10.2118/163583-MS

[39] Appleyard J, Appleyard M, Wakefield A, Desitter A. Accelerating reservoir simulators using GPU technology. In: Paper SPE 141265, Presented at the SPE Reservoir Simulation Symposium. Woodlands, TX, US; 21-23 February 2011. DOI: 10.2118/141402-MS

[40] Hayder M, Baddourah M. Challenges in high performance computing for reservoir simulation. In: Paper SPE 152414, Presented at the SPE Europe. Copenhagen, Denmark; 4-7 June 2012. DOI: 10.2118/152414-MS

[41] Michelsen M. Simplified flash calculations for cubic equations of state. Industrial & Engineering Chemistry Process Design and Development. 1986; **25**:184-188. DOI: 10.1021/i200032a029

[42] Hendriks E, van Bergen A. Application of a reduction method to phase equilibria calculations. Fluid Phase Equilibria. 1992;**74**:17-34. DOI: 10.1016/0378-3812(92)85050-I

[43] Firoozabadi A, Pan H. Fast and robust algorithm for compositional modeling: Part I – Stability analysis testing. SPE Journal. 2002;7:78-89. DOI: 10.2118/77299-PA

[44] Pan H, Firoozabadi A. Fast and robust algorithm for compositional modeling: Part II – Two phase flash calculations. SPE Journal. 2002;**12**: 380-391. DOI: SPE-87335-PA

[45] Nichita D, Graciaa A. A new reduction method for phase equilibrium calculations. Fluid Phase Equilibria. 2011;**302**:226-233. DOI: 10.1016/ j.fluid.2010.11.007

[46] Gaganis V, Varotsis N. An improved BIP matrix decomposition method for reduced flash calculations. Fluid Phase Equilibria. 2013;**340**:63-76. DOI: 10.1016/ j.fluid.2012.12.011

[47] Haugen K, Beckner B. Are reduced methods in EoS calculations worth the effort? In: Paper SPE 141399, Presented at the SPE Reservoir Simulation Symposium. Woodlands, TX, US; 21-23 February 2011. DOI: 10.2118/141399-MS

[48] Petitfrere M, Nichita V. A comparison of conventional and reduction approaches for phase equilibrium calculations. Fluid Phase Equilibria. 2015;**386**:30-46. DOI: 10.1016/j.fluid.2014.11.017

[49] Bishop C. Pattern Recognition and Machine Learning. NY, USA: Springer; 2006. p. 137

[50] Voskov D, Tchelepi H. Tie-simplex based mathematical framework for thermodynamic equilibrium computations of mixtures with an arbitrary number of phases. Fluid Phase Equilibria. 2009;**283**:1-11. DOI: 10.1016/ j.fluid.2009. 04.018

[51] Belkadi A, Michelsen M, Stenby E. Comparison of two methods for speeding up flash calculations in

compositional simulations. In: Paper SPE 142132, Presented at the SPE Reservoir Simulation Symposium. Woodlands, TX, US; 21-23 February 2011. DOI: 10.2118/142132-MS

[52] Gaganis V, Varotsis N. Machine learning methods to speed up compositional reservoir simulation. In: Paper SPE 154505, Presented at the SPE Europec. Copenhagen, Denmark; 4-7 June 2012. DOI: 10.2118/154505-MS

[53] Gaganis V, Varotsis N. An integrated approach for rapid phase behavior calculations in compositional modeling. Journal of Petroleum Science and Engineering. 2014;**118**:74-87. DOI: 10.1016/j.petrol.2014.03.011

[54] Burges C. A tutorial on support vector machines for pattern recognition. Data Mining and Knowledge Discovery. 1998;**2**:121-167. DOI: 10.1023/ A: 1009715923555

[55] Gaganis V. Rapid phase stability calculations in fluid flow simulation using simple discriminating functions. Computers and Chemical Engineering. 2018;**108**:112-127. DOI: 10.1016/j. compchemeng.2017.09.006

Chapter 4

Life Is Not on the Edge of Chaos but in a Half-Chaos of Not Fully Random Systems. Definition and Simulations of the Half-Chaos in Complex Networks

Andrzej Gecow

Abstract

The research concerns the dynamics of complex autonomous Kauffman networks. The article defines and shows using simulation experiments half-chaotic networks, which exhibit features much more similar to typically modeled systems like a living, technological or social than fully random Kauffman networks. This represents a large change in the widely held view taken of the dynamics of complex systems. Current theory predicts that random autonomous systems can be either ordered or chaotic with fast phase transition between them. The theory uses shift of finite, discrete networks to infinite and continuous space. This move loses important features like e.g. attractor length, making description too simplified. Modeled adapted systems are not fully random, they are usually stable, but the estimated parameters are usually "chaotic", they place the fully random networks in the chaotic regime, far from the narrow phase transition. I show that among the not fully random systems with "chaotic parameters", a large third state called half-chaos exists. Half-chaotic system simultaneously exhibits small (ordered) and large (chaotic) reactions for small disturbances in similar share. The discovery of half-chaos frees modeling of adapted systems from sharp restrictions; it allows to use "chaotic parameters" and get a nearly stable system more similar to modeled one. It gives a base for identity criterion of an evolving object, simplifies the definition of basic Darwinian mechanism and changes "life on the edge of chaos" to "life evolves in the half-chaos of not fully random systems".

Keywords: Kauffman networks, complex networks, chaos, edge of chaos, damage spreading, connectivity

1. Introduction

This is empirical work[1] using simulation. It concerns dynamics in complex autonomous Kauffman networks that are finite and discrete, shows that current theory used for them, based on Lyapunov coefficient in infinite, continuous space, implies false expectations.

Kauffman [5, 6] has considered as a general model of a system an autonomous, dynamic, deterministic, complex, random Boolean networks (known as RBN). The discovery of chaos, order, and phase transition between them in such the networks, allowed to look into this very complex world. Lot of works are based on this model [5, 7–17]. Now, slowly become aware of new important aspects that we have not yet considered adequately. In this paper, such the way is developed, but considered networks are not fully random and use more signal variants than only two. A new obtained vision is clearly different and more adequate for description of adapted objects than now widely accepted. The discovery of half-chaos is the main new element here, which above all frees from strong limitations in systems modeling imposed by the contemporary vision. The statistical properties of the systems are easiest to investigate for fully random systems and from these, we should have started (like Kauffman did), but the systems we model are usually suited to some tasks and are certainly not completely random.

Lyapunov exponents are the most widely used measures to describe chaotic behavior of dynamical systems, however, to check an adequateness of theory by observation of the behavior of finite dynamical networks it must be defined using its main features expected by the theory. The main characteristic of the chaotic behavior of dynamic systems is a high sensitivity to initial conditions, leading to maximally different effects for very similar initial conditions. A small disturbance is a small change in initial conditions. An effect of such small disturbance is called damage. Distribution of damage size [6] ("size distribution of avalanches" in [17]) is then the main feature observed in experiment and expected by the theory that may be compared. It is original; theories using Lyapunov exponents or percolation are derivative. The term 'chaos' is used here in such the meaning, similarly as Kauffman does (see ch.2.3). To fit the current theory the damage distributions should fit a Derrida's annealed approximation model [18] and for chaotic systems - an equilibrium level found in it.

It is commonly believed, that system can be only either chaotic or ordered, but not simultaneously both of them - this is shown to be false. On this believing, a "criticality hypothesis" (critical regions have also been said to be "at the edge of

---

[1] The description of the investigation and the arguments for introducing the half-chaos given in this article is necessarily shortened and simplified. A much more extensive description is available in supplement [1, 2] to this article. Earlier, simpler versions of the article are available in preprints [3, 4]. The data (programs and its sources, results of simulations) analyzed during the current study are available from the author on any request.
Wider list of abbreviations and new terms is placed on the end.
In this work a few abbreviations that are not standard are used: ***s*** - number of equally probable signal variants.
Network types: ***sf*** - scale-free; ***ss*** - single scale; ***er*** - Erdős-Rényi "random"; ***sh*** and ***si*** are respectively ***sf*** and ***ss*** with 30% removal of nodes.
tmx - maximum number of counting steps of time *t*;
dmx - mean maximal damage *d*, i.e. Derrida equilibrium for chaotic behavior;
q - degree of order, fraction of damage which are a small change of network functioning at tmx, capacity of left peak of $P(d)$.

chaos" in parameter space of systems) is formulated (see e.g. [17]). Evolution needs small changes which practically occur only in critical regions in such the systems. Current theory and this believing (see e.g. [17]) are based on the assumption that networks are fully random. However, interesting phenomena concerning life occur in not fully random networks due to natural selection. The current theory of chaos was built for functions in infinite and continuous space, but it is used for finite discrete networks [7, 8], such a method is an approximation. It loses a few important phenomena present in such the networks, but absent in the infinite and continuous space. Due to such the reasons, expectations of the theory that life is on the edge of chaos can be and are inadequate. Here such phenomena are shown; they need much more complex theory which will not use the assumption of full randomness of network and infinite continuous space, but to build such theory is the next step, which is the task for mathematicians. The description of this experiment in the language of mathematical equations seems to me inadequate and unattainable, and in my opinion useless, but mathematicians may have a different opinion. Programming languages are a natural and appropriate tool for describing such issues. I can share the program, but it is complex. It is not true that the below description of experiment is not exact enough to be repeated by every IT specialist. Therefore it is enough exact to understand by mathematicians too.

Indication of adequate ranges of parameters of a complex "purposeful" (adapted) system describing living, technological or social object is a key for modeling their processes. An important parameter is a connectivity [19], which current theory strongly limits. The system can be any, e.g. the solar system is also a system, but usually, in human intuition, the system has to somehow work (therefore above "purposeful"), and despite some changeability, it has to keep its identity. The evolution of the system is a term that reconciles two adversities - variability that is the essence of evolution, and the identity of the evolving object. This is not a philosophical problem, but a particular problem for modeling. In this work a base for solving this problem is found. A good approximation of the system description is a dynamic complex network, although it undoubtedly has many important simplifications. We are just entering this subject and it is difficult for our intuitions to operate on more complex, more adequate descriptions, such as process algebras [20].

Half-chaos is a state of the system that is not fully random, with parameters that make the random system strongly chaotic (hereinafter we will call them "chaotic parameters", such the parameters are usually estimated for real systems), however small disturbances give an ordered reaction (small damage) with a similar probability to a chaotic reaction (damage near the Derrida balance [18], **Figure 1c,d**). Acceptance of changes that trigger ordered reactions preserves the half-chaotic state allowing for a long evolution of the slowly changing system (the system retains identity), but acceptance of one change that gives a chaotic reaction leads to practically irreversible entry into normal chaos (the system works completely different, ceases to be itself). Thus, the basic Darwinian mechanism emerges - this has large interpretational consequences.

The assessment of whether a given system is chaotic or ordered is currently based on parameters that in the case of a half-chaotic system indicate chaos for the fully random system (I call them "chaotic parameters"), but the behavior of the system turns out to be inconsistent with such prediction. This work presents half-chaotic systems and simple ways to obtain such systems. The experimental results are unambiguous and easy to repeat. The constraints forming the half-chaotic system are small, which means that there are a lot of such systems, though undoubtedly significantly less than of fully random.

The practical result of this work is the realignment of the acceptable range of parameters for system modeling. This is a fundamental change. First and foremost,

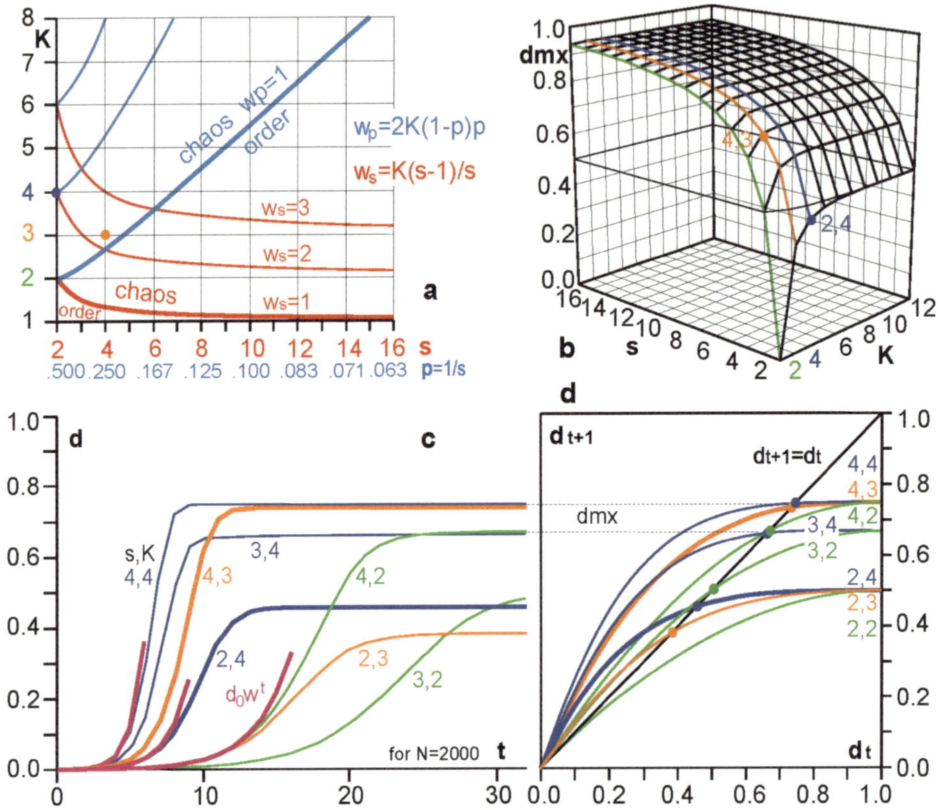

Figure 1.

Comparison of models based on p *and* s*; - of influence of* s *and* K *on Derrida equilibrium, Derrida plot;* d(t). ***a*** *- Comparison of models based on probability* p *of one Boolean signal variant and on* s *equally probable signal variants in dependency on* K*. As the basic argument* s *is taken. For it* p *is added as* 1/s*, it is for the case if in reality there are* s *different equally probable signal variants, but we are interested only in one of them, and rest we collect to the second one. Values of the coefficient of damage propagation* w_s *for* s,K *and* w_p *for* p *and* K *are used. The equation for* w_p *is taken from [5, 14]. Both models give very different results, it means, that they cannot replace each other. (See also ([21]* ***Fig. 4****)).* ***b*** *- Derrida equilibrium* (dmx) *for chaotic response in the system of* s,K*. Kauffman using Boolean networks has considered only* K *as the most interesting variable, but* s *influences* dmx *more hardly. However, he cannot use* s *other than 2, because for each* $s > 2$ *the chaos is present* ($dmx > 0$ *exists), like for any* $K > 2$*. Among sensible* s,K*, only for 2,2 exist order, it is an especially extreme case.* ***c,d*** *- theoretical damage spreading calculated using the Derrida's annealed approximation model.* ***d*** *- The change of damage in one step of the time in synchronous calculation known as the 'Derrida plot', extended [21] for the case* $s > 2$*. The crossing of curves* $d_{t+1}(d_t, s, K)$ *with diagonal* $d_{t+1} = d_t$ *shows equilibrium levels* dmx *up to which damage can grow. Case* s,K = 2,2 *has a damage equilibrium level in* d = 0*. These levels are reached on the left which shows damage size in time dependency. For* $s > 2$ *they are significantly higher than for Boolean networks. All cases with the same* K *have the same color to show the influence of* s*.* ***c*** *- In this plot expected* d(t) *for* N = 2000 *is shown. It is an effect of 'Derrida plot' shown in d. A simplified expectation* $d(t) = d_0 w_s^t$ *based on coefficient* w_s *is shown for the first critical period when* d *is still small - three short curves to the left of the longer curves reaching equilibrium. Parameter* **s,K** *(treated as a vector) is the main variables in the simulations. Most of the studies are made for* s,K = 4,3*, also sometimes for* s,K = 2,4 *(that is, for Boolean network). They provide highly chaotic random systems - 'coefficient* w *of damage propagation' is significantly higher than one.*

"chaotic parameter" for the Kauffman (Boolean) network - connectivity is included in this scope, but also a larger number than two of signal variants, also omitted due to the effect in the form of a chaotic system (for fully random systems). The need to introduce a larger number of signal variants for statistical investigations was already explained in [21]. The maintenance of the name of the 'Kauffman network' for such a network was there postulated, to be no longer synonymous with Boolean networks. However, these postulates acquire practical significance only after demonstrating half-chaos.

2. Main assumptions

2.1 Variables *K, k, t, N, A* and *d* in Kauffman networks

The considerations concern the statistical stability of the deterministic discrete Kauffman networks [5, 6, 22] (a little bit extended). The network consists of ***N*** nodes. A node in such a network receives signals at the ***K*** inputs, converts them uniquely using its function to the output signal called the state of the node, and then sends it to other nodes by ***k*** output links. States of all *N* nodes together creates a state of the network. The calculation of function takes a time step. Up to now, 2 (logical) signal states (variants) have been used. In the simplest case, it was assumed the same probability of signal variants and full randomness of connections, functions, and initial states of each node, such networks were called **RBN** (Random Boolean Networks). Here, deviation from this full randomness is made[2] by assuming **short attractor** (a small number of time-steps until meeting the same network state), especially – **point attractor** (next network state is the same). In other here described investigations (met7, ch.3.5) – by controlled construction of in-ice-modular network (ch.3.3) or (met1-4b, ch.4) – by an increase of the fraction of negative feedbacks or classic modularity. *K* (called "connectivity", see [19]) was the basic variable for Kauffman.

Synchronous computing is used, i.e., the states of nodes from the discrete time ***t*** are input signals and arguments of the function of other nodes, and the results of these functions are nodes states at the next moment ($t + 1$). Variable *t* – is the number of time steps from a disturbance initiation. As the disturbance a permanent change in the value of the function of the node for its input state is used at the time $t = 0$; in method '8' (**met8**) it was an addition or removing a node. Parameter ***tmx*** - the maximum number of calculated time steps is chosen arbitrarily, but it is checked whether its increase does not change the results (**Figures 2**, **3** and 5).

Considerations have been limited to autonomous systems – they do not take signals from the environment. Determining the states and functions of all nodes and the connections between nodes uniquely determines the trajectory - consecutive states of the whole network (sets of states of all nodes). We simulated the process of transformation of the disturbed system on the section *tmx*, then we compared the resulting state of the system with the undisturbed system. It is also looked after the node functions are correctly random, but this assumption cannot always be fully met, so the impact of the derogations is checked.

The size of a change in a network function at time *t* after a small disturbance is measured by the number ***A*** (from Avalanche [23]) of the nodes, which have a different state in the pattern network – identical network, but without disturbance. The value $\boldsymbol{d} = A/N$ is called damage. The distribution of damage size at the time *tmx* as ***P(d)*** or ***P(A)*** is an especially important result (**Figure 6**).

For random networks, this result creates two system states – ordered and chaotic. In system parameters space they occupy areas which Kauffman calls 'solid' and 'gas' respectively. Between them, there is a fairly quick transition (near $K = 2$, if Boolean signals are equally probable) treated as a phase transition. Only in systems in the vicinity of this transition (Kauffman calls it 'liquid' - the area between 'solid' and 'gas') changes in the system function (damage) often enough are small, therefore suitable for biological evolution. This is the main basis for the Kauffman's hypothesis: life on the edge of chaos. However, this conclusion aroused doubts [21], therefore, it has been subjected to a deeper analysis presented here.

[2] It is made using few method, in short: 'met'. Each of them is called using digit on the end and, if need be, some letter for its variant. In this case they are: met4c, met4d, met5, met6, met8, described in ch.3.1–4.

a

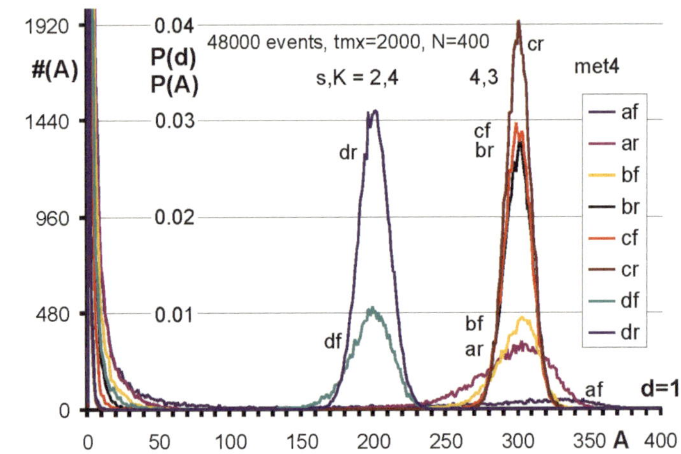

b

peak	af	ar	bf	br	cf	cr	df	dr
Left	43270	29258	32889	16851	15234	6922	31017	8345
Right	4730	18742	15111	31149	32766	41078	16983	39655

c

d

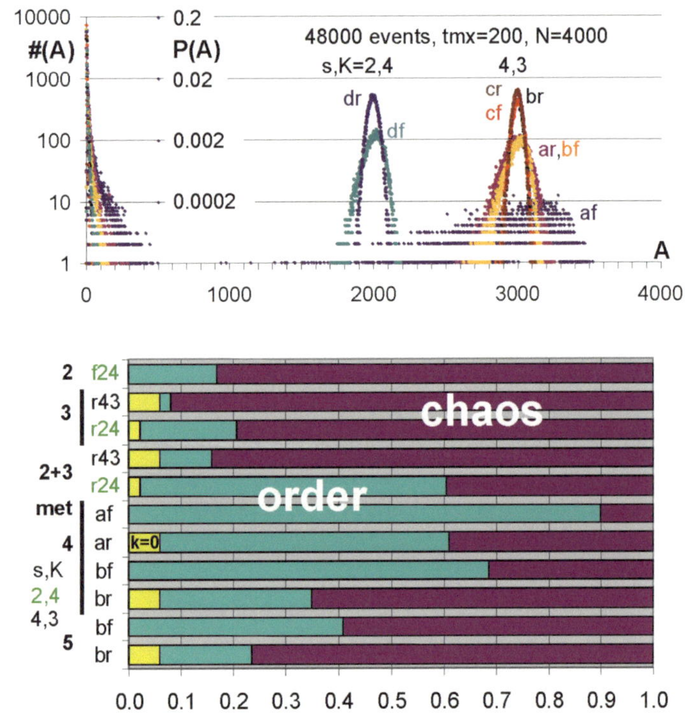

Figure 2.

Half-chaos and chaos in the presentation of A(t) *for a full set of initiations on the example of met7b* J *and* X *for network* ss. *This is a presentation observed dynamically during a simulation on the screen pixels. The details should be watched in enough magnification. In met7b* N = 800, tmx = 2000 *was used. A rectangle has the dimension of 400*1000 pixels, so on each axis, one pixel shows 2 values. In* ***Figure*** *3, for which this figure is a description of form,* N = 400 *and* tmx = 1000 *is used, so the there unit on the axes corresponds to a pixel. The vertical axis is originally scaled in the* A *- number of the nodes states different than in the pattern. The horizontal axis is the number of steps* t *of simulation of network functioning. After each initiation by small permanent change, the state* A(t) *was drawn with a continuous line on the screen after every step of the calculation. In case of initiation of a node in the in-ice-module black color is used and for initiation in the walls between in-ice-modules - purple. In met5 shown in* ***Figure*** *3, this distinction was not known and always black was used. To optimize the simulation a counting after 70 steps from the explosion to chaos (crossing over the threshold, here* = 300, *marked in red on the left) was stopped - there the process has no chance to return. As can be seen, the transition to chaos in the vicinity Derrida balance is not slow, but rapid in several to over a dozen steps, where A increases drastically, so - "explosion." after deflection from a small value to say* A = 80 *no longer the returns happened (as checked without optimization, see [1]). After the end of initiation set, the red curve* q(t) *was added to the figure. In met7 it is originally scaled by the A as the number of initiation, which does not exceed the threshold = 300, but there are 3* N = 2400 *of initiations. In met5 in a* ***Figure*** *3* q(t) *is divided by the number = 3 of initiation in node, so that* q = 1 *for* A = N. *The red description in the left has been added for readability and here* q(t) *is the share of processes that in the time* t *did not pass the threshold.* ***a*** *– Half-chaos, experiment* J *for network* ss, *model b. There were 600 of such simulations for each type of networks* sf, ss, er *and models a and b of met7. The red curve* q(t) *quickly stabilizes at a high level* q = 0.22. *In the lower part of the graph, many trajectories are visible (there are* L = 532 *of 2400) that a little over* t = 200 *no longer explode. So* R = 1868 *processes from the very beginning went to chaos - a Derrida balance.* ***b*** *- Chaos on the example of experiment* X *performed immediately after the measurement of the* J *illustrated above in the a. There were 300 of such simulations for each type of networks* sf, ss, er *and models a and b of met7 and for each experiment of* X, S, T, F. *Here,* q(t) *is steadily decreased until all the processes are not 'exploded'. At the end there is exact LX = 0 of them, means* q = 0. *Blue points describe the number of processes that currently have* A = 0, *i.e., damage fade out, but for the* X *the secondary initiations lead to their explosion.*

The conflict [7, 8] of a size of K in the Kauffman model and K estimated from nature [19] is a problem solved here. Kauffman postulates that the natural property of the random ordered systems (order for free [10]) is the source of stability, but then K should be extremely small ($K \leq 2$) [18]. The attempts to prove that the real genetic network (using model **GRN** – Gene Regulatory Network) is ordered [14, 15, 17, 23] assume such a source of stability. Different circumstances allowing system with greater K to be in the ordered phase were indicated (p.48 in [7]), such as a significant difference in probabilities of logical states [18], or deviation from the randomness of the function (canalizing [11]), but these and other suggestions are not satisfactory for many reasons [21]. The model GRN has disappointed many expectations, mainly due to restrictions arising from the range of 'liquid region', it was replaced by the more attractive Banzchaf model [24], but GRN is still being studied [25].

For investigation shown here, as typically, the same K for all N nodes of the network are taken.

2.2 More than two signal variants $s \geq 2$

According to my previous [21] suggestions, here I also study a larger number of s (>2, usually 4) of equally probable signal states, which in random networks for every sensible K (≥ 2) always gives chaos (**Figure 1**). In the range of sensible parameters s and K, the order appears only for $s = 2$ and $K = 2$, it is absolutely exceptional (**Figure 1b,d**). Attempts to introduce more signal states already exist [12, 16], but they assume the possibility of an ordered phase for the random network therefore these states cannot be equally probable.

I repeat here briefly my basic arguments given in ([21]; ch.2) for using $s \geq 2$ in Kauffman networks for statistical investigations:

1. Using Boolean network we can describe each complex relationship (mechanism), but bringing to two-value description frequent cases where significant signals take more than two variants, we generate unrealistic situations, presumably - to skip. In the statistical analysis, however, they are not skipped and give a false picture. Or we simplify something which we do not want to simplify. In both cases the statistical investigation is false. It was shown on the example of the thermostat ([21]; **Fig. 3**, p. 292). The only way is to use a real number of signal variants and not limit ourselves to only two Boolean alternatives. Because such case is frequent, then in one system it should appear for a lot of signals and the investigations based on $s = 2$ must be false (extreme).

2. Two variants are often subjective ([21] ch.2.1.2). There are typically lots of real alternatives, but we are watching one of them and all the remaining we collect into the second one. Typically our interesting variant has much lower

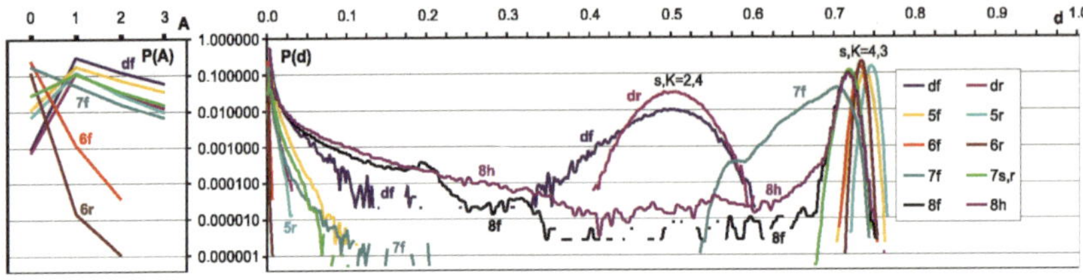

Figure 3.
Simulations met5 (changes accumulation) in the presentation of A(t). *Except for red description* q *on the left, each drawing was created dynamically on the screen during the simulation of one full set of initiation without blocking of reverse initial changes. It is accurate to the pixel. Description of the presentation elements in* ***Figure** 2. **a** - Full typical image for the* M13 *met5c (met5 in other figures, model c from met4), network* sf. *Almost an immediate end of the explosions to the chaos can be seen. At the top - the state of chaos in the Derrida balance (short due to optimization by interrupting the counting after 70 steps, as in **Figure** 2). At the bottom - a repeating pattern in accordance with the global attractor marked on the top frame (pattern network state as in* tmx *before the first initiation of the set). Here L and R under the lower frame is the sum from the beginning of the evolution simulation of this network. In this set 383 of initial changes were accumulated of 1200 tested, but accepted changes defining* q *(not exceeding the threshold = 150) were a little bit more (with global attractor < 7). **b** - Typical image of network* er *simulation in met5c. The upper part of the almost identical to a is cut. The level of* q(t) *is lower, the belt at the bottom - clearly thinner, the time of the latest explosion to chaos - shorter. **c, d** – The lower part of the image for met5b (with minimal regulation). Here the level of* q(t) *was much higher than in a. In the model b, the width of the lower belt is greater due to the possibility of regulation. Simulations slightly different model than in **Figure 4d** - here without blocking of reverse changes, but with the condition non-decreasing of global attractor and accumulation of changes not less than* A = 3, *the shift of beginning = 2, but not 50. In these simulations, a distribution of damage size for ordered cases* (A < 150) *was studied on the section from* t = 600 *to* tmx *for a given set of initiations (purple curve on the right frame) and the sum of the sets in the final set* M20 *(blue curve in c). It is one of several ways to look for proof of the in-ice-modules existence. As can be seen, in both (c,d) shown cases in these distributions the significant peaks are visible. They indicate an existence of one (in the c* M20) *or two (in d* M1) *hypothetical in-ice-modules. Under the scope of these peaks, there is a clear gap in the minimum of distribution. An interpretation of these peaks can vary, they are not proof of the in-ice-modules existence, which was shown later watching nodes states repeating, but they are a strong premise. The* q *level here is high: in c* q = 0.46 *and in d* q = 0.55. *In c the attractor was not found at the beginning of the set (attr ≥900), and because it could not decrease, no one accumulation happened (not.PAS saved = 0). It does not mean, however, that there is no here acceptable* (A < 150) *cases (there are 220), which indicates* q *and wide black belt below the* A = 150.

probability, resulting from the similar probability of each one, but description basing on ***p*** – the probability of signal variant leads to much different results (**Figure 1a**) than for the case, when we consider all variants. Adding the parameter p to the two-value description does not solve the problem here. Adoption of $\boldsymbol{s} \geq 2$ equally probable signal variants is an alternative method of model realignment. However, it seems to be often more adequate. Both methods give different results which significantly increases the importance of the correct choice of description.

3. Parameter s is more important in the description of damage spreading than K treated as the most important – see **Figure 1b**. Value ***dmx*** – mean maximal damage, i.e. Derrida equilibrium for chaotic behavior (**Figure 1c,d**), much stronger depends on s than on K.

2.3 Criteria of chaos, coefficient of damage propagation *w*

The main characteristic of the chaotic behavior of dynamic systems is high sensitivity to initial conditions, leading to maximally different effects for very similar initial conditions. It is original, theories using Lyapunov exponents or percolation are derivative. I use the term 'chaos' in such the meaning, similarly as Kauffman [6] does. For chaotic Kauffman networks a small initiation of damage

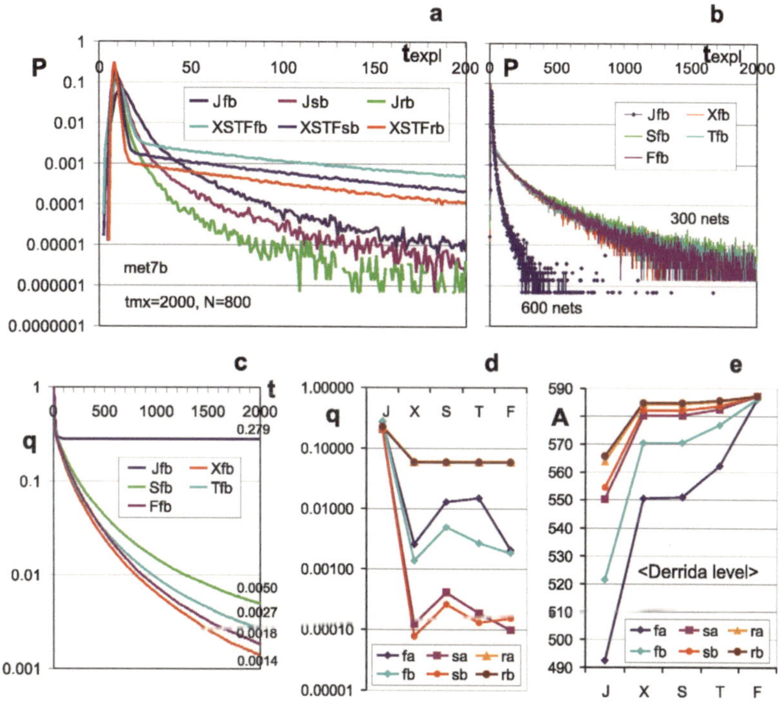

Figure 4.
Increasing regulation or another factor - the point attractor. The primary result of the met4. In the met4 removing a presumed cause of the poor performance of the met2, we start with the non-random system with extremely short attractor – a point attractor: initially, all states are set to 0 and $f(0) = 0$ (f – node function). The models were tested in the sequence a, b, c (s,K = 4,3) and d (s,K = 2,4) starting from strong regulation and ending with the lack of regulation in the models c and d. Care was taken that each signal has the same probability in the function of each node. ***Model a*** *contains a negative feedback with a positive (1) and negative (3) deflection from equilibrium (0) in each of the three input signals. It contains also the leaving of homeostasis into the area of randomness (when deflection is too great or one of the input signals = 2, then the node function is defined randomly). A more exact description of this formula can be found in the text of ch.4.3 and is available in [1].* ***Model b*** *has a minimal regulation: the condition of the point attractor $f(0,0,0) = 0$ is supplemented only by condition $f(0,0,1) = f(0,1,0) = f(1,0,0) = 0$ that there is no in* ***model c. Model d*** *of Boolean network (s, K = 2,4) has only condition $f(0,0,0) = 0$ similar to model c. Each model is simulated for three combinations of N,tmx = 400,200; 400,2000; 4000,200 for networks sf and er, so as to always number of initiations was 48,000 in the series. The threshold of small change for N = 400 was set to 100, and for N = 4000 to 800. Each initiation by definition of met4 is made for node state = 0 and for input state = (0,0,0). So only in the model c 3 other function values may be used for initiation. For model a the only one value 2 remains, for b only two values: 2 and 3, which are new states of a node without the mandatory fade out of damage at the destination.* ***a,c*** *- The counts #(A) of processes ending in tmx with value A (changed states of nodes in tmx) are shown. Also, the scale of the P(A) or P(d) are added. The results showed here in the linear plot a (N = 400) for models c and d are also in* ***Figure 6*** *in log scale. The series showed in c contains 10 times fewer networks, which gave peaks much narrower (in damage d scale instead of A) than in a. The right peak for models b, a is becoming smaller due to increased regulation, which is reflected in the diagram d as less participation of chaos. Place of the right peak in a and c are well designated by Derrida balance (****Figure 1d****) (different for s = 2 and s = 4), which is the property of a mature chaos.* ***b*** *- The table of results #(A) for tmx = 200 for the same networks as in a for which tmx = 2000. The counts differ only for af by 140 and for df by 2 (less for left peak).* ***d*** *- A complementary for* ***Figure 8*** *juxtaposition of a fraction of ordered cases (q) and chaotic cases (1-q) for minor experiments discussed in the article. While* ***Figure 8*** *lists only the study of impact of small attractor, it is here - the impact of increasing the share of regulation in met2 (only sf 2,4 can be considered in met2 as entry into half-chaos, see* ***Figure 5a,b****); of modularity in met3; assembling of met3 and met2 (****Figure 9****); assembling of point attractor and regulations in met4ab and met5b. Among them only met5b examined the evolutionary stability included in the definition of half-chaos. As can be seen, the assembling is more effective than approach alone and should be expected of such a strategy in biological evolution. The case af shows that the way evolution can lead to a state where the half-chaotic system may seem as ordered. Evolution met5b decreased q comparing met4b when met5 (****Figure 8****) worked in the opposite direction relative to met4c (these are uncertain trends), but the expected strategies of biological evolution its creative aspect is important, not modeled in the presented simulations, too simplified to such a task.*

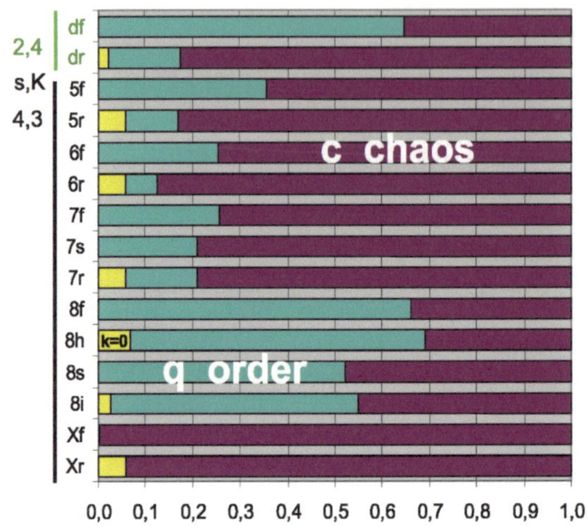

Figure 5.
Ordered fraction (q) *as a function of time* (t) *after raising the share of negative feedbacks (met2) and the classic modularity (met3). The upper row of all part -* s,K = 2,4 *(Boolean network), lower -* s,K = 4,3. ***A*** *– For some moments* t *the shares of mechanisms: Wild - without interference met2; function narrowing as a side effect of the method; the increased participation of negative feedback by met2. For network* er, *the level of q resulting from participation* k = 0 *(nodes without outputs) is indicated by the green line. In the right column as a wild the modular system resulting from met3 is used, further described in (c) as a curve a. the type of networks* sf, ss, er *is described by a second letter. As can be seen, the results for the simulation parameters* s,K = 2,4 *and* 4,3, *and network types, differ significantly. For* s,K = 2,4 *the function narrowing is of utmost importance to increase* q, *but for* s,K = 4,3 *the importance of feedback turns out to be essential. For small* t *the effect of increase* q *is significant. From these data it can be suspected to achieve half-chaos for:* sf 2,4 *- the result of functions narrowing and increase of the share of regulatory feedback, and for the assembly of modularity met3 with met2 using nets* er *- for 2,4 mainly due to the functions narrowing, but for 4,3 due to the met2. In the remain 5 presented cases the effect practically disappears already for* tmx = 1000, *the use of it by living entities require very rapid multiplication in comparison to the transformation of the construction and metabolism, which seems unattainable. Here evolutionary stability (included in the definition of half-chaos in the result of further studies restricting fundamental factors to a short attractor* ***Figures 6–8****) was not examined. The degree of entry into the plateau can be better assessed in* ***b*** *and* ***c****. the network* ss *gives a similar effect to the network* er, *but without the confounding effect of* k = 0. ***b*** *– Net* sf *4,3 (350 nets) not reached a plateau even at* t = 20,000, *where* q *is negligible, but* sf *2,4 (700 nets) is almost on plateau* q *at* t = 5000, *and this level is high (compare* ***Figure 4d****).* ***c*** *– The result of modularity (met3) and assembling it with met2. Result of met2 for network* er *is added, such as in* ***b****, omitting, however, the share of function narrowing enough presented in* ***a****. it can be seen that the wild system (without forced modularity, 700 nets for* s = 2, *350 nets for* s = 4*) of network* er *very quickly descends to the level of* q *resulting only from* k = 0. *Also curve b - the result of the met2 quickly closer to that level, which can also be seen in* ***a****. forced modularity (curve a, 100 nets) gives a clear stable increase of* q, *and met2 help it (curve ab) to radically increase* q, *but for* s,K = 4,3 *appears to fall within the plateau above* t = 20,000. *For* s,K = 2,4, *almost all large and stable met2 effect results from the function narrowing only (curve afb). The network has* N = 400 *nodes assembled of* N2 = 50 *modules each of* N1 = 8 *nodes.*

typically causes a large avalanche of damage which spreads onto a big part (percolates) of the discrete and finite system and ends at a Derrida equilibrium level *dmx* [18, 21, 26] (**Figure 1c,d**), which is a maximal loss of information about the previous system. Such distance cannot be infinite in the finite network witch finite discrete *s*. The existence of this limitation is the main difference between this 'chaos' and the more commonly taken definition [27] used for continuous variables on infinite space, where Lyapunov description works. **The term 'chaos' is not reserved for one of those separate areas. The distribution of damage size is the experimental base to classify a particular system of Kauffman network as chaotic or ordered using levels of damage equilibrium calculated from Derrida's annealed approximation (Figure 1d)**. In Derrida's model only case $s, K = 2{,}2$ (I use s,K as a vector) is ordered – it has no other cross of diagonal than in $d_{t+1} = d_t = 0$. In any other cases, such cross called here ***dmx*** exists; it is Derrida equilibrium level of chaotic reaction for the disturbance.

In a typical case, the chaos is indicated by Lyapunov exponent, which describe the growth of distance for two, near, initial states. For finite discrete networks, it corresponds to "coefficient of damage propagation" $\boldsymbol{w}$ described in ([21] ch.2.2.1) and earlier, or eq. 4.8 in [23]. $w = \langle k \rangle (s\text{-}1)/s$. It can be treated as damage multiplication coefficient on one node if only one input signal is changed. It indicates how many output signals of a node will be changed on average. For an autonomous network with fixed K, $\langle k \rangle = K$ and we can use $w = K(s\text{-}1)/s$. It is easy to see that for $w > 1$ damage grows, for $w < 1$ it disappears and $w = 1$ is critical – for $s = 2$ it gives known critical $K_c = 2$. In [7] similar eq. (6.2): $K_c(s\text{-}1)/s = 1$ is given which is a case for the condition $w = 1$. Coefficient w is a simplification for the beginning of damage growth, later a case of more than one changed input signal happens more and more often, but this first period is crucial (**Figure 1c**).

Note, we are going to know: is a particular network chaotic, ordered, or something else, therefore we test it by statistical experiment. We make small disturbance (perturbation) and look how great is a change of a function (damage d) of this deterministic network comparing to undisturbed network. Damage is an effect of this small disturbance. We make a lot of such small disturbances (see **Figure 2**), each in the same network being tested, and we get distribution $P(d)$ for one, tested network. For a chaotic network, the $P(d)$ contains one peak near dmx, for ordered – one narrow peak near $d = 0$. If there are both the peaks in the distribution for one particular network, then it is neither chaotic nor ordered network, it may be half-chaotic.

2.4 Types of networks

Several **types** of networks are considered. They differ in the rules of their creation (for *sf* and *ss* see **Figure 2** in [21]) and distributions of k (output links), (K – input links is fixed for all nodes of particular network): ***sf*** (scale-free [28]), ***er*** (classic Erdős-Rényi [29] "random"), and ***ss*** (single-scale). In the figures, the second letter of these shortcuts indicates the network type. In studies **met8** (denoted in figures by '8') the network grew - an addition or removal of the node was the disturbution. There networks ***sh*** and ***si*** are respectively ***sf*** and ***ss*** with 30% removal.

Parameters: network **type** together with $\boldsymbol{s,K}$ (treated as a vector) are the main variables in the simulations. In a wider description [1, 2] of here presented investigation, I used more network types.

2.5 The main results

At the beginning, in ch.2.1 there is the statement: 'the distribution of damage size at the time *tmx* as $\boldsymbol{P(d)}$ or $\boldsymbol{P(A)}$ is an especially important result'. It is shown in **Figure 6** for the main range of investigation and in **Figure 4** for mechanisms supporting half-chaos. However, it is the base for more important conclusions.

In obtained here distribution of damage size for the particular system there are two peaks: the left of small changes (ordered behavior) and the right of big changes (chaotic, near Derrida balance). Sharp boundaries of these peaks, supported by a clear gap between them define a "**small change**".

The main result, however, is a "**degree of order**" $\boldsymbol{q}$ – a fraction of effects (damage) of small perturbations which fit into the range of the "**small change**" of the functioning at the time *tmx*. It is summarized in **Figures 8** and **4d**. This q corresponds to the contents of the left peak or probability of acceptance of changes in the modeled evolution (lack of elimination).

The **degree of order** $\boldsymbol{q}$ is the base (see ch.2.3) to state, that we found half-chaos using definition given in the Introduction: **Half-chaos is a state of a system that is not fully random, with parameters that the random system make strongly**

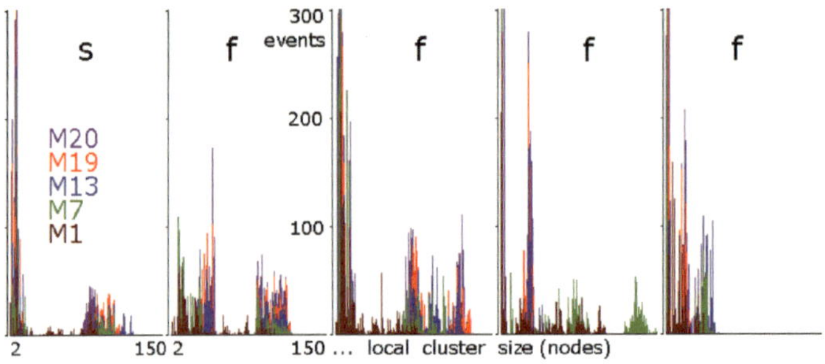

Figure 6.
The main result – distribution of damage size. Symbol of the method begins a signature. The methods: 'd', '5', '8' start from point attractor; 'd' (met4d, see ch.4.3) is the only with s,K = 2,4 *and without evolution, remain* s, K = 4,3 *with evolution; '6' (met6) starts from small attractors; '7' (met7ea) starts from constructed in-ice-modular system. After the method the second letter of network type ends signature. Results presented here (except 'd') are a sum from 4 already stabilized sets of initiation (see* ***Figure 7****). The gap between the peaks - left (ordered) and right (chaotic, near Derrida balance, different for* s = 2 *and 4) is not empty only for not really small disturbations by adding or removing a node ('8' – met8). The share of the left peak as* q *– degree of order is summarized in* ***Figure 8****. It is the basic result of this study; it allows to introduce half-chaos. Collecting only permanent changes which give damage from the left peak (i.e. small changes) is sufficient to keep half-chaos in the evolution (****Figure 7****). The shape of the left peak is important for the modeling an evolution of adapted systems. It is shown (without '8') in more details on the left for variable* A = d*N *where* N *is* = 400*. In the experiment '6' there is practically only* A = 0 *due to lack of in-ice-modularity. Network* sf *of '7' differs from the others in the left slope of the right peak, (see also* ***Figure 7c****) mechanism of this is unknown.*

chaotic, but small disturbances give the ordered reaction with a similar probability to the chaotic reaction. Such state is contrary to the current view, but the current view is based on the assumption of full randomness of the network which typically is not fitted.

The "**small change**" is a criterion of the acceptance of perturbing permanent changes creating the evolution, which is enough (**Figure 7**) to stay in half-chaos. It is the **evolutionary stability of half-chaos.** It **was included in the half-chaos definition. Acceptance of one** perturbing permanent change that gives a **big change** of the functioning at the time *tmx* (chaotic reaction) **leads to practically irreversible entry into normal chaos** (elimination). Note that in such great change of behavior only states of network nodes differ before and after, but in both cases they have the same, random–look distribution. Nothing has changed for currently used methods to define: is this network chaotic or ordered, but the behavior is absolutely different.

3. Half-chaotic systems, construction and mechanisms

3.1 Short attractor and secondary initiation as the main mechanism

The preliminary search (met1-4ab described in ch.4) of mechanisms enlarging stability for chaotic systems allowed for a deeper look at the process and its determinants. However, it turned out that they concern mechanisms of secondary importance which only support the main mechanism based on a short attractor effected from the phenomenon of secondary initiation. The first initiation does not have to lead to quick explosion to chaos, it even could fade out. Secondary initiation - the cases of re-appear at the inputs of disturbed node its initial inputs state for which the function has been permanently changed are responsible for the decline of $q(t)$ with increasing t (**Figure 2b** and 5). Such a secondary initiation takes place in different conditions than the previous one and can also lead to entering chaos or fade out. After a round of attractor new such cases are no longer present (see **Figure 9a,b**). If up to this point explosion to the chaos does not take place, then it will not appear later. For short attractor, it can happen with not a negligible probability. To check it, *tmx* must be greater than the sum of length (in time steps) of attractor and path to the attractor. In below-described researches, it turned out that global attractor can be large if it is assembled of few independent short local attractors, which is a typical case for in-ice-modularity (ch.3.3).

3.2 System with point attractor is half-chaotic

The study of **the systems with a point attractor** (further – 'point attractor system', system state is not changed over time *t*, attractor is extremely short, length = 1), with parameters s,K = 4,3 (**met4c**) and 2,4 (**met4d**), (see more in ch.3.3, ch.4.3 and description of **Figure 4**) which make random systems highly chaotic, gave clear results - such systems **are neither ordered, nor chaotic**. Both reaction variants on a small initial perturbation (ordered - a small change in the functioning and chaotic - a big change nearby of Derrida equilibrium – **Figure 6**) appear in similar proportions (**Figure 8**). **This state was named "half-chaos"**. In this state, the resultant change in the functioning (damage) can be either very small or very large (explosions to the chaos **Figure 2**), but almost no intermediate changes (**Figures 2–4** and **6**). This defines a **small change** in a natural way. There remains the problem of the length and condition of the evolution of the half-chaotic system.

Obtaining a point attractor is simple, just after the random generation of networks (nodes connections and functions) and the states, it is enough to take that for the current state of the node inputs a node function gives the current node state. For the remain states of the input - functions stay random. The point attractor system in Kauffman terms is a completely frozen system – there is only "ice" (nothing changes). The predominance of the ice is a spontaneous property of ordered systems. Obtaining small change after disturbation of half-chaotic, point attractor system, we can expect "a small lake of activity in the ice," (originally [5]: "unfrozen islands"), which is the essence of the 'liquid' area of random systems, where Kauffman sees place for life. But such a system ceases to be a point attractor system. It turns out that the vast majority (typically over 99%) of "small changes of functioning" gives also point attractor systems. Therefore, evolution may be long, however, such the model is quite extreme and unattractive.

Simulation studies and their analysis include many important details that are unfeasible to include in this article. They are described in more than 170 pages of the report [1], Only basic ones will be listed here. The particular system is calculated at *tmx* discrete time steps *t* after disturbation, and then at $t = tmx$, more adequate value for the final results (**Figures 6** and **8**) is recorded as averaged A over the last 50 counting steps *t*. Due to the strong influence of various factors often sporadic, **formal errors in the obtained results are not calculated, judging such a calculation as clearly inadequate and misleading**. This problem is limited to the similarity of results from the similar simulations and the visual evaluation of fluctuations. Given here a number of networks in described series of simulations concern showed results, but often experiments were repeated in a similar way, giving a much greater certainty.

3.3 The evolution from the point attractor

Next, for models b (see ch.4.3) and c of the met4 started from point attractors we checked how long can be evolution if it accumulates small disturbances caused

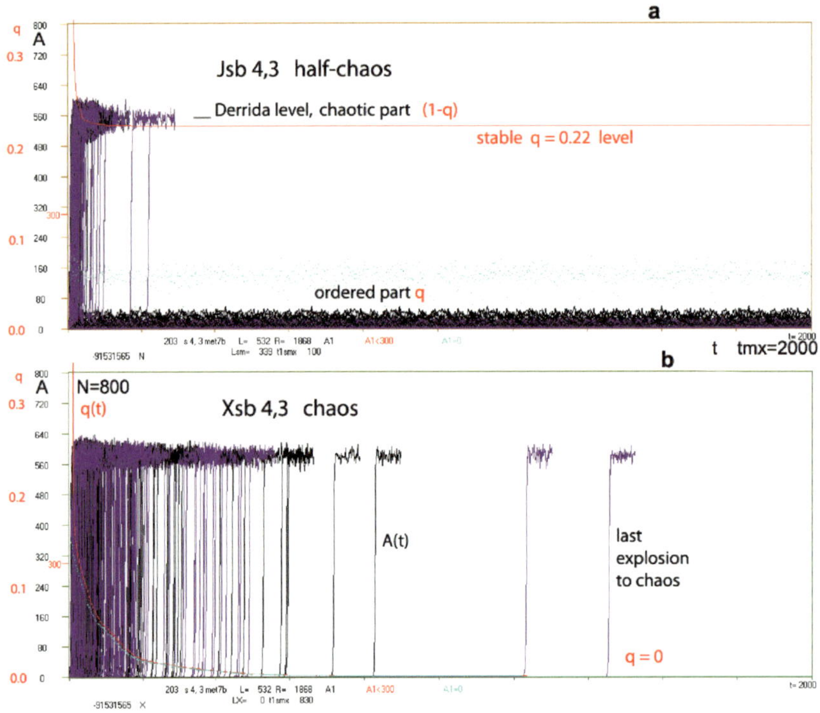

Figure 7.
The variability of basic parameters during evolution. The similarity of results for these 4 methods shows the similarity of obtained half-chaos, mainly its evolutionary stability, despite the differences in the way of obtaining. In a-c only met5c and met7ea are shown. ***a*** *- Stability of parameter* q *(degree of order of the system, the contents of the left peak in* ***Figure 6****) shows lack of moving towards the chaos during the evolution - accepting permanent changes which give small changes in the functioning (in the range of left peak, additionally excluded global attractors less than 7, and in the* M20 *of met5-7 also smaller than the already obtained).* ***b*** *- The average time of five latest explosions to the chaos (see also* ***Figure 5a,b****) does not grow in spite of the above indicated conditions on attractor's length. In the chaotic networks such explosions (see* ***Figure 4****) happen almost until the not yet exploded processes exist.* ***c*** *- The average size of local clusters (in met8 they are not checked) and the ice. It makes sense for in-ice-modularity, so not for the met6 where a single local cluster covers the whole network* (N = 400)*. In met7e network* sf *has a specific derogation. A mechanism of it has not been elucidated (see also* ***Fig. 1****, wider recognition in [1]).* ***d*** *- The average number of global clusters. In the met7 it also stabilizes from the* M7*. In the initial set of initiation* (J)*, still without accumulation, it is sometimes even greater than the number generated in-ice-modules, which shows that few so defined clusters may arise within one constructed in-ice-module.* ***e*** *- The average number of local clusters.*

small changes of functioning (small damage), but it does not allow new point attractors (**met5**). We received (also in **met6-8**) that it allows to any length of maintenances of the half-chaotic state and stabilizes its parameters (**Figure 7**). It is the **evolutionary stability** of half-chaos which **was included in the half-chaos definition**. The system still has a **significant prevalence of ice** (**Figure 7c**), and there are usually some **"small lakes of activity" forming "in-ice-modules[3]"**. Among the methods used to check the presence and properties of the in-ice-modules (see also **Figures 3c,d** and **10**), the most effective was to **track periods of node states**. The set of nodes with the same period in the process ended of accumulation was treated as a **local cluster corresponding with in-ice-module**. On average, at the same time occurred about 2 local clusters (**Figure 7e**). In the evolution, sometimes after many in the meantime accumulated changes, there appeared local clusters very similar in terms of nodes composition - a **collection of such local clusters is treated as a global cluster**. Methods to identify global clusters are very complex due to the wealth of different circumstances, including merger and disintegration of global clusters during evolution. However, we can say that they are generally quite

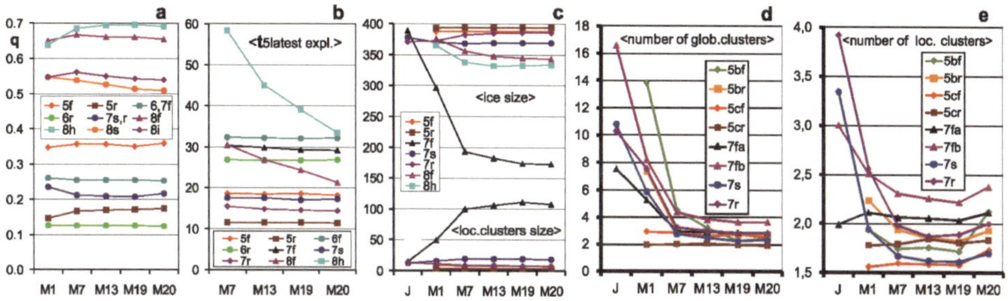

Figure 8.
Half-chaos – fractions of ordered events (q) *and chaotic* (c = 1-q)*. Experiments described as in* ***Figure 6****. In the range of* q*, an order resulting from the absence of output in some nodes* (k = 0) *in the network* er, sh *and* si *is isolated as yellow. All results presented here concern only the effects of length limitation of global attractors ('6' - met6) or length limitation of local attractors through in-ice-modularity. For '8' local attractors are not detected, but the level of ice (****Figure 7c****) shows, that local clusters cannot be large. For 'd' and 'X' (met7a) there is no evolution, the results concern the network immediately after generation of half-chaos, but for 'X' also after acceptance of one chaotic change, which gives a typical chaos (see also* ***Figure 7f****). In the remaining methods ('5', '6', '7' and '8') result is a sum of the results of 4 stable complete* M*, as in* ***Figure 6****, (see* ***Figure 7****). Except 'd' where* s,K = *2,4, in remain cases* s,K = *4,3. See also* ***Figure 4d****.*

stable formations, though they often disappear (freeze) and reappear, often in the other company of remaining global clusters, often changing period. Their average number for a set of initiations presents **Figure 7d**.

It should be emphasized that the **structure of the nodes connections in the investigated networks was constant and random**, although the randomness had various formulas that define the type of the network. **In-ice-modules are also the classic modules**, however this is only one, supporting, but less important factor. **The main property of the in-ice-modules is the activity** - changes of the states of nodes forming the in-ice-module. **The ice** (the area where the nodes do not change their states) surrounds them and **isolates from the other in-ice-modules**. In-ice-modules are the result of the functioning defined by the functions and states of the nodes in a given structure. Despite the selection of functions for obtaining initial point attractor state, **functions and states of nodes had truly random characteristics**.

Simulations met4, met5 and met8 start from the system with point attractor. **In the met4** (see also ch.4.3) networks *sf* and *er* were tested. Number of nodes N = 400 and 4000, section tmx = 200 and 2000 (no variant N = 4000, tmx = 2000). One set of initialization was tested - for s = 2 (met4d) each node is able to one initiation, for s = 4 there was 3 of the remaining function values. There were gained 48,000 events for each of the three variants of (N,tmx). The differences in the results of these variants were not significant (**Figures 4** and **6**), for further research in met5 we used N = 400, tmx = 1000.

We limited **met5** (and next met6, 7, 8) to s,K = 4,3, but these studies were much more complex. For a long process of evolution (accumulation of initiating permanent changes, which give small damage) we were studied many full sets of initiations, therefore the same change in function as an initiation has been repeated, but it was separated by many accumulations. Full evolution of the particular network is a collection of 20 sets (M) of initiations after one initial (J in **Figures 2**, 7, and **9**) set. In most of these sets, **retrogressive changes were blocked**. This results in the exclusion of a large number of initiations from the measurements and leads to a significant slowdown of evolution. After several such sets, the reversal is allowed (*M1, M7, M13, M19, M20*), assuming that the change has already another circumstance. It also allows to correctly measure of various phenomena that illustrate evolution (**Figure 7**). Since the attractor is decreasing spontaneously, making it

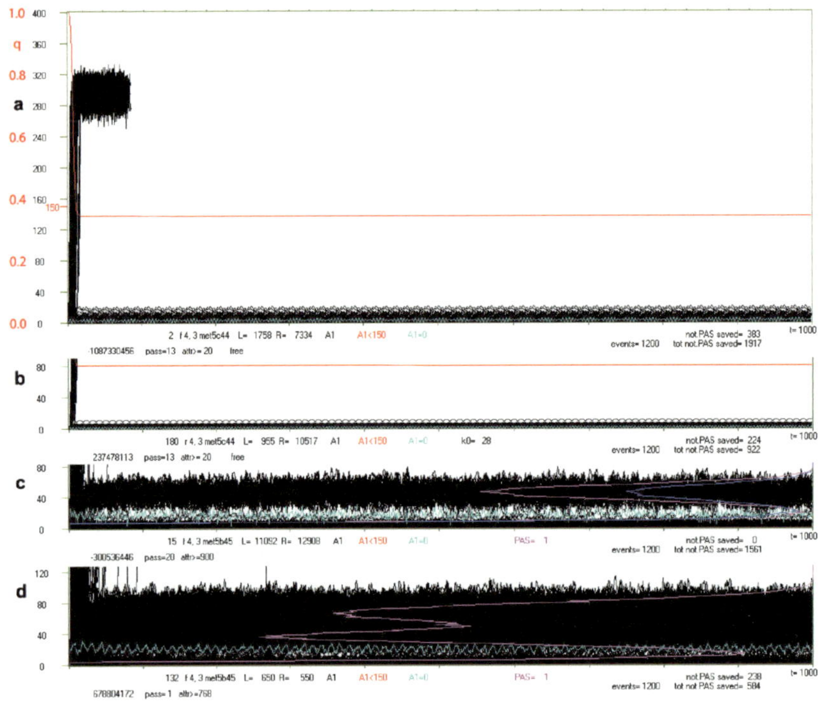

Figure 9.

The difference between half-chaos and chaos in met7a and b. Experiments met7a and b (without evolution) were supposed to deeper and more accurately demonstrate the distinctiveness of the achieved half-chaotic state and chaos. In comparison to studying the evolution, an elevated N = 800 *and* tmx = 2000 *were used. Variant b, over the conditions used in variant a, is forcing small attractors in in-ice-modules, and limitations: local attractor* ≤ 100 *and global attractor* >200 *also a shift to the latest start of local attractor* <500*. Experiment* J *- immediately after generation in-ice-modularity (600 networks), and after* J *further experiments* X, S, T, F *(300 networks).* X *- after acceptance of one chaotic change,* S *- after changing the node states to be random,* T *- the shift of functions to other nodes,* F *- after a generation of random functions for nodes. Despite the lacking possibility of the meaningful designation of measurement errors, the reproducibility of the results and the radical behavior otherness of* J *experiment clearly shows that the obtained state strongly differs from chaos.*
a,b *- Probability of time of explosion to chaos for met7b. This aspect is shown in the graphs of* A(t) *shown in **Figures** 2 and 3 where late explosions resemble the image to the chaotic and increase the uncertainty of the appropriate selection of* tmx. ***a*** *-* J *and* X *for network* sf, ss, *and* er. *For* J *the probability smoothly decreases with time increasing, for* X *appears the collapse near* t = 22 *and the transition to a much slower decline associated with the presence of chaotic explosion after the secondary initiations. None of the collapses for the* J *results from the completion of the first round of short local attractor. After this moment there is no explosion as a result of secondary initiation inside the in-ice-module, which would be happened in the new circumstances. This mechanism is an approximation since initiations are also held in the icy walls between in-ice-modules, but there damage spreads more difficult, and after penetration into in-ice-module already subjects to the indicated mechanism. There was a clear difference in the behavior of the tested types of networks -* sf *has later explosions, in this aspect it is the most similar to the chaos;* er *has the least of late explosions.* ***b*** *-* J, X, S, T, F *for network* sf. *Apart from the half-chaotic* J, *the remaining chaotic* X, S, T, F *practically overlap.* X *protrudes somewhat from below, and the* S *and* T *– from above. Very late explosions also occur in half-chaos, but they are rare. These are usually cases of especially large global attractors, sometimes not at all found in the range of* tmx, *furthermore, most initiations appear in the ice between in-ice-modules, where damage normally builds up slowly.* ***c*** *- Average* q(t) *for* fb *(network* sf *in met7b) in experiments* J, X, S, T, F. *Half-chaos in the* J *is clearly different and quickly stabilizes* q, *but* X, S, T, F *drop up to* tmx *and probably further and are a little bit different. In this measurement the difference may be within a measurement error, which is practically impossible to determine due to the multiplicity of factors, but in d at least the* S *and* T *seem to consistently differ from the* X *and* F. *Reviewing diagrams* A(t) *as in **Figure 2b** similarity is noted in the range of* X, S, *and* F, *but in the case of* T *there are frequent derogation of different nature, particularly for* fa, *where the result is strongly disturbed for a few special cases.* ***d*** *- Average* q *for all the tested types of networks (*sf, ss, er*) and models (a, b) in all the five experiments* J, X, S, T, F. *network* er *in chaotic cases hides differences due to presence* k = 0. *See also the discussion of differences in the description c above.* ***e*** *- Average position for the right peak of chaotic Derrida balance. Particularly large deviation for the* Jfa *and* Jfb *is shown in more detail in **Figure 6** and **Figure 7c**.* X, S, *and* T *behave here the nature of the derogation and the statistical derogation from the randomness of functions, which suggests such a source of visible here differences and determines the magnitude of the impact of non-randomness of functions on the results.* X *and* S *retain a correlation of non-randomness of functions with node place in the structure of the network, which* T *breaks.*

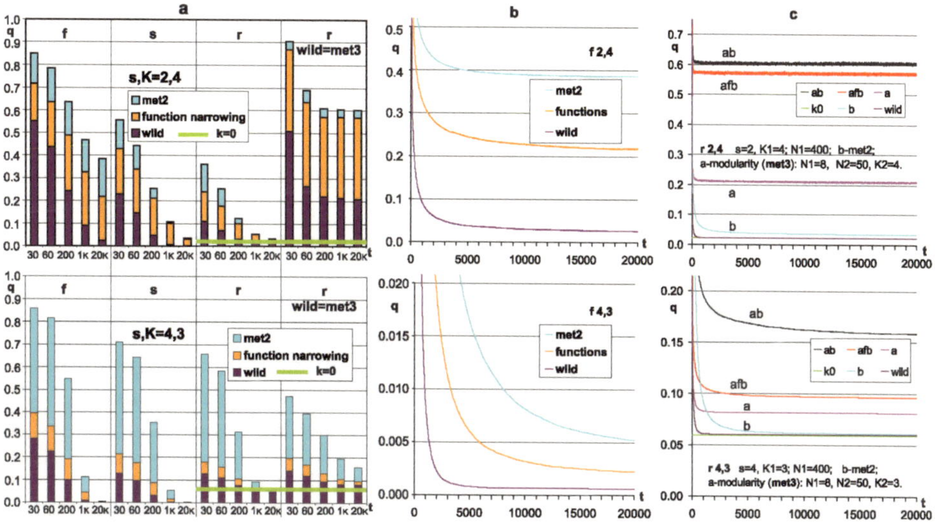

Figure 10.
Dynamical size distribution of local clusters and their stability through evolution met7eb. Distributions at end of M20 *of the size of local clusters in the range of up to 150 (of* N = *400) nodes collected only in indicated sets. It should be analyzed on the greatly enlarged picture in pixels – one pixel up means one event, to right – one node more in the cluster. Dynamically observed increases are significantly more uneven than this is due to randomness, it can be assessed painstakingly analyzing the size of the growth of a specific color assigned to the particular set* M, *but it does not reflect the image of a dynamic inside the set. This non-uniformity is associated with the presence of different in-ice-modules also changing during the accumulation. Such results are practically identical in met7ea for* er *and* ss, *only* sf *clusters are there typically larger. In the* er, *larger local clusters are very rare. Presented image, especially in the dynamical form in part reproduced through colors, is a strong, eye-argument for the existence and functioning of in-ice-modules. As can be seen, in-ice-modules may even be quite large.*

difficult to move away from the point attractor, it is also forbidden to reduce the global attractor to less than 7, and in the *M20* (in met5, 6, 7) to reduce the attractor.

Parameters q and average time of five the latest "explosions to chaos" are the most important, they demonstrate in **Figure 7a,b** lack of converging into chaos. They stabilize starting from set *M7*, despite a slightly elevated length of global attractor was forced. There were happen that the conditions for the attractor size block further evolution. Such processes were interrupted, however, in the main series (of met5 and met6) 100 networks were obtained, which reached the end of *M20*.

It turned out that the amount of a shift (in the range of 2-50) of the point of process start (place of the initiation) after each accumulation is an important factor. We assumed a shift of 50 steps. The study was much broader and deeper, their wide description can be found in [1]. Additional attempts of evolution referral more towards the boundaries of chaos gave no noticeable nearing - a condition of acceptance of a small change is enough for any long evolution - gives evolutionary stability of half-chaos.

3.4 Controlled design of the system with short-attractor

Point attractor, as extremely short, gave sought half-chaos. However, extreme is specific and in the evolution (met5) half-chaos was maintained even when attractor was not found in the range of *tmx* (**Figure 3c**). It should be checked whether the alone condition of a short attractor, but significantly greater than 1, is sufficient. For that, simulations **met6** causing in the random system a global attractor (of the whole network) = 21 was performed. From t = 21 for the unused input states of the node, the function value was changed to state 20 steps backward. **We obtained the evolutionarily stable half-chaos** even with a high q (**Figure 8**) for the same parameters and rules of the evolution simulation as in the met5. **The primary difference is the shape of the resulting left peak** (of small changes) in the distribution of damage size - there are practically only changes of a magnitude $A = 0$, but $A = 1$ and $A = 2$ are present in negligible amounts (**Figure 6**). This means that practically there are no changes in the functioning and in spite of the acceptance of permanent changes in the functions of nodes, nothing is changed. Such a process **is not suitable for modeling of adaptive biological evolution**, only for neutral evolution. A total **lack of in-ice-modules** was found, but the classic modules are present like in met5. In half-chaos based on in-ice-modularity as in the met5, the peak of a small damage contains a significant amount of change in the range $A = 1$ to 4, and also larger changes occur markedly frequent (**Figure 6**). In-ice-modularity in met5 explains achieved stability for the larger global attractors - they are assembled of small local attractors (in in-ice-modules), but this solution was checked in met7.

3.5 Controlled design of the in-ice-modular system

To determine the sufficiency of the in-ice-modular state to obtain stable half-chaos, we have attempted to controlled create it without booting from the point attractor (**met7**). Networks *sf, ss,* and *er*, $s,K = 4,3$ was studied. First, a network of N nodes and their states are randomly generated (dependently on network type). Next, analyzing of the node connections, a collection of 'in-ice-modules' was created and everyone node was assigned to an in-ice-module or separating them ice. Node created new in-ice-module when none of its link (input and output) was connected to a node belonging to an already existing in-ice-module. When it was connected to nodes belonging to only one in-ice-module, it was assigned to this in-ice-module. When it was connected to the nodes belonging to several in-ice-modules or if the limit of in-ice-modules (= 10) or the size of the in-ice-module (= 100 nodes for $N = 800$, 25 nodes for the study of evolution) was exhausted, the node was assigned to the ice.

Next, a trajectory was calculated by appropriately functions selecting. For the current input state, if it was not previously defined, nodes of ice get the value of the function equal to 0, but nodes belonging to in-ice-modules – random value.

A number of additional conditions and adjustments was applied, documentation [1] contain a full description, their details are not important here. **Initially, short attractor was forced in each in-ice-module** and using this assumption basic investigations were made: (**b**) – of the in-ice-modularity state (series with N **= 800 and *tmx* = 2000 without evolution** roughly corresponding to the met4) and (**eb**) - **the evolution** as in the met5 and met6 (series with N **= 400, tmx = 1000**). In the end, the necessity of **this assumption** was verified and surprisingly it **occurs unnecessary**. So the two most important research without the forcing of the short attractor in in-ice-modules were repeated (called **a** and **ea** - as logically simpler).

Examination (*J*) of the in-ice-modularity with N **= 800** mainly relied on checking the q and the distributions of damage size. In the versions b, we demanded the global attractor to be greater than 200 when the local attractor could not exceed 100 - the result was in line with the tested vision which explains the admissibility of larger global attractors. **In both versions (a and b) it was verified that the statistical properties of non-randomly selected functions are not responsible for the increase of stability**, namely - how such a system behaves after: the acceptance of one large change (***X***), randomly changing of node states (***S***), moving the functions to other nodes (***T***), and the random generation of new functions (***F***).

In the experiments *X, S, T* functions retained their statistics. In all these experiments chaos yielded (like *X* in **Figures 2b** and **8**), but it systematically slightly differed from the full version of chaos *F* (**Figure 9**).

Comparing with the met5, particular for network sf, both peaks of the distribution of damage size have been a little bit changed (**Figure 6**). Also in distributions of the ice size and the local clusters size the blur arise what caused a marked decreasing of average ice and increasing average size of local clusters (**Figure 7c**). This shows getting a **slightly different state of in-ice-modularity**. Like in the met5 and met6, system parameters stabilize from the *M7* and **the small change as a condition of acceptance is sufficient to any long maintain of half-chaos in the version of such the in-ice-modularity.**

3.6 Growing half-chaotic networks

Much more complicated and stronger is the **disturbation of a system through adding or removing a node (met8)** [2]. There are problems with the comparison to the undisturbed system and the interpretation of secondary initiation. These simulations **start from a small system (*N* = 50) with a point attractor**. The network grows in 5 successive stages *M* by 100 nodes and reached *N* = 550 at the end. The overall picture was very close to met5 and met7. Also, **half-chaos** (**Figure 6** right, **Figures** 7a,b and 8) **with evolutionary stability and stable presence of large ice share are obtained** (**Figure** 7c). It suggests similarity of mechanisms of increased stability to in-ice-modularity. In this case, the network grows by evolving under the control of a small change. The gap between the right and left peaks is not so empty here (**Figure 6**), probably because adding or removing a node is not a very small disturbation.

4. Supports for stability

4.1 More of negative feedbacks in a random system, function narrowing

It is generally believed that the stability of the various systems results from homeostasis based on regulation by negative feedbacks. Kauffman pointed instead to the property of the ordered phase (order for free) [10] as the most important reason, but for it extremely small *K* should be expected. **The regulatory feedbacks** are generally considered the basis for the stability of living entities and their **concentration is considered to be significantly increased** in relation to the random one. **However, the complex structure of the feedbacks for this statistical surplus has been replaced in the Kauffman model by their proper effect (ice)** and it remains only in random share. So much simplified model is not able to give a proper statistical picture of a system failure and conclusions for a stability mechanism can (and seem) significantly differ from reality.

This doubt was the main reason for undertaking the research, which initially aimed to strong raise the share of regulatory mechanisms.

In the presented study **we transform part of the feedbacks in random structure into negative feedback**. It is done by changing the random function when the state on the inputs was not used yet. It was the first method (**met1**) of correction of a random chaotic system. The similar, stronger **met2** has iterative change the pattern. Network *s,K* = 2,4 and 4,3 were investigated. **Figure 1c** suggests that Derrida chaotic balance is achieved even before the 15-th time step. **Initial research for *tmx* = 60 steps yielded very promising results** (**Figure 5a**) - *q* was significantly increased (especially for *s,K* = 4,3), the distribution of damage size already contained two peaks separated by a gap. A large part of this effect (especially for *s*, *K* = 2,4) was the result of deviation from the randomness of node functions (function narrowing), which also may be included [11] to evolution tools. But it turned out (**Figure 5**) that obtained in met2 stability of ***q*** **usually significantly decreases with the elongation of *tmx***, practically disappears already for *tmx* = 1000, only in the case of Boolean networks *sf* 2,4 this method could be considered to be effective to achieve half-chaos (not tested for evolutionary stability). As it can be seen in **Figure 5b,c**, *tmx* = 20,000 was used. Simulation series contains 700 nets for *s* = 2 and 350 nets for *s* = 4.

These studies demonstrated a **high range of results dependence on the network type** - the network *sf* is more ordered [9]; network *ss* and *er* are more chaotic, similar to the reaction, but *er* has part *k* = 0 (**Figures 4**, **5** and **8**) obstructing observation. **The parameters *s,K* = 2,4 and 4,3 also give a very different picture**. The simulation allowed for a deeper look at the process and its determinants, which **pointed to the short attractor** (ch.3.1).

4.2 Modularity

It seemed that the most natural way to get short attractors is modularity. In met1 and 2 no modular effects were observed, although modules exist in practically every network. It was assumed that in a random network the modules are too 'weak', then it was pre-checked, what stronger modularity gives for stability (**met3**). Here it turned out that sufficiently small spontaneous attractors can be expected only in so small modules that the consideration a state of chaos in them losing meaning. Consideration of chaos in the modules network has been postponed.

In the study, the network has ***N*** **= 400 nodes. It was assembled of *N2* = 50 modules, each of *N1* = 8 nodes**. Connectivity *K1* between nodes inside modules *K1* = *K2* connectivity between modules. The rule of connection is taken like in type *er*. Simulation series consists of 100 nets.

The modularity also gave raise *q* (**Figure 5c**), **especially when met2**, which increased the share of negative feedback, **is used at the same time**, however, evolutionary stability was not checked. In the distribution of damage size, the typical for the half-chaos radical **gap between peaks was not observed, only the clear minimum**. An increase of *q* in the experiment met3 + met2 with *s,K* = 2,4, almost entirely resulted from non-randomness of functions (**function narrowing**). Both of these methods and their associated factors (such as function narrowing) belong to the most important methods of producing desired stability by biological evolution, but in both the short attractor is an important factor.

As was described in ch.3.3, classic modules cooperate with in-ice-modules. The theme of classic modularity and its role in system stability was here recognized only provisionally and requires much deeper research. However, it is one more source of modularity, than was found in [30], where the role of modularity is studied in depth in evolution.

4.3 Regulation in system with point attractor

Lack of expected radical effect of regulatory mechanisms in the met2 was found in the system starting from a random network, then we introduced strong regulation in a system with a radically short attractor – point attractor (**met4a**). This time the result was surprisingly strong (**Figure 4**), so we decreased the regulation to the minimum (**met4b**, see also **met5b**, **Figure 4d**) and next, regulation was rejected at all (**met4c,d** and later), which showed that the point attractor is sufficient to achieve half-chaos.

In met4 point attractor starts with all node states equal 0. It was not permitted in met8, where states are random. Model **met4c** for $s,K = 4,3$ used later in met5 is defined as f(0,0,0) = 0. **Model d** for $s,K = 2,4$ – as f(0,0,0,0) = 0. They are based only point attractor, without regulation. **Model b** with minimal regulation ($s,K = 4,3$) also used later in met5, has in addition to c also: f(0,0,1) = f(0,1,0) = f(1,0,0) = 0. For signal value 1 interpretation was taken: deviation from proper state '0', but still in the range of homeostasis. For **model a** ($s,K = 4,3$) also f(0,0,0) = 0, but description is much more complicated. Here direction of deviation in homeostasis range: 1 – positive; 3 – negative. The deviation of one of 3 input signals gives 0. The function also gives 0 if 2 signals are deviated, but in the opposite direction and third is 0. If 2 signals deviate in the same direction but third is 0 or 3 signals deviate, but they are not equal, then function result is deviated, i.e., is 1 or 3. If 3 signals deviate and are equal or at least one is 2, then the result of the base function is 2, but such value for a particular node is converted into random value in the way that share of each function value be equal. Other parameters of simulation in met4 are described in ch.3.3.

The result of met4a shows how strong may be the effect of the regulation in the half-chaotic system – right peak almost disappears, that is the probability of entry into chaos as a result of a small system failure (internal cause) is small. **This gives a deceptive picture of the ordered phase** [14, 23]. There remain external causes, which model of the autonomous network does not take into account from assumption. However, adaptation is to the environment, which can vary and the evolution should be tested using open systems as in [31].

5. False assumptions of Kauffman's model – summary

The Kauffman's widely known hypothesis "life on the edge of chaos and order" [5, 6], pointed out an important factor in modeling of biological evolution, processes in social organizations, and technical constructions, however, it was based on too simple model, even – on few false assumptions:

1. Any network of conditions can be described as Boolean, then it is sufficient to study the Random Boolean Networks (RBNs). Such complex networks are finite, discrete, deterministic, and fully random.

   The assumption that the statistical properties of Boolean networks are general is false [21]. The number s of equally probable signal variants should be also considered higher than only two.

2. RBNs can be either ordered or chaotic, which is observed and confirmed by the current mathematical theory of chaos

   The current mathematical theory defines chaos by Lyapunov coefficients in **infinite, continuous space**. High sensitivity to initial conditions, leading to maximally different effects for very similar initial conditions is the main characteristic of the chaotic behavior of dynamical systems. Kauffman [6] uses such the term 'chaos' to describe **finite, discrete networks. The term 'chaos' is not reserved for just one of those separate areas**. This theory is used for finite discrete networks (e.g. [19]), but such a method **is an approximation, which loses a few important phenomena**, e.g., repeating the same argument for a function, the path length to the attractor and attractor length (in steps of a process). Analog of Lyapunov exponent for networks (coefficient of damage propagation [21], eq. 4.8 in [23] or eq. 6.2 in [7] in the case of half-chaos turns out to be misleading.

Model-forced strong limitations on parameters are not compatible with estimations from nature [7, 12, 16, 19]. For the evolution of life, the model allows only extreme $K = 2$ (connectivity, K—number of node inputs) and $s = 2$. Higher values of K or s lead to useless chaos.

3. Random networks contain all possible networks, then it is not important that **living organisms are not random** in the aspect of stability due to natural selection. Many works have assumed that this stability is explained by natural properties of the ordered system known as "order for free" [10]. These are false assumptions. Such a picture was not very consistent with the observed delicacy of living entities, not emphasized of regulatory structures and did not contain a model of death necessary for the Darwinian elimination. Kauffman [6] considered negative feedbacks, but practically [21] he left them on a random level.

In this work, it is experimentally shown that among discrete and finite systems that are not fully random, with parameters s and K which for fully random system result in chaos, there is a third state of systems I call **half-chaos.** The not fully random networks where half-chaos is found are obtained considering the specific correlation of parameters which Kauffman simplifying took as random. The analogy to the phase transition is more complex here - it is rather the "superheating".

The particular half-chaotic system exhibits small and large damage. Current theory does not foresee such the possibility, but it is easy to show examples using computer simulation – system with point attractor is half-chaotic. The modeled objects (like living or administrative units, technological processes, and technical constructions) are certainly neither infinite nor continuous. Half-chaotic systems better describe the modeled objects, **freeing modeling from difficult theoretical limitations** (see point 2 above), which until now are the typical basis of many considerations [7, 10, 12, 13, 16, 17, 23]. This opens the door to adequate models with complex networks.

The large gap between small and large damage defines in a natural way a small change, which is very important for interpretation. The peak of great changes (of functioning—damage) well model a death and elimination. After the great change, the system becomes forever simply chaotic, but a small change retains half-chaos and identity of the system, then evolution can go on. This feature as 'evolutionary stability of half-chaos' was included in the half-chaos definition. Half-chaos together with given initializing changeability completed by the multiplication of evolving system resulting from the demand of long evolution offers the full basic Darwinian mechanism.

The Kauffman model is trying to describe living systems and similar ones using several easy to show, and it would appear that the main parameters, the rest of them simplifies assuming their randomness, but natural selection works on all possible parameters, which may be easier and more important for selection and its effect. Indeed, it is difficult to imagine the possibility of the existence of half-chaotic systems from Kauffman point of view. In fact, after the system is drawn, it is either chaotic or ordered (ad. 'observed' in point 2 above) and the set of random systems contains all the possible ones (point 3 above). In the interpretation of the results of this approach, it has not been seen that the statistical absence of intermediate systems does not imply a small number of such systems. There are a lot of half-chaotic systems, but their share is negligible because there are radically more chaotic systems with given parameters (e.g., K) - for larger N not imaginable many. Model GRN based on RBN is not false, but its assumptions are too simple. Each model is a simplification, for some applications it can be useful, but if it gives a false expectation of important parameter, then some simplifications must be rejected which is the next step of approximation. It cannot be found without a previous step.

Regulatory feedbacks (misinterpreted and practically included only on the random level in the Kauffman model [21]) also the classic modularity and narrowing of the function significantly increase the stability, which was noticed, but the main and the new condition is the short attractor. They take over the role of explaining the experience [14, 15, 17] from "order for free", which in the half-chaos lost importance. The reached a deeper interpretation of Kauffman hypothesis gives a picture much more consistent with the observation and indicates systems more adequate to the modeling of biological evolution. This significantly alters the existing basis of many considerations and probably their conclusions. Likewise, the description of the systems from 'liquid' region [5], where Kauffman saw living objects - "small lakes of activity in the ice" (originally [5]: "unfrozen islands") remains valid for the primary and the most appropriate form of the half-chaos for the evolution - in-ice-modularity discovered in these studies. The base of the in-ice-modularity is an activity of nodes (they change their states) in the ice (where nodes do not change their states), however, in-ice-modularity is supported by classic modularity, which is always present.

6. Conclusion

This work examines the systems described by networks that are: autonomous, complex, finite, discreet, directed, functioning, deterministic, and designed as the Kauffman network (Boolean), but with the admission of more than two ($s \geq 2$) equally probable signal variants. The number K of the node inputs in a given network is fixed. Parameters s and K have values ($s = 2$, $K = 4$ or $s = 4$, $K = 3$), which in the case of fully random networks give unambiguous chaos.

Half-chaos is the state of such a system in which small disturbances cause both small and large 'damage' (changes in the system's functioning) occurring statistically similarly often. As a reminder, in the chaotic system there are only large damages and in the ordered - only very small. The studied half-chaotic systems are not fully random, but they have typical characteristics indicating a full randomness.

The evidence presented in the work indicates that half-chaos is an experimental fact. Its basic mechanism is based on a short attractor, but it is too weak a condition for modeling adaptive evolution. Much more adequate (to describe the purposeful systems) the half-chaos variant depends on in-ice-modularity. The simplest way to obtain such a state is starting from an easily attainable system with a point attractor, but it has been shown that it is possible to build such a state based on its description recognized in the evolution started from the point attractor. These are 'small lakes of (nodes) activity' in 'ice' (the area of the network where nodes do not change their states) - a picture similar to the one described by Kauffman (in the network parameters space) of the 'liquid' area at the boundary of the ordered ('frozen / solid') area and chaotic ('gas'). The Kauffman model is the basis of the famous hypothesis "life on the edge of chaos", however, this model strongly limits the parameters allowed for modeling life to $K = 2$ and $s = 2$ and their immediate vicinity called phase transition. Half-chaos allows a much larger range of these parameters, many estimates indicate such a need.

The experiments used a constant random structure (mainly scale-free and Erdős-Rényi random networks, but also others), random initial states of nodes and random functions, but despite the maintenance of characteristics indicating the randomness of the function, they were non-randomly correlated with states. Evolutionary variability concerned node functions. However, half-chaos was also observed in experiments, where the network grew by random addition and removing of nodes. This testifies to the more general nature of the discovered phenomenon of half-chaos.

Acceptance (as an evolutionary change) of a disturbance that gives great damage leads to ordinary chaos, which practically does not return to half-chaos. This is the elimination model - death. On the other hand, acceptance of a disturbance that gives small damage is enough to remain in half-chaos. This feature is: 'evolutionary stability of half-chaos', it is one of the most important, added to the definition of half-chaos. It creates a natural criterion of identity of the evolving object. The distinction between small and large damage is natural because they create separate peaks in the distribution of damage size separated by a large gap, in which there are practically no counts.

The discovery of half-chaos radically changes the vision of the dynamics of the studied systems. The famous Kauffman's hypothesis 'life on the edge of chaos' is strongly reinterpreted to 'life evolve in half-chaos of not fully random systems' and the analogy to phase transition is substituted by the comparison of half-chaos to 'superheated liquid'. Strong limitations contrary to the observation, on the parameters of modeling of purposeful systems are removed.

List of abbreviations and new terms

Abbreviations used in figure descriptions are defined in those descriptions.

N	network consists of ***N*** nodes.
K	connectivity. A node in a network receives signals at the ***K*** inputs.
k	number of output links of the node.
types of networks	***sf*** - scale-free (Barabási-Albert), ***er*** - classic "random" Erdős-Rényi, ***ss*** - single-scale, ***sh*** and ***si*** are respectively ***sf*** and ***ss*** with 30% removal.
Parameters	network **type** together with ***s,K*** (treated as a vector) are the main variables in the simulations.
RBN	Random (classic Erdős-Rényi) Boolean Network.
GRN	Gene Regulatory Network proposed by Kauffman, based on **RBN**.
t	is the number of time steps from a disturbance initiation.
tmx	the maximum number of counting steps.
A	Avalanche. The size of a change in a network function at time *t* after a small disturbance is measured by the number ***A*** of the nodes, which have a different state from the pattern network – identical, but without disturbance.
d	damage ***d*** = *A*/*N*.
dmx	maximal damage, i.e., Derrida equilibrium for chaotic behavior.
P(d) or ***P(A)***	the distribution of damage size at the time *tmx*, an especially important result.
w	coefficient of damage propagation. $w = <k> (s-1)/s$. For an autonomous network with fixed K, $<k> = K$ and we can use $w = K(s-1)/s$.
small change	in obtained here the distribution of damage size ***P(d)*** for the particular system there are two peaks and the

	clear gap between them: the left of **small changes** (ordered behavior) and the right of big changes (chaotic, near Derrida balance).
***q* - degree of order**	a fraction of damage in the range of the "**small change**", the capacity of the left peak of ***P(d)***.
chaotic parameters	they make random system strongly chaotic.
Half-chaos	state of a not fully random system, with **chaotic parameters**, but small disturbances give the ordered reaction with a similar probability to the chaotic reaction.
evolutionary stability of half-chaos	the "**small change**" as a criterion of the acceptance of perturbing permanent changes creating the evolution is enough to stay in **half-chaos**. It was included in the **half-chaos** definition. (Acceptance of one change that gives a chaotic reaction leads to practically irreversible entry into normal chaos).
in-ice-module	the set of connected active nodes surrounded by ice – inactive nodes (with the constant state).
local cluster	the set of nodes with the same period of their states in the process ended of accumulation. It corresponds with **in-ice-module**.
global cluster	a collection of **local clusters** very similar in terms of nodes composition in the evolution of one network.
met#	method #, where # is a digit 1 to 8. Separate experiments with different rules described in this article.

Author details

Andrzej Gecow[†]

[*]Address all correspondence to: andrzejgecow@gmail.com; gecow@op.pl

[†]My personal page is: https://sites.google.com/site/andrzejgecow/home
This research did not receive any specific grant from funding agencies in the public, commercial, or not-for-profit sectors.

References

[1] Gecow A (2016) Report of simulation investigations, a base of statement that life evolves in the half-chaos. http://vixra.org/abs/1603.0220

[2] Gecow A (2017) Report of simulation investigations, part II, a growth of half-chaotic autonomous networks. http://viXra.org/abs/1711.0467

[3] Gecow A (2016) Life evolves in half-chaos of not fully random systems. http://vixra.org/abs/1612.0390

[4] Gecow A (2017) Experimentally confirmed half-chaos removes the strong limitations in modeling using dynamic complex networks. 2017. http://arxiv.org/abs/1712.09609 v1

[5] Kauffman SA (1990) Requirements for Evolvability in Complex Systems - Orderly Dynamics and Frozen Components, *Physica D* 42; pp. 135-152

[6] Kauffman SA (1993) The Origins of Order: Self-Organization and Selection in Evolution. New York: Oxford University Press

[7] Aldana M, Coppersmith S, Kadanoff LP. (2003) Boolean Dynamics with Random Couplings. in Perspectives and Problems in Nonlinear Science, Applied Mathematical Sciences Series, ed. Kaplan E, Marsden JE, Sreenivasan KR, Berlin: Springer-Verleg

[8] Aldana M (2003) Dynamics of Boolean Networks with Scale Free Topology. *Physica D* 185; pp. 45-66

[9] Iguchi K, Kinoshita S, Yamada H (2007) Boolean dynamics of Kauffman models with a scale-free network. *J Theor Biol* 247; pp. 138-151

[10] Kauffman SA. (1996) At Home in the Universe. Oxford University Press USA;

[11] Kauffman SA, Peterson C, Samuelsson B, Troein C (2004) Genetic networks with canalyzing Boolean rules are always stable. *PNAS* vol. 101 no. 49; pp. 17102-7

[12] Luque B, Ballesteros FJ (2004) Random walk networks. *Physica A* 342; pp. 207-213

[13] Nghe P, Hordijk W, Kauffman SA, Walker SI, Schmidt FJ, Kemble H, Yeates JAM, Lehman N (2015) Prebiotic network evolution: Six key parameters. *Molecular BioSystems*, 11(12); pp. 3206-17. DOI: 10.1039/c5mb00593k

[14] Shmulevich I, Kauffman SA, Aldana M. (2005) Eukaryotic cells are dynamically ordered or critical but not chaotic. *PNAS* 102 (38); pp. 13439-44

[15] Serra R, Villani M, Semeria A (2004) Genetic network models and statistical properties of gene expression data in knock-out experiments. *J Theor Biol* 227; pp. 149-157

[16] Sole RV, Luque B, Kauffman S. (2000) Phase transitions in random networks with multiple states". *Technical Report* 00-02-011, Santa Fe Institute

[17] Villani M, La Rocca L, Kauffman SA, Serra R, (2018) Dynamical Criticality in Gene Regulatory Networks. *Complexity* Hindawi https://doi.org/10.1155/2018/5980636

[18] Derrida B, Pomeau Y (1986) Random Networks of Automata: A Simple Annealed Approximation", *Europhys. Lett.*, 1(2); pp. 45-49

[19] Turnbull L, Hütt MT, Ioannides AA, Kininmonth S, Poeppl R, Tockner K, Bracken LJ, Keesstra S, Liu L, Masselink R, Parsons AJ, (2018) Connectivity and complex systems: learning from a multi-disciplinary

perspective, *Applied Network Science*; 3: 11 https://doi.org/10.1007/s41109-018-0067-2

[20] Nowostawski M, Gecow A. (2011) Identity criterion for living objects based on the entanglement measure, ICCCI 2011, Studies in Computational Intelligence 381, Radosław Katarzyniak, Tzu-Fu Chiu, Chao-Fu Hong, Ngoc Thanh Nguyen (Eds.) Semantic Methods for Knowledge Management and Communication, Springer; pp. 159-170

[21] Gecow A (2011) Emergence of Matured Chaos During Network Growth, Place for Adaptive Evolution and More of Equally Probable Signal Variants as an Alternative to Bias p. In: Chaotic Systems, E. Tlelo-Cuautle (ed.); pp. 280-310, ISBN: 978-953-307-564-8, http://www.intechopen.com

[22] Kauffman SA (1969) Metabolic stability and epigenesis in randomly constructed genetic nets. *J Theor Biol* 22; pp. 437-467

[23] Serra R, Villani M, Graudenzi A, Kauffman SA (2007) Why a simple model of genetic regulatory networks describes the distribution of avalanches in gene expression data. *J Theor Biol* 246; pp. 449-460

[24] Banzhaf W (2003) On the Dynamics of an Artificial Regulatory Network. In *Advances in Artificial Life, 7th European Conference, ECAL'03, LNAI Springer*, 2801, pp. 217-227

[25] Kinoshita S, Yamada H (2019) Role of Self-Loop in Cell-Cycle Network of Budding Yeast. *Open Journal of Biophysics*, 9, pp. 10-20. https://doi.org/10.4236/ojbiphy.2019.91002

[26] Derrida B, Weisbuch G (1986) Evolution of Overlaps Between Configurations in Random Boolean Networks. *Journal De Physique* 47; pp. 1297-1303

[27] Schuster H (1984) Deterministic Chaos: An Introduction. *Physik-Verlag*

[28] Barabási A-L, Albert R, Jeong H (1999) Mean-field theory for scale-free random networks. *Physica A* 272; pp. 173-187

[29] Erdős P, Rényi A (1960) On the evolution of random graphs. Publication of the Mathematical Institute of the Hungarian Academy of Science, pp. 17-61

[30] Altenberg L (2005) Modularity in Evolution: Some Low-Level Questions. In: *Modularity: understanding the development and evolution of natural complex systems.* W. Callebaut and D. Rasskin-Gutman (eds), The Vienna series in theoretical biology, MIT Press; pp. 99-128

[31] Gecow A (2009) Emergence of Growth and Structural Tendencies During Adaptive Evolution of System. In: From System Complexity to Emergent Properties. M.A. Aziz-Alaoui & Cyrille Bertelle (eds), Springer, Understanding Complex Systems Series; pp. 211-241

Chapter 5

Perturbation Methods to Analysis of Thermal, Fluid Flow and Dynamics Behaviors of Engineering Systems

Gbeminiyi M. Sobamowo

Abstract

This chapter presents the applications of perturbation methods such as regular and homotopy perturbation methods to thermal, fluid flow and dynamic behaviors of engineering systems. The first example shows the utilization of regular perturbation method to thermal analysis of convective-radiative fin with end cooling and thermal contact resistance. The second example is concerned with the application of homotopy perturbation method to squeezing flow and heat transfer of Casson nanofluid between two parallel plates embedded in a porous medium under the influences of slip, Lorentz force, viscous dissipation and thermal radiation. Additionally, the dynamic behavior of piezoelectric nanobeam embedded in linear and nonlinear elastic foundations operating in a thermal-magnetic environment is analyzed using homotopy perturbation method which is presented in the third example. It is believed that the presentation in this chapter will enhance the understanding of these methods for the real world applications.

Keywords: perturbation method, thermal analysis, fluid flow behavior, dynamic response, engineering systems

1. Introduction

The descriptions of the behaviors of the real world phenomena and systems through the use of mathematical models often involve developments of nonlinear equations which are difficult to solve exactly and analytically. Consequently, recourse is always made to numerical methods as alternative methods in solving the nonlinear equations. However, the developments of analytical solutions are obviously still very important. Analytical solutions for specified problems are also essential and required to show the direct relationship between the models parameters. When analytical solutions are available, they provide good insights into the significance of various system parameters affecting the phenomena. Such solutions provide continuous physical insights than pure numerical or computation methods. Indisputably, analytical solutions are convenient for parametric studies, accounting for the physics of the problem and appear more appealing than the numerical solutions. Also, they help in reducing the computation and simulation costs as well as the task involved in the analysis of real-life problems.

Although, there is no general exact analytical method to solve all nonlinear problems, over the years, the nonlinear problems have been solved using different approximate analytical methods such as regular perturbation, singular perturbation method, homotopy perturbation method, homotopy analysis method, methods of weighted residual, variational iterative method, differential transformation method, variation parameter method, Adomian decomposition method, etc. The non-perturbative approximate analytic methods present explicit approximate analytical solutions which often involve complex mathematical analysis leading to analytic expressions involving large number terms. Furthermore, the methods are inherently with high computational cost and time accompanied with the requirement of high skills in mathematics. Moreover, in practice, analytical solutions with large number of terms and conditional statements for the solutions are not convenient for use by designers and engineers. Also, in these methods, there are always search for particular value(s) that will satisfy the end boundary condition(s). This always necessitates the use of software and such could result in additional computational cost in the generation of solution to the problem. Also, the quests involve applications of numerical schemes to determine the required value(s) that will satisfy the end boundary condition(s). This fact renders most of the approximate analytical methods to be taken as more of semi-analytical methods than total approximate analytical methods. Moreover, these methods have their own operational restrictions that severely narrow their functioning domain and when they are routinely implemented, they can sometimes lead to erroneous results. Specifically, the transformation of the nonlinear equations and the development of equivalent recurrence equations for the nonlinear equations using differential transformation method proved somehow difficult in some nonlinear system such as in rational Duffing oscillator, irrational nonlinear Duffing oscillator, finite extensibility nonlinear oscillator. There is difficulty in the determination of Adomian polynomials for the application of Adomian decomposition method for nonlinear problems. There are lack of rigorous theories or proper guidance for choosing initial approximation, auxiliary linear operators, auxiliary functions, and auxiliary parameters in the use of homotopy analysis method. Therefore, the need for comparatively simple, flexible, generic and high accurate total approximate analytical solutions is well established. One of the techniques that can be applied for such quest is the perturbation method. Perturbation method, although comparably old, as a pioneer method for finding approximate analytical solutions to nonlinear problems, it offers an alternative approach to solving certain types of nonlinear problems. In the limit of small parameter, perturbation method is widely used for solving many heat transfer, vibration, fluid mechanics and solid mechanics problems. It is capable of solving nonlinear, inhomogeneous and multidimensional problems with reasonable high level of accuracy. The most significant efforts and applications of the method were focused on celestial mechanics, fluid mechanics, and aerodynamics. Although, the solutions reported for other sophisticated methods to difference problems have good accuracy, they are more complicated for applications than perturbation method. Therefore, over the years, the relative simplicity and high accuracy especially in the limit of small parameter have made perturbation method an interesting tool among the most frequently used approximate analytical methods. Although, the perturbation method provides in general, better results for small perturbation parameters, besides having a handy mathematical formulation, it has been shown to have a good accuracy, even for relatively large values of the perturbation parameter [1–5].

2. Example 1: regular perturbation method to thermal analysis of convective-radiative fin with end cooling and thermal contact resistance

Consider a convective-radiative fin of temperature-dependent thermal conductivity $k(T)$, length L and thickness δ, exposed on both faces to a convective environment at temperature T_∞ and a heat transfer co-efficient h subjected to magnetic field shown in **Figure 1**. The dimension x pertains to the length coordinate which has its origin at the tip of the fin and has a positive orientation from the fin tip to the fin base. In order to analyze the problem, the following assumptions are made. The following assumptions were made in the development of the model

i. The heat flow in the fin and its temperatures remain constant with time.

ii. The temperature of the medium surrounding the fin is uniform.

iii. The temperature of the base of the fin is uniform.

iv. The fin thickness is small compared with its width and length, so that temperature gradients across the fin thickness and heat transfer from the edges of the fin is negligible compared with the heat leaving its lateral surface.

Applying thermal energy balance on the fin and using the above model assumptions, the following nonlinear thermal model is developed

$$\frac{d}{dx}\left[[1+\lambda(T-T_a)]\frac{dT}{dx}\right]-\frac{h}{k_a\delta}(T-T_a)-\frac{\sigma\varepsilon}{k_a\delta}\left(T^4-T_a^4\right)-\frac{\sigma B_o^2u^2}{k_aA_{cr}}(T-T_a)=0 \tag{1}$$

The boundary conditions are

$$x=0,\quad -k(T)\frac{\partial T}{\partial x}=h_e(T-T_a)+\sigma\in\left(T^4-T_a^4\right) \tag{2}$$

$$x=L,\quad -k(T)\frac{\partial T}{\partial x}=h_c(T_b-T)+\sigma\in\left(T^4-T_a^4\right) \tag{3}$$

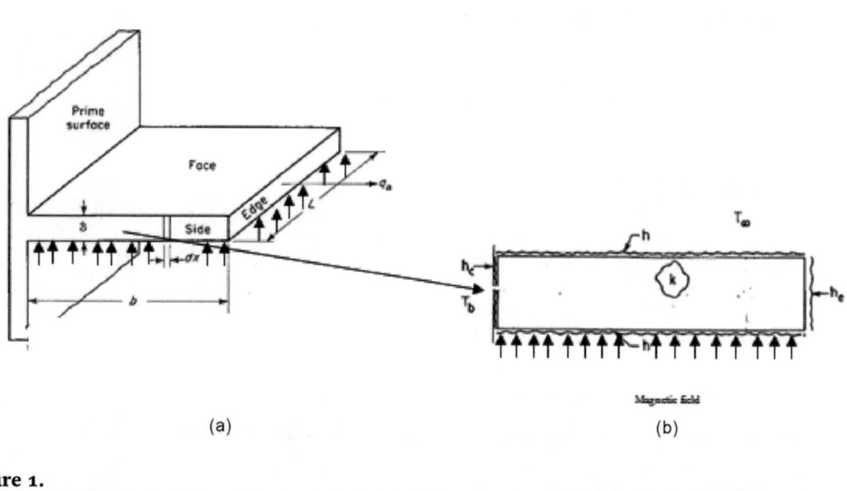

Figure 1.
(a) Schematic of the convective-radiative longitudinal straight fin with magnetic field. (b) Schematic of the longitudinal straight fin geometry showing thermal contact resistance and boundary conditions.

Considering a case when a small temperature difference exists within the material during the heat flow. This actually necessitated the use of temperature-invariant physical and thermal properties of the fin. Also, it has been established that under such scenario, the term T^4 can be expressed as a linear function of temperature. Therefore, we have

$$T^4 = T_a^4 + 4T_a^3(T - T_a) + 6T_a^2(T - T_a)^2 + \dots \cong 4T_a^3 T - 3T_a^4 \tag{4}$$

On substituting Eq. (4) into Eq. (1), one arrives arrived at

$$\frac{d}{dx}\left[[1 + \lambda(T - T_\infty)]\frac{dT}{dx}\right] - \frac{h}{k_a\delta}(T - T_a) - \frac{4\sigma\varepsilon T_a^3}{k_a\delta}(T - T_a) - \frac{\sigma B_o^2 u^2}{k_a A_{cr}}(T - T_a) = 0 \tag{5}$$

The boundary conditions

$$x = 0, \quad -k(T)\frac{\partial T}{\partial x} = h_e(T - T_a) + 4\sigma \in T_a^3(T - T_a) \tag{6}$$

$$x = L, \quad -k(T)\frac{\partial T}{\partial x} = h_c(T_b - T) + 4\sigma \in T_a^3(T - T_a) \tag{7}$$

On introducing the following dimensionless parameters in Eq. (8) into Eq. (5),

$$X = \frac{x}{L}, \; \theta = \frac{T - T_a}{T_b - T_a}, \; Ra = \frac{gk\beta(T_b - T_a)b}{\alpha v k_r}, \; N = \frac{4\sigma_{st} b T_a^3}{k_a}, \; Ha = \frac{\sigma B_0^2 u^2}{k_a A_{cr}}. \tag{8}$$

$$Bi_e = \frac{h_e b}{k_a}, \quad Bi_c = \frac{h_c b}{k_a}, \quad M^2 = \frac{hb^2}{k_a\delta}, \quad \varepsilon = \lambda(T_b - T_a) Bi_{e,eff} = \frac{(h_e + \sigma\varepsilon)b}{k_a},$$

$$Bi_{ceff} = \frac{(h_c + \sigma\varepsilon)b}{k_a}$$

The dimensionless form of the governing Eq. (5) is arrived at as

$$\frac{d}{dX}\left[(1 + \varepsilon\theta]\frac{d\theta}{dX}\right] - M^2\theta - Nr\theta - Ha\theta = 0 \tag{9}$$

On expanding Eq. (9), one has

$$\frac{d^2\theta}{dX^2} + \varepsilon\theta\frac{d^2\theta}{dX^2} + \varepsilon\left(\frac{d\theta}{dX}\right)^2 - M^2\theta - Nr\theta - Ha\theta = 0 \tag{10}$$

The boundary conditions are

$$X = 0, \quad (1 + \varepsilon\theta)\frac{d\theta}{dX} = -Bi_{e,eff}\theta \tag{11}$$

$$X = 1, \quad (1 + \varepsilon\theta)\frac{d\theta}{dX} = -Bi_{c,eff}(1 - \theta) \tag{12}$$

3. Method of solution using regular perturbation method

It is very difficult to develop closed-form solution for the above non-linear Eq. (10). Therefore, in this work, recourse is made to apply a relatively simple and accurate method approximate analytical method, the perturbation method.

Perturbation theory is based on the fact that the equation(s) describing the phenomena or process under investigation contain(s) a small parameter (or several small parameters), explicitly or implicitly. Therefore, the perturbation method is applicable to very small magnitudes of ε where the nonlinearity is slightly effective. Although, it has been shown to have a good accuracy, even for relatively large values of the perturbation parameter, ε [1, 2].

In solving Eq. (10), one needs to expand the dimensionless temperature as

$$\theta = \theta_0 + \varepsilon\theta_1 + \varepsilon^2\theta_2 + \ldots \tag{13}$$

Substituting Eq. (13) into Eq. (10), up to first order approximate, we have

$$\frac{d^2\theta_0}{dX^2} - (M^2 + Nr + Ha)\theta_0 + \varepsilon\left[\frac{d^2\theta_1}{dX^2} + \theta_0\frac{d^2\theta_0}{dX^2} + \left(\frac{d\theta_0}{dX}\right)^2 - (M^2 + Nr + Ha)\theta_1\right]$$
$$+\varepsilon^2\left[\frac{d^2\theta_2}{dX^2} + \theta_1\frac{d^2\theta_0}{dX^2} + \theta_0\frac{d^2\theta_1}{dX^2} + 2\left(\frac{d\theta_1}{dX}\right)\left(\frac{d\theta_0}{dX}\right) - (M^2 + Nr + Ha)\theta_2\right] = 0 \tag{14}$$

Leading order and first order equations with the appropriate boundary conditions are given as:

Leading order equation:

$$\frac{d^2\theta_o}{dx^2} - (M^2 + Nr + Ha)\theta_o = 0 \tag{15}$$

Subject to:

$$X = 0, \quad \frac{d\theta_o}{dX} = -Bi_{e,eff}\theta_o \tag{16}$$

$$X = 1, \quad \frac{d\theta_o}{dX} = Bi_{c,eff}(\theta_o - 1) \tag{17}$$

First-order equation:

$$\frac{d^2\theta_1}{dX^2} - (M^2 + Nr + Ha)\theta_1 = -\left(\frac{d\theta_o}{dX}\right)^2 - \theta_0\frac{d^2\theta_o}{dX^2} \tag{18}$$

Subject to:

$$X = 0, \quad \theta_0\frac{d\theta_0}{dX} + \frac{d\theta_1}{dX} = -Bi_{e,eff}\theta_1 \tag{19}$$

$$X = 1, \quad \theta_0\frac{d\theta_0}{dX} + \frac{d\theta_1}{dX} = Bi_{c,eff}\theta_1 \tag{20}$$

Second-order equation

$$\frac{d^2\theta_2}{dX^2} - (M^2 + Nr + Ha)\theta_2 = -\theta_1\frac{d^2\theta_0}{dX^2} - \theta_0\frac{d^2\theta_1}{dX^2} - 2\left(\frac{d\theta_1}{dX}\right)\left(\frac{d\theta_0}{dX}\right) \tag{21}$$

The boundary conditions

$$X = 0, \quad \theta_1\frac{d\theta_o}{dX} + \theta_0\frac{d\theta_1}{dX} + \frac{d\theta_2}{dX} = -Bi_{e,eff}\theta_2 \tag{22}$$

$$X = 1, \quad \theta_1\frac{d\theta_0}{dX} + \theta_0\frac{d\theta_1}{dX} + \frac{d\theta_2}{dX} = Bi_{c,eff}\theta_2 \tag{23}$$

It can be shown from Eq. (15), (18) and (21) with the corresponding boundary conditions of Eqs. (16), (19) and (22) that the:

Leading order solution for θ_o is

$$\theta_o = \frac{Bi_c\left\{\sqrt{(M^2+Nr+Ha)}\cosh\left(\sqrt{(M^2+Nr+Ha)}X\right) - Bi_e\sinh\left(\sqrt{(M^2+Nr+Ha)}X\right)\right\}}{\left\{\begin{matrix} Bi_c\left\{\left(\sqrt{(M^2+Nr+Ha)}\right)\cosh\left(\sqrt{(M^2+Nr+Ha)}\right) - Bi_e\sinh\left(\sqrt{(M^2+Nr+Ha)}\right)\right\} \\ +\sqrt{(M^2+Nr+Ha)}\left\{Bi_e\cosh\left(\sqrt{(M^2+Nr+Ha)}\right) - \left(\sqrt{(M^2+Nr+Ha)}\right)\sinh\left(\sqrt{(M^2+Nr+Ha)}\right)\right\}\end{matrix}\right\}} \tag{24}$$

While the first order solution θ_1 is

$$\begin{aligned}\theta_1 = {} & \frac{-Bi_c^2Bi_e}{3}\left[\frac{Bi_e^{2}\left\{\begin{matrix}Bi_c\cosh\left(\sqrt{(M^2+Nr+Ha)}\right)\\ -\left(\sqrt{M^2+Nr+Ha}\right)\sinh\left(\sqrt{M^2+Nr+Ha}\right)\end{matrix}\right\} + \left\{\begin{matrix}[Bi_c(M^2+Bi_e^2)+4MBi_c^2Bi_e]\cosh\left(2\sqrt{(M^2+Nr+Ha)}\right)\\ +[M(M^2+Bi_e^2)-2MBi_cBi_e]\sinh\left(2\sqrt{M^2+Nr+Ha}\right)\end{matrix}\right\}}{\left[\begin{matrix}\left\{\begin{matrix}\left(\sqrt{(M^2+Nr+Ha)}\right)\left\{\begin{matrix}Bi_c\cosh\left(\sqrt{(M^2+Nr+Ha)}\right)\\ -\left(\sqrt{M^2+Nr+Ha}\right)\sinh\left(\sqrt{M^2+Nr+Ha}\right)\end{matrix}\right\}\\ -Bi_e\left\{\begin{matrix}\left(\sqrt{(M^2+Nr+Ha)}\right)\cosh\left(\sqrt{(M^2+Nr+Ha)}\right)\\ -Bi_c\sinh\left(\sqrt{(M^2+Nr+Ha)}\right)\end{matrix}\right\}\end{matrix}\right\}\left\{\begin{matrix}Bi_c\left(\begin{matrix}\left(\sqrt{(M^2+Nr+Ha)}\right)\cosh\left(\sqrt{(M^2+Nr+Ha)}\right)\\ -Bi_e\sinh\left(\sqrt{(M^2+Nr+Ha)}\right)\end{matrix}\right)\\ +\left(\sqrt{(M^2+Nr+Ha)}\right)\left(\begin{matrix}Bi_e\cosh\left(\sqrt{(M^2+Nr+Ha)}\right)\\ -\left(\sqrt{(M^2+Nr+Ha)}\right)\sinh\left(\sqrt{(M^2+Nr+Ha)}\right)\end{matrix}\right)\end{matrix}\right\}\end{matrix}\right]^2}\right]\cosh\left(\sqrt{(M^2+Nr+Ha)}\right)X \\ & +\frac{Bi_c^2}{3}\left[\frac{\left(\sqrt{(M^2+Nr+Ha)}\right)\left\{\begin{matrix}[Bi_c(M^2+Bi_e^2)+4MBi_c^2Bi_e]\cosh\left(2\sqrt{(M^2+Nr+Ha)}\right)\\ +[M(M^2+Bi_e^2)-2MBi_cBi_e]\sinh\left(2\sqrt{M^2+Nr+Ha}\right)\end{matrix}\right\} + Bi_e^{3}\left\{\begin{matrix}\left(\sqrt{(M^2+Nr+Ha)}\right)\cosh\left(\sqrt{(M^2+Nr+Ha)}\right)\\ -Bi_e\sinh\left(\sqrt{M^2+Nr+Ha}\right)\end{matrix}\right\}}{\left[\begin{matrix}\left\{\begin{matrix}\left(\sqrt{(M^2+Nr+Ha)}\right)\left\{\begin{matrix}Bi_c\cosh\left(\sqrt{(M^2+Nr+Ha)}\right)\\ -\left(\sqrt{M^2+Nr+Ha}\right)\sinh\left(\sqrt{M^2+Nr+Ha}\right)\end{matrix}\right\}\\ -Bi_e\left\{\begin{matrix}\left(\sqrt{(M^2+Nr+Ha)}\right)\cosh\left(\sqrt{(M^2+Nr+Ha)}\right)\\ -Bi_c\sinh\left(\sqrt{(M^2+Nr+Ha)}\right)\end{matrix}\right\}\end{matrix}\right\}\left\{\begin{matrix}Bi_c\left(\begin{matrix}\left(\sqrt{(M^2+Nr+Ha)}\right)\cosh\left(\sqrt{(M^2+Nr+Ha)}\right)\\ -Bi_e\sinh\left(\sqrt{(M^2+Nr+Ha)}\right)\end{matrix}\right)\\ +\left(\sqrt{(M^2+Nr+Ha)}\right)\left(\begin{matrix}Bi_e\cosh\left(\sqrt{(M^2+Nr+Ha)}\right)\\ -\left(\sqrt{(M^2+Nr+Ha)}\right)\sinh\left(\sqrt{(M^2+Nr+Ha)}\right)\end{matrix}\right)\end{matrix}\right\}\end{matrix}\right]^2}\right]\sinh\left(\sqrt{(M^2+Nr+Ha)}\right)X \\ & -\frac{Bi_c^2}{3}\left[\frac{\left\{\begin{matrix}[Bi_c(M^2+Bi_e^2)+4MBi_c^2Bi_e]\cosh\left(2\sqrt{(M^2+Nr+Ha)}\right)X\\ +[M(M^2+Bi_e^2)-2MBi_cBi_e]\sinh\left(2\sqrt{M^2+Nr+Ha}\right)X\end{matrix}\right\}}{\left\{Bi_c\left(\begin{matrix}\left(\sqrt{(M^2+Nr+Ha)}\right)\cosh\left(\sqrt{(M^2+Nr+Ha)}\right)\\ -Bi_e\sinh\left(\sqrt{(M^2+Nr+Ha)}\right)\end{matrix}\right)+\left(\sqrt{(M^2+Nr+Ha)}\right)\left(\begin{matrix}Bi_e\cosh\left(\sqrt{(M^2+Nr+Ha)}\right)\\ -\left(\sqrt{(M^2+Nr+Ha)}\right)\sinh\left(\sqrt{(M^2+Nr+Ha)}\right)\end{matrix}\right)\right\}^2}\right]\end{aligned} \tag{25}$$

The second-order solution θ_2 is too huge to be included in the manuscript.

On substituting Eqs. (24) and (25) into Eq. (13) up to the first order (i.e. neglecting the higher orders), one arrives at

$$\begin{aligned}\theta(X) = {} & \frac{Bi_c\left\{\sqrt{(M^2+Nr+Ha)}\cosh\left(\sqrt{(M^2+Nr+Ha)}X\right) - Bi_e\sinh\left(\sqrt{(M^2+Nr+Ha)}X\right)\right\}}{\left\{\begin{matrix} Bi_c\left\{\left(\sqrt{(M^2+Nr+Ha)}\right)\cosh\left(\sqrt{(M^2+Nr+Ha)}\right) - Bi_e\sinh\left(\sqrt{(M^2+Nr+Ha)}\right)\right\} \\ +\sqrt{(M^2+Nr+Ha)}\left\{Bi_e\cosh\left(\sqrt{(M^2+Nr+Ha)}\right) - \left(\sqrt{(M^2+Nr+Ha)}\right)\sinh\left(\sqrt{(M^2+Nr+Ha)}\right)\right\}\end{matrix}\right\}} \\ & \frac{-\varepsilon Bi_c^2Bi_e}{3}\left[\frac{Bi_e^{2}\left\{\begin{matrix}Bi_c\cosh\left(\sqrt{(M^2+Nr+Ha)}\right)\\ -\left(\sqrt{M^2+Nr+Ha}\right)\sinh\left(\sqrt{M^2+Nr+Ha}\right)\end{matrix}\right\} + \left\{\begin{matrix}[Bi_c(M^2+Bi_e^2)+4MBi_c^2Bi_e]\cosh\left(2\sqrt{(M^2+Nr+Ha)}\right)\\ +[M(M^2+Bi_e^2)-2MBi_cBi_e]\sinh\left(2\sqrt{M^2+Nr+Ha}\right)\end{matrix}\right\}}{\left[\begin{matrix}\left\{\begin{matrix}\left(\sqrt{(M^2+Nr+Ha)}\right)\left\{\begin{matrix}Bi_c\cosh\left(\sqrt{(M^2+Nr+Ha)}\right)\\ -\left(\sqrt{M^2+Nr+Ha}\right)\sinh\left(\sqrt{M^2+Nr+Ha}\right)\end{matrix}\right\}\\ -Bi_e\left\{\begin{matrix}\left(\sqrt{(M^2+Nr+Ha)}\right)\cosh\left(\sqrt{(M^2+Nr+Ha)}\right)\\ -Bi_c\sinh\left(\sqrt{(M^2+Nr+Ha)}\right)\end{matrix}\right\}\end{matrix}\right\}\left\{\begin{matrix}Bi_c\left(\begin{matrix}\left(\sqrt{(M^2+Nr+Ha)}\right)\cosh\left(\sqrt{(M^2+Nr+Ha)}\right)\\ -Bi_e\sinh\left(\sqrt{(M^2+Nr+Ha)}\right)\end{matrix}\right)\\ +\left(\sqrt{(M^2+Nr+Ha)}\right)\left(\begin{matrix}Bi_e\cosh\left(\sqrt{(M^2+Nr+Ha)}\right)\\ -\left(\sqrt{(M^2+Nr+Ha)}\right)\sinh\left(\sqrt{(M^2+Nr+Ha)}\right)\end{matrix}\right)\end{matrix}\right\}\end{matrix}\right]^2}\right]\cosh\left(\sqrt{(M^2+Nr+Ha)}\right)X \\ & +\frac{\varepsilon Bi_c^2}{3}\left[\frac{\left(\sqrt{(M^2+Nr+Ha)}\right)\left\{\begin{matrix}[Bi_c(M^2+Bi_e^2)+4MBi_c^2Bi_e]\cosh\left(2\sqrt{(M^2+Nr+Ha)}\right)\\ +[M(M^2+Bi_e^2)-2MBi_cBi_e]\sinh\left(2\sqrt{M^2+Nr+Ha}\right)\end{matrix}\right\} + Bi_e^{3}\left\{\begin{matrix}\left(\sqrt{(M^2+Nr+Ha)}\right)\cosh\left(\sqrt{(M^2+Nr+Ha)}\right)\\ -Bi_e\sinh\left(\sqrt{M^2+Nr+Ha}\right)\end{matrix}\right\}}{\left[\begin{matrix}\left\{\begin{matrix}\left(\sqrt{(M^2+Nr+Ha)}\right)\left\{\begin{matrix}Bi_c\cosh\left(\sqrt{(M^2+Nr+Ha)}\right)\\ -\left(\sqrt{M^2+Nr+Ha}\right)\sinh\left(\sqrt{M^2+Nr+Ha}\right)\end{matrix}\right\}\\ -Bi_e\left\{\begin{matrix}\left(\sqrt{(M^2+Nr+Ha)}\right)\cosh\left(\sqrt{(M^2+Nr+Ha)}\right)\\ -Bi_c\sinh\left(\sqrt{(M^2+Nr+Ha)}\right)\end{matrix}\right\}\end{matrix}\right\}\left\{\begin{matrix}Bi_c\left(\begin{matrix}\left(\sqrt{(M^2+Nr+Ha)}\right)\cosh\left(\sqrt{(M^2+Nr+Ha)}\right)\\ -Bi_e\sinh\left(\sqrt{(M^2+Nr+Ha)}\right)\end{matrix}\right)\\ +\left(\sqrt{(M^2+Nr+Ha)}\right)\left(\begin{matrix}Bi_e\cosh\left(\sqrt{(M^2+Nr+Ha)}\right)\\ -\left(\sqrt{(M^2+Nr+Ha)}\right)\sinh\left(\sqrt{(M^2+Nr+Ha)}\right)\end{matrix}\right)\end{matrix}\right\}\end{matrix}\right]^2}\right]\sinh\left(\sqrt{(M^2+Nr+Ha)}\right)X \\ & -\frac{\varepsilon Bi_c^2}{3}\left[\frac{\left\{\begin{matrix}[Bi_c(M^2+Bi_e^2)+4MBi_c^2Bi_e]\cosh\left(2\sqrt{(M^2+Nr+Ha)}\right)X\\ +[M(M^2+Bi_e^2)-2MBi_cBi_e]\sinh\left(2\sqrt{M^2+Nr+Ha}\right)X\end{matrix}\right\}}{\left\{Bi_c\left(\begin{matrix}\left(\sqrt{(M^2+Nr+Ha)}\right)\cosh\left(\sqrt{(M^2+Nr+Ha)}\right)\\ -Bi_e\sinh\left(\sqrt{(M^2+Nr+Ha)}\right)\end{matrix}\right)+\left(\sqrt{(M^2+Nr+Ha)}\right)\left(\begin{matrix}Bi_e\cosh\left(\sqrt{(M^2+Nr+Ha)}\right)\\ -\left(\sqrt{(M^2+Nr+Ha)}\right)\sinh\left(\sqrt{(M^2+Nr+Ha)}\right)\end{matrix}\right)\right\}^2}\right]\end{aligned} \tag{26}$$

4. Example 2: homotopy perturbation method to analysis of squeezing flow and heat transfer of Casson nanofluid between two parallel plates embedded in a porous medium under the influences of slip, Lorentz force, viscous dissipation and thermal radiation

Consider a Casson nanofluid flowing between two parallel plates placed at time-variant distance and under the influence of magnetic field as shown in the **Figure 2**. It is assumed that the flow of the nanofluid is laminar, stable, incompressible, isothermal, non-reacting chemically, the nanoparticles and base fluid are in thermal equilibrium and the physical properties are constant. The fluid conducts electrical energy as it flows unsteadily under magnetic force field. The fluid structure is everywhere in thermodynamic equilibrium and the plate is maintained at constant temperature.

Following the assumptions, the governing equations for the flow are given as

$$\frac{\partial u}{\partial x}+\frac{\partial v}{\partial y}=0 \tag{27}$$

$$\rho_{nf}\left(\frac{\partial u}{\partial t}+u\frac{\partial u}{\partial x}+v\frac{\partial u}{\partial y}\right)=-\frac{\partial p}{\partial x}+\mu_{nf}\left(1+\frac{1}{\beta}\right)\left(2\frac{\partial^2 u}{\partial x^2}+\frac{\partial^2 u}{\partial x \partial y}+\frac{\partial^2 u}{\partial y^2}\right)-\sigma B_{v}^{2}u-\frac{\mu_{nf}u}{K_p} \tag{28}$$

$$\rho_{nf}\left(\frac{\partial v}{\partial t}+u\frac{\partial v}{\partial x}+v\frac{\partial v}{\partial y}\right)=-\frac{\partial p}{\partial y}+\mu_{nf}\left(1+\frac{1}{\beta}\right)\left(2\frac{\partial v}{\partial x^2}+\frac{\partial^2 v}{\partial x \partial y}+\frac{\partial^2 v}{\partial y^2}\right)-\frac{\mu_{nf}v}{K_p} \tag{29}$$

$$\begin{aligned}\frac{\partial T}{\partial t}+u\frac{\partial T}{\partial x}+u\frac{\partial T}{\partial y}&=\frac{k_{nf}}{\left(\rho C_p\right)_{nf}}\left(\frac{\partial^2 T}{\partial x^2}+\frac{\partial^2 T}{\partial y^2}\right)\\&+\frac{\mu_{nf}}{\left(\rho C_p\right)_{nf}}\left(1+\frac{1}{\beta}\right)\begin{pmatrix}2\left(\frac{\partial^2 u}{\partial x^2}\right)^2+\left(\frac{\partial^2 u}{\partial y^2}+\frac{\partial^2 v}{\partial x^2}\right)^2\\+2\left(\frac{\partial^2 v}{\partial y^2}\right)^2\end{pmatrix}-\frac{1}{\left(\rho C_p\right)_{nf}}\frac{\partial q_r}{\partial x}\end{aligned} \tag{30}$$

where

$$\rho_{nf}=\rho_f(1-\phi)+\rho_s\phi \tag{31}$$

$$\mu_{nf}=\frac{\mu_f}{(1-\phi)^{2.5}} \tag{32}$$

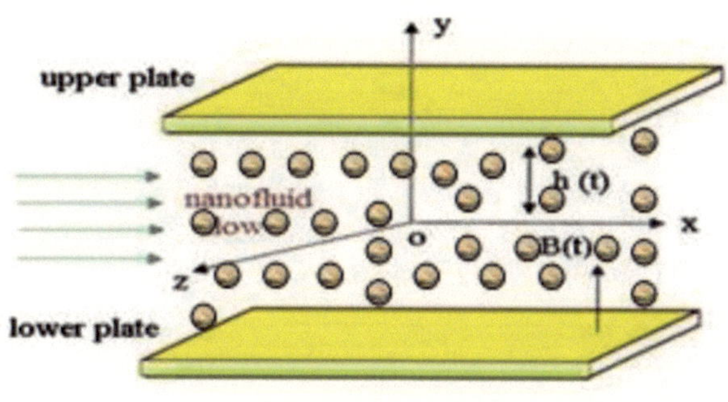

Figure 2.
Model diagram of MHD squeezing flow of nanofluid between two parallel plates embedded in a porous medium.

and the magnetic field parameter is given as

$$B(t) = \frac{B_0}{\sqrt{1-\alpha t}} \tag{33}$$

$$\sigma_{nf} = \sigma_f \left[1 + \frac{3\left\{\frac{\sigma_s}{\sigma_f} - 1\right\}\phi}{\left\{\frac{\sigma_s}{\sigma_f} + 2\right\}\phi - \left\{\frac{\sigma_s}{\sigma_f} - 1\right\}\phi} \right], \tag{34}$$

$$k_{nf} = k_f \left[\frac{k_s + (m-1)k_f - (m-1)\left(k_f - k_s\right)\phi}{k_s + (m-1)k_f + \left(k_f - k_s\right)\phi} \right], \tag{35}$$

The Casson fluid parameter, $\beta = \mu_B \sqrt{2\pi / P_y}$ and k is the permeability constant. The radiation term is given as

$$\frac{\partial q_r}{\partial y} = -\frac{4\sigma}{3K}\frac{\partial T^4}{\partial y} \cong -\frac{16\sigma T_s^3}{3K}\frac{\partial^2 T}{\partial y^2} \text{ (using Rosseland's approximation)} \tag{36}$$

The appropriate boundary conditions are given as

$$u = 0, \quad v = v_w = \frac{dh}{dt}, \quad T = T_H \quad at \ y = h(t) = H\sqrt{1-\alpha t}, \tag{37}$$

$$\frac{\partial u}{\partial y} = 0, \quad \frac{\partial T}{\partial y} = 0, \quad v = 0, \quad at \quad y = 0, \tag{38}$$

On introducing the following dimensionless and similarity variables

$$\begin{gathered} u = \frac{\alpha H}{2\sqrt{1-\alpha t}} f'(\eta, t), \quad v = -\frac{\alpha H}{2\sqrt{1-\alpha t}} f(\eta, t), \quad \eta = \frac{y}{H\sqrt{1-\alpha t}}, \quad \theta = \frac{T - T_0}{T_H - T_0}, \quad Ec = \frac{1}{C_p}\left(\frac{\alpha H}{2(1-\alpha t)}\right)^2 \\ Re = -SA(1-\phi)^{2.5} = \frac{\rho_{nf} H V_w}{\mu_{nf}}, \quad S = \frac{\alpha H^2}{2v_f}, \quad Da = \frac{K_p}{H^2}, \quad A_1 = (1-\phi) + \phi\frac{\rho_s}{\rho_f}, \quad \Pr = \frac{\mu C_p}{k}, \quad \delta = \frac{H}{x}, \\ B_1 = \left[\frac{(\sigma_s + (m-1)\sigma_f) + (m-1)(\sigma_s - \sigma_f)\phi}{(\sigma_s + (m-1)\sigma_f) - (m-1)(\sigma_s - \sigma_f)\phi} \right], \quad A_2 = (1-\phi) + \phi\frac{(\rho C_p)_s}{(\rho C_p)_f}, \quad A_3 = \frac{k_{nf}}{k_f}, \quad R = \frac{4\sigma T_\infty^3}{3kK} \end{gathered} \tag{39}$$

One arrives at the dimensionless equations

$$\left(1 + \frac{1}{\beta}\right) f^{iv} - SA_1(1-\phi)^{2.5}\left(\eta f''' + 3f'' + ff''' - f'f''\right) - Ha^2 f'' - \frac{1}{Da} f'' = 0 \tag{40}$$

$$\left(1 + \frac{4}{3}R\right)\theta'' + PrS\left(\frac{A_2}{A_3}\right)(\theta' f - \eta\theta') + \frac{\text{Pr}Ec}{A_3(1-\phi)^{2.5}}\left(\left(f''\right)^2 + 4\delta^2\left(f'\right)^2\right) = 0 \tag{41}$$

with the boundary conditions as follows

$$f = 0, \quad f'' = 0, \quad \theta' = 0, \quad when \quad \eta = 0, \tag{42}$$

$$f = 1, \quad f' = 0, \quad \theta = 1, \quad when \quad \eta = 1, \tag{43}$$

where m in the above Hamilton Crosser's model in Eq. (35).

5. Method of solution by homotopy perturbation method

The comparative advantages and the provision of acceptable analytical results with convenient convergence and stability coupled with total analytic procedures of homotopy perturbation method compel us to consider the method for solving the system of nonlinear differential equations in Eqs. (40) and (41) with the boundary conditions in Eq. (42).

5.1 The basic idea of homotopy perturbation method

In order to establish the basic idea behind homotopy perturbation method, consider a system of nonlinear differential equations given as

$$A(U) - f(r) = 0, \quad r \in \Omega, \tag{44}$$

with the boundary conditions

$$B\left(u, \frac{\partial u}{\partial \eta}\right) = 0, \qquad r \in \Gamma, \tag{45}$$

where A is a general differential operator, B is a boundary operator, $f(r)$ a known analytical function and Γ is the boundary of the domain Ω.

The operator A can be divided into two parts, which are L and N, where L is a linear operator, N is a non-linear operator. Eq. (44) can be therefore rewritten as follows

$$L(u) + N(u) - f(r) = 0. \tag{46}$$

By the homotopy technique, a homotopy $U(r,p) : \Omega \times [0,1] \to R$ can be constructed, which satisfies

$$H(U,p) = (1-p)[L(U) - L(U_o)] + p[A(U) - f(r)] = 0, \qquad p \in [0,1], \tag{47}$$

or

$$H(U,p) = L(U) - L(U_o) + pL(U_o) + p[N(U) - f(r)] = 0. \tag{48}$$

In the above Eqs. (47) and (48), $p \in [0,1]$ is an embedding parameter, u_o is an initial approximation of equation of Eq. (44), which satisfies the boundary conditions.

Also, from Eq. (47) and Eq. (48), one has

$$H(U,0) = L(U) - L(U_o) = 0, \tag{49}$$

or

$$H(U,0) = A(U) - f(r) = 0. \tag{50}$$

The changing process of p from zero to unity is just that of $U(r,p)$ from $u_o(r)$ to $u(r)$. This is referred to homotopy in topology. Using the embedding parameter p as a small parameter, the solution of Eqs. (47) and Eq. (48) can be assumed to be written as a power series in p as given in Eq. (51)

$$U = U_o + pU_1 + p^2U_2 + \ldots \tag{51}$$

It should be pointed out that of all the values of p between 0 and 1, $p = 1$ produces the best result. Therefore, setting $p = 1$, results in the approximation solution of Eq. (42)

$$u = \lim_{p \to 1} U = U_o + U_1 + U_2 + \ldots \tag{52}$$

The basic idea expressed above is a combination of homotopy and perturbation method. Hence, the method is called homotopy perturbation method (HPM), which has eliminated the limitations of the traditional perturbation methods. On the other hand, this technique can have full advantages of the traditional perturbation techniques. The series Eq. (29) is convergent for most cases.

5.2 Application of the homotopy perturbation method to the fluid flow problem

According to homotopy perturbation method (HPM), one can construct an homotopy for Eq. (36)–(39) as

$$H_1(p,\eta) = (1-p)\left[\left(1+\frac{1}{\beta}\right)f^{iv}\right] + p\left[\begin{matrix}\left(1+\frac{1}{\beta}\right)f^{iv} - SA_1(1-\phi)^{2.5}\begin{pmatrix}\eta f''' + 3f'' \\ +ff''' - f'f''\end{pmatrix} \\ -Ha^2 f'' - \frac{1}{Da}f''\end{matrix}\right] = 0, \tag{53}$$

$$H_2(p,\eta) = (1-p)\left[\left(1+\frac{4}{3}R\right)\theta''\right] + p\left[\begin{matrix}\left(1+\frac{4}{3}R\right)\theta'' + PrS\left(\frac{A_2}{A_3}\right)(\theta' f - \eta\theta') \\ +\frac{\mathrm{PrEc}}{A_3(1-\phi)^{2.5}}\left(\left(f''\right)^2 + 4\delta^2\left(f'\right)^2\right)\end{matrix}\right] = 0, \tag{54}$$

Taking power series of velocity, temperature and concentration fields, gives

$$f = f_0 + pf_1 + p^2 f_2 + p^3 f_3 + \ldots \tag{55}$$

and

$$\theta = \theta_0 + p\theta_1 + p^2\theta_2 + p^3\theta_3 + \ldots \tag{56}$$

Substituting Eqs. (55) and (56) into Eq. (53) and (54) as well as the boundary conditions in Eq. (42), and grouping like terms based on the power of p, the fluid flow velocity equation is given as:

Zeroth-order equations

$$p^0: \quad f_0^{iv}(\eta) + \frac{1}{\beta}f_0^{iv}(\eta) = 0, \tag{57}$$

$$p^0: \quad \left(1+\frac{4}{3}R\right)\theta_0'' = 0, \tag{58}$$

First-order equations

$$p^1: \ \frac{1}{\beta}f_1^{iv}(\eta) + f_1^{iv}(\eta) - SA_1(1-\phi)^{2.5}\eta f_0(\eta) - \frac{1}{Da}f_0''(\eta) - Ha^2 f_0''(\eta)$$
$$-3SA_1(1-\phi)^{2.5}f_0''(\eta) - SA_1(1-\phi)^{2.5}f_0'(\eta)f_0''(\eta) + SA_1(1-\phi)^{2.5}f_0(\eta)f_0'''(\eta) = 0, \tag{59}$$

$$p^1: \ \left(1+\frac{4}{3}R\right)\theta_1'' + PrS\left(\frac{A_2}{A_3}\right)(\theta_0' f_0 - \eta\theta_0') + \frac{\text{Pr}Ec}{A_3(1-\phi)^{2.5}}\left(\left(f_0''\right)^2 + 4\delta^2\left(f_0'\right)^2\right) = 0 \tag{60}$$

Second-order equations

$$p^2: \ \frac{1}{\beta}f_2^{iv}(\eta) + f_2^{iv}(\eta) - SA_1(1-\phi)^{2.5}\eta f_1(\eta) - \frac{1}{Da}f_1''(\eta) - Ha^2 f_1''(\eta) - 3SA_1(1-\phi)^{2.5}f_2''(\eta)$$
$$-SA_1(1-\phi)^{2.5}f_1'(\eta)f_0''(\eta) - SA_1(1-\phi)^{2.5}f_0'(\eta)f_1''(\eta) + SA_1(1-\phi)^{2.5}f_1(\eta)f_0'''(\eta)$$
$$+SA_1(1-\phi)^{2.5}f_0(\eta)f_1'''(\eta) = 0, \tag{61}$$

$$p^2: \ \left(1+\frac{4}{3}R\right)\theta_2'' + PrS\left(\frac{A_2}{A_3}\right)(\theta_1' f_0 + \theta_0' f_1 - \eta\theta_1') + \frac{2\text{Pr}Ec}{A_3(1-\phi)^{2.5}}\left(f_0'' f_1'' + 4\delta^2 f_0' f_1'\right) = 0 \tag{62}$$

the boundary conditions are

$$f_0 = f_1 = f_2 = 0, \quad f_0'' = f_1'' = f_2'' = 0, \quad \theta_0' = \theta_1' = \theta_2' = 0, \ when \quad \eta = 0,$$
$$f_0 = 1, \ f_1 = f_2 = 0, \quad f_0' = f_1' = f_2' = 0, \ \theta_0 = 1, \ \theta_1 = \theta_2 = 0, \ when \quad \eta = 1, \tag{63}$$

In a similar way, the higher orders problems are obtained.

On solving Eqs. (57), (61) and (64) with their corresponding boundary conditions, we arrived at

$$f_0(\eta) = \frac{1}{2}(3\eta - \eta^3) \tag{64}$$

$$f_1(\eta) = -\frac{1}{6720(1+\beta)}\left(\begin{array}{l} \left(168\left(\frac{1}{Da}\right)\beta + 168Ha^2\beta + 419SA_1(1-\phi)^{2.5}\beta\right)\eta \\ -\left(336\left(\frac{1}{Da}\right)\beta + 336Ha^2\beta + 873SA_1(1-\phi)^{2.5}\beta\right)\eta^3 \\ +\left(168\left(\frac{1}{Da}\right)\beta + 168Ha^2\beta + 504SA_1(1-\phi)^{2.5}\beta\right)\eta^5 \\ -28SA_1(1-\phi)^{2.5}\beta\eta^6 - 24SA_1(1-\phi)^{2.5}\beta\eta^7 \\ +2SA_1(1-\phi)^{2.5}\beta\eta^8 \end{array}\right) \tag{65}$$

$$f_2(\eta) = -\frac{1}{9686476800(1+\beta)^2}\left(\begin{array}{l}
\left(\begin{array}{l}-12684672\left(\frac{1}{Da}\right)^2\beta^2 - 25369344\left(\frac{1}{Da}\right)Ha^2\beta^2 - 12684672Ha^4\beta^2 - 92692600\left(\frac{1}{Da}\right)SA_1(1-\phi)^{2.5}\beta^2\\ -92692600Ha^2A_1(1-\phi)^{2.5}\beta^2 - 154163807S^2A_1^2(1-\phi)^5\beta^2\end{array}\right)\eta\\
+\left(\begin{array}{l}31135104\left(\frac{1}{Da}\right)^2\beta^2 + 62270208\left(\frac{1}{Da}\right)Ha^2\beta^2 + 31135104Ha^4\beta^2 + 205741536\left(\frac{1}{Da}\right)SA_1(1-\phi)^{2.5}\beta^2\\ +205741536Ha^2A_1(1-\phi)^{2.5}\beta^2 + 324472661S^2A_1^2(1-\phi)^5\beta^2\end{array}\right)\eta^3\\
+\left(\begin{array}{l}-24216192\left(\frac{1}{Da}\right)^2\beta^2 - 48432384\left(\frac{1}{Da}\right)Ha^2\beta^2 - 24216192Ha^4\beta^2 - 135567432\left(\frac{1}{Da}\right)SA_1(1-\phi)^{2.5}\beta^2\\ -135567432Ha^2SA_1(1-\phi)^{2.5}\beta^2 - 188756568S^2A_1^2(1-\phi)^5\beta^2\end{array}\right)\eta^5\\
+\left(672672\left(\frac{1}{Da}\right)SA_1(1-\phi)^{2.5}\beta^2 + 672672Ha^2SA_1(1-\phi)^{2.5}\beta^2 + 1677676S^2A_1^2(1-\phi)^5\beta^2\right)\eta^6\\
+\left(\begin{array}{l}5765760\left(\frac{1}{Da}\right)^2\beta^2 + 11531520\left(\frac{1}{Da}\right)Ha^2\beta^2 + 5765760Ha^4\beta^2 + 24216192\left(\frac{1}{Da}\right)\beta^2\\ +24216192Ha^2SA_1(1-\phi)^{2.5}\beta^2 + 17976816S^2A_1^2(1-\phi)^5\beta^2\end{array}\right)\eta^7\\
-\left(1009008\left(\frac{1}{Da}\right)SA_1(1-\phi)^{2.5}\beta^2 + 1009008Ha^2SA_1(1-\phi)^{2.5}\beta^2 - 332946S^2A_1^2(1-\phi)^5\beta^2\right)\eta^8\\
-\left(1441440\left(\frac{1}{Da}\right)SA_1(1-\phi)^{2.5}\beta^2 + 1441440Ha^2SA_1(1-\phi)^{2.5}\beta^2 + 1441440S^2A_1^2(1-\phi)^5\beta^2\right)\eta^9\\
+\left(80080\left(\frac{1}{Da}\right)SA_1(1-\phi)^{2.5}\beta^2 + 80080Ha^2SA_1(1-\phi)^{2.5}\beta^2\beta^2\right)\eta^{10} - 109928S^2A_1^2(1-\phi)^5\beta^2\eta^{11}\\
+12376S^2A_1^2(1-\phi)^5\beta^2\eta^{12} + 168S^2A_1^2(1-\phi)^5\beta^2\eta^{13}
\end{array}\right), \tag{66}$$

In the same manner, the energy equations are solved. Following the definition of the homotopy perturbation method as presented in Eq. (52), one could write the solution of the fluid flow equation as

$$f(\eta) = \frac{1}{2}(3\eta-\eta^3) - \frac{1}{6720(1+\beta)}\left(\begin{array}{l}\left(168\left(\frac{1}{Da}\right)\beta + 168Ha^2\beta + 419SA_1(1-\phi)^{2.5}\beta\right)\eta - \left(336\left(\frac{1}{Da}\right)\beta + 336Ha^2\beta + 873SA_1(1-\phi)^{2.5}\beta\right)\eta^3\\ +\left(168\left(\frac{1}{Da}\right)\beta + 168Ha^2\beta + 504SA_1(1-\phi)^{2.5}\beta\right)\eta^5 - 28SA_1(1-\phi)^{2.5}\beta\eta^6 - 24SA_1(1-\phi)^{2.5}\beta\eta^7 + 2SA_1(1-\phi)^{2.5}\beta\eta^8\end{array}\right)$$
$$-\frac{1}{9686476800(1+\beta)^2}\left(\begin{array}{l}
\left(\begin{array}{l}-12684672\left(\frac{1}{Da}\right)^2\beta^2 - 25369344\left(\frac{1}{Da}\right)Ha^2\beta^2 - 12684672Ha^4\beta^2 - 92692600\left(\frac{1}{Da}\right)SA_1(1-\phi)^{2.5}\beta^2-\\ 92692600Ha^2A_1(1-\phi)^{2.5}\beta^2 - 154163807S^2A_1^2(1-\phi)^5\beta^2\end{array}\right)\eta\\
+\left(\begin{array}{l}31135104\left(\frac{1}{Da}\right)^2\beta^2 + 62270208\left(\frac{1}{Da}\right)Ha^2\beta^2 + 31135104Ha^4\beta^2 + 205741536\left(\frac{1}{Da}\right)SA_1(1-\phi)^{2.5}\beta^2\\ +205741536Ha^2A_1(1-\phi)^{2.5}\beta^2 + 324472661S^2A_1^2(1-\phi)^5\beta^2\end{array}\right)\eta^3\\
+\left(\begin{array}{l}-24216192\left(\frac{1}{Da}\right)^2\beta^2 - 48432384\left(\frac{1}{Da}\right)Ha^2\beta^2 - 24216192Ha^4\beta^2 - 135567432\left(\frac{1}{Da}\right)SA_1(1-\phi)^{2.5}\beta^2-\\ 135567432Ha^2SA_1(1-\phi)^{2.5}\beta^2 - 188756568S^2A_1^2(1-\phi)^5\beta^2\end{array}\right)\eta^5\\
+\left(672672\left(\frac{1}{Da}\right)SA_1(1-\phi)^{2.5}\beta^2 + 672672Ha^2SA_1(1-\phi)^{2.5}\beta^2 + 1677676S^2A_1^2(1-\phi)^5\beta^2\right)\eta^6\\
+\left(\begin{array}{l}5765760\left(\frac{1}{Da}\right)^2\beta^2 + 11531520\left(\frac{1}{Da}\right)Ha^2\beta^2 + 5765760Ha^4\beta^2 + 24216192\left(\frac{1}{Da}\right)\beta^2+\\ 24216192Ha^2SA_1(1-\phi)^{2.5}\beta^2 + 17976816S^2A_1^2(1-\phi)^5\beta^2\end{array}\right)\eta^7\\
-\left(1009008\left(\frac{1}{Da}\right)SA_1(1-\phi)^{2.5}\beta^2 + 1009008Ha^2SA_1(1-\phi)^{2.5}\beta^2 - 332946S^2A_1^2(1-\phi)^5\beta^2\right)\eta^8\\
-\left(1441440\left(\frac{1}{Da}\right)SA_1(1-\phi)^{2.5}\beta^2 + 1441440Ha^2SA_1(1-\phi)^{2.5}\beta^2 + 1441440S^2A_1^2(1-\phi)^5\beta^2\right)\eta^9\\
+\left(80080\left(\frac{1}{Da}\right)SA_1(1-\phi)^{2.5}\beta^2 + 80080Ha^2SA_1(1-\phi)^{2.5}\beta^2\beta^2\right)\eta^{10} - 109928S^2A_1^2(1-\phi)^5\beta^2\eta^{11}+\\
12376S^2A_1^2(1-\phi)^5\beta^2\eta^{12} + 168S^2A_1^2(1-\phi)^5\beta^2\eta^{13}
\end{array}\right) \tag{67}$$

6. Example 3: homotopy perturbation method to dynamic behavior of piezoelectric nanobeam embedded in linear and nonlinear elastic Foundation in a thermal-magnetic environment

Consider a nanobeam embedded in linear and nonlinear elastic media as shown in **Figure 3**. The nanobeam is subjected to stretching effects and resting on Winkler, Pasternak and nonlinear elastic media in a thermo-magnetic environment as depicted in the figure.

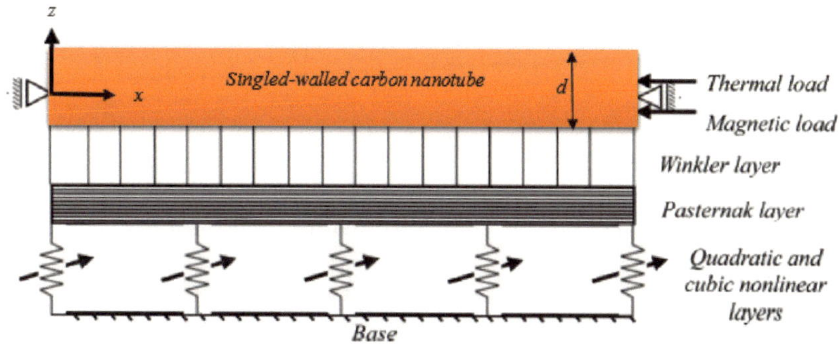

Figure 3.
A nanobeam embedded in linear and nonlinear elastic media (note: Only the bottom side of the elastic media is shown).

Following the nonlocal theory and Euler-Bernoulli theorem, the governing equation of the structure is developed as

$$\begin{aligned}&EI\left(\frac{\partial^4\overline{w}}{\partial\overline{x}^4}\right)+\rho A_c\frac{\partial^2}{\partial\overline{t}^2}\left[\overline{w}-(e_0a)^2\frac{\partial^2\overline{w}}{\partial\overline{x}^2}\right]+k_w\left[\overline{w}-(e_0a)^2\frac{\partial^2\overline{w}}{\partial\overline{x}^2}\right]-k_p\frac{\partial^2}{\partial\overline{x}^2}\left[\overline{w}-(e_0a)^2\frac{\partial^2\overline{w}}{\partial\overline{x}^2}\right]\\&+k_2\left[\overline{w}^2-(e_0a)^2\frac{\partial^2(\overline{w}^2)}{\partial\overline{x}^2}\right]+k_3\left[\overline{w}^3-(e_0a)^2\frac{\partial^2(\overline{w}^3)}{\partial\overline{x}^2}\right]-\eta A_cH_{\overline{x}}^2\frac{\partial^2}{\partial\overline{x}^2}\left[\overline{w}-(e_0a)^2\frac{\partial^2\overline{w}}{\partial\overline{x}^2}\right]\\&+\left(EA_c\frac{\alpha_{\overline{x}}\Delta T}{1-2\upsilon}\right)\frac{\partial^2}{\partial\overline{x}^2}\left[\overline{w}-(e_0a)^2\frac{\partial^2\overline{w}}{\partial\overline{x}^2}\right]-\left[\left(\frac{EA_c}{2L}\int_0^L\left(\frac{\partial\overline{w}}{\partial\overline{x}}\right)^2d\overline{x}\right)\left(\frac{\partial^2\overline{w}}{\partial\overline{x}^2}-(e_0a)^2\frac{\partial^4\overline{w}}{\partial\overline{x}^4}\right)\right]=0\end{aligned}\tag{68}$$

It is assumed that the midpoint of the nanobeam is subjected to the following initial conditions

$$\overline{w}(\overline{x},0)=\overline{w}_o,\quad\frac{\partial\overline{w}(\overline{x},0)}{\partial\overline{t}}=0\tag{69}$$

The following boundary conditions for the multi-walled nanotubes for simply supported nanotube is given,

$$\overline{w}(0,\overline{t})=0,\quad\frac{\partial^2\overline{w}(0,\overline{t})}{\partial^2\overline{x}}=0,\quad\overline{w}(L,\overline{t})=0,\quad\frac{\partial^2\overline{w}(L,\overline{t})}{\partial^2\overline{x}}=0.\tag{70}$$

Using the following adimensional constants and variables

$$\begin{aligned}&x=\frac{\overline{x}}{L};\ w=\frac{\overline{w}}{r};\ t=\sqrt{\frac{EI}{\rho A_cL^4}};\ r=\sqrt{\frac{I}{A_c}};h=\frac{e_0a}{L};\ \alpha_t^d=\frac{N_{thermal}L^2}{EI};\ A=\frac{\overline{w}_o}{r}\\&K_w=\frac{k_wL^4}{EI};\ K_p=\frac{k_pL2}{EI};\ Ha_m=\frac{\eta A_cH_{\overline{x}}^2L^2}{EI};\ K_2^d=\frac{k_2rL^4}{EI};\ K_3^d=\frac{k_3r^2L^4}{EI}.\end{aligned}\tag{71}$$

The adimensional form of the governing equation of motion for the nanobeam is given as

$$\begin{aligned}&\left[1+K_ph^2+Ha_mh^2-\alpha_t^dh^2+\frac{h^2}{2}\int_0^1\left(\frac{\partial w}{\partial x}\right)^2dx\right]\frac{\partial^4w}{\partial x^4}+\left[\alpha_t^d-K_wh^2-K_p-Ha_m-\frac{1}{2}\int_0^1\left(\frac{\partial w}{\partial x}\right)^2dx\right]\frac{\partial^2w}{\partial x^2}\\&+K_ww+\frac{\partial^2w}{\partial t^2}-h^2\frac{\partial^4w}{\partial x^2\partial t^2}+K_2^d\left[w^2-h^2\frac{\partial^2(w^2)}{\partial x^2}\right]+K_3^d\left[w^3-h^2\frac{\partial^2(w^3)}{\partial x^2}\right]=0\end{aligned}\tag{72}$$

And the boundary conditions become

$$w(0,t) = 0, \quad \frac{\partial^2 w(0,t)}{\partial^2 x} = 0, \quad w(1,t) = 0, \quad \frac{\partial^2 w(1,t)}{\partial^2 x} = 0. \tag{73}$$

6.1 Solution methodology: Galerkin decomposition and homotopy perturbation methods

The method of solution for the governing equation includes Galerkin decomposition and homotopy perturbation methods. As the name implies the Galerkin decomposition method is used to decompose the governing partial differential equation of motion can be separated into spatial and temporal parts. The resulting temporal equations are solved using homotopy perturbation method.

The procedures for the analysis of the equations are given in the proceeding sections as follows:

6.1.1 Galerkin decomposition method

With the application of Galerkin decomposition procedure, the governing partial differential equations of motion can be separated into spatial and temporal parts of the lateral displacement function as

$$w(x,t) = \phi(x)q(t) \tag{74}$$

Using one-parameter Galerkin decomposition procedure, one arrives at

$$\int_0^1 R(x,t)\phi(x)dx = 0 \tag{75}$$

where $R(x,t)$ is the governing equation of motion for nanobeam i.e.

$$\begin{aligned} R(x,t) = {} & \left[1 + K_p h^2 + Ha_m h^2 - \alpha_t^d h^2 + \frac{h^2}{2}\int_0^1 \left(\frac{\partial w}{\partial x}\right)^2 dx\right]\frac{\partial^4 w}{\partial x^4} + \left[\alpha_t^d - K_w h^2 - K_p - Ha_m - \frac{1}{2}\int_0^1 \left(\frac{\partial w}{\partial x}\right)^2 dx\right]\frac{\partial^2 w}{\partial x^2} \\ & + K_w w + \frac{\partial^2 w}{\partial t^2} - h^2\frac{\partial^4 w}{\partial x^2 \partial t^2} + K_2^d\left[w^2 - h^2\frac{\partial^2 (w^2)}{\partial x^2}\right] + K_3^d\left[w^3 - h^2\frac{\partial^2 (w^3)}{\partial x^2}\right] = 0 \end{aligned} \tag{76}$$

where $\phi(x)$ is the basis or trial or comparison function or normal function, which must satisfy the boundary conditions in Eq. (73), and $q(t)$ is the temporal part (time-dependent function).

Substituting Eqs. (75) into (74), then multiplying both sides of the resulting equation by $\phi(x)$ and integrating it for the domain of (0,1), we have

$$\frac{d^2 q(t)}{dt^2} + \lambda_1 q(t) + \lambda_2 q^2(t) + \lambda_3 q^3(t) = 0 \tag{77}$$

where

$$\lambda_1 = \frac{\bar{\lambda}_1}{\bar{\lambda}_0}; \lambda_2 = \frac{\bar{\lambda}_2}{\bar{\lambda}_0}; \lambda_3 = \frac{\bar{\lambda}_3}{\bar{\lambda}_0}; \tag{78}$$

$$\bar{\lambda}_0 = \int_0^1 \left(\phi^2 - h^2\phi\frac{\partial^2\phi}{\partial x^2}\right)dx \tag{79}$$

$$\bar{\lambda}_1 = \int_0^1 \left(K_w\phi^2 + (1 + K_ph^2 + Ha_mh^2 - \alpha_t^dh^2)\phi\frac{\partial^4\phi}{\partial x^4} + (\alpha_t^d - K_wh^2 - K_p - Ha_m)\phi\frac{\partial^2\phi}{\partial x^2}\right)dx \tag{80}$$

$$\bar{\lambda}_2 = \int_0^1 K_2^d\left(\phi^3 - h^2\phi\frac{\partial^2(\phi^2)}{\partial x^2}\right)dx \tag{81}$$

$$\bar{\lambda}_3 = \int_0^1 K_3^d\left(\phi^4 - h^2\phi\frac{\partial^2(\phi^4)}{\partial x^2}\right)dx + \frac{h^2}{2}\int_0^1\left(\frac{\partial\phi}{\partial x}\right)^2dx\int_0^1\phi\frac{\partial^2\phi}{\partial x^2}dx - \frac{1}{2}\int_0^1\left(\frac{\partial\phi}{\partial x}\right)^2dx\int_0^1\phi\frac{\partial^4\phi}{\partial x^4}dx \tag{82}$$

The initial conditions are given as

$$q(0) = A, \quad \frac{dq(0)}{dt} = 0 \tag{83}$$

A is the maximum vibration amplitude of the structure.

From the initial conditions in Eq. (83), one can write the initial approximation, u_o as

$$u_o = Acos(\omega t) \tag{84}$$

Eq. (22) satisfies the initial conditions in Eq. (83).

The homotopy perturbation representation of Eq. (77) is

$$H(q,p) = \left[\frac{d^2q}{dt^2} + \lambda_1 q\right] - \left[\frac{d^2u_o}{dt^2} + \lambda_1 u_o\right] + p\left[\frac{d^2u_o}{dt^2} + \lambda_1 u_o\right] + p\left(\lambda_2 q^2 + \lambda_3 q^3\right) = 0 \tag{85}$$

From the procedure of homotopy perturbation method, assuming that the solution of Eq. (77) takes the form of:

$$q = q_0 + pq_1 + p^2q_2 + p^3q_3 + \ldots, \tag{86}$$

On substituting Eqs. (86) into the homotopy Eq. (85)

$$H(q,p) = \left[\frac{d^2(q_0 + pq_1 + p^2q_2 + p^3q_3 + \ldots)}{dt^2} + \lambda_1(q_0 + pq_1 + p^2q_2 + p^3q_3 + \ldots)\right]$$
$$-\left[\frac{d^2u_o}{dt^2} + \lambda_1 u_0\right] + p\left[\frac{d^2u_o}{dt^2} + \lambda_1 u_0\right] + p\begin{pmatrix}\lambda_2(q_0 + pq_1 + p^2q_2 + p^3q_3 + \ldots)^2 \\ +\lambda_3(q_0 + pq_1 + p^2q_2 + p^3q_3 + \ldots)^3\end{pmatrix} = 0 \tag{87}$$

rearranging the coefficients of the terms with identical powers of p, one obtains series of linear differential equations as.

Zero-order equation

$$p^0:\left[\frac{d^2q_0}{dt^2}+\lambda_1 q_0\right]-\left[\frac{d^2u_o}{dt^2}+\lambda_1 u_o\right]=0 \tag{88}$$

with the conditions

$$q_0(0)=A \text{ and } \frac{dq_0(0)}{dt}=0 \tag{89}$$

First-order equation

$$p^1:\frac{d^2q_1}{dt^2}+\lambda_1 q_0+\frac{d^2u_o}{dt^2}+\lambda_1 u_o+\lambda_2 q_0^2+\lambda_3 q_0^3=0 \tag{90}$$

with corresponding initial conditions

$$q_1(0)=0 \text{ and } \frac{dq_1(0)}{dt}=0 \tag{91}$$

Second-order equation

$$p^2:\frac{d^2q_2}{dt^2}+\lambda_1 q_2+2\lambda_2 q_0 q_1+3\lambda_3 q_0^2 q_1=0 \tag{92}$$

with corresponding initial conditions

$$q_2(0)=0 \text{ and } \frac{dq_2(0)}{dt}=0 \tag{93}$$

The solution of the zero-order is given by.
From Eq. (27), we have

$$q_0=Acos(\omega t) \tag{94}$$

On substituting Eq. (94) into Eq. (90) and using trigonometric identities, after the colllection of like terms, one arrives at

$$\frac{d^2q_1}{dt^2}+\lambda_1 q_1+A\left(\lambda_1-\omega^2+\frac{3}{4}A^2\lambda\right)cos\,(\omega t)+\frac{A^2\lambda_2}{2}cos\,(2\omega t)+\frac{A^3\lambda_3}{4}cos\,(3\omega t)+\frac{A^2\lambda_2}{2}=0 \tag{95}$$

The solution of the above Eq. (95) provides

$$q_1(t)=\begin{bmatrix} A\left(\lambda_1-\omega^2+\frac{3}{4}A^2\lambda\right)\left(\frac{\lambda_1}{\omega^2-\lambda_1^2}\right)cos\,(\omega t)+\frac{A^2\lambda_2}{2}\left(\frac{\lambda_1}{4\omega^2-\lambda_1^2}\right)cos\,(2\omega t) \\ +\frac{A^3\lambda_3}{4}\left(\frac{\lambda_1}{9\omega^2-\lambda_1^2}\right)cos\,(3\omega t)+\frac{A^2\lambda_2}{2} \end{bmatrix}$$
$$+\left[A\left(\lambda_1-\omega^2+\frac{3}{4}A^2\lambda\right)\left(\frac{\lambda_1}{\lambda_1^2-\omega^2}\right)+\frac{A^2\lambda_2}{2}\left(\frac{\lambda_1}{\lambda_1^2-4\omega^2}\right)+\frac{A^3\lambda_3}{4}\left(\frac{\lambda_1}{\lambda_1^2-9\omega^2}\right)+\frac{A^2\lambda_2}{2\lambda_1}\right]cos\,(\alpha t) \tag{96}$$

Based on the procedure of HPM, setting $p = 1$,

$$q(t) = \lim_{p\to1} q(t) = \lim_{p\to1} \left[q_0 + pq_1 + p^2q_2 + p^3q_3 + \ldots\right] = q_0 + q_1 + q_2 + q_3 + \ldots \tag{97}$$

On substituting Eqs. (94) and (96) into Eq. (97), the result is

$$q(t) = A\cos(\omega t) + \left[\begin{array}{l} A\left(\lambda_1 - \omega^2 + \frac{3}{4}A^2\lambda\right)\left(\frac{\lambda_1}{\omega^2-\lambda_1^2}\right)\cos(\omega t) + \frac{A^2\lambda_2}{2}\left(\frac{\lambda_1}{4\omega^2-\lambda_1^2}\right)\cos(2\omega t) \\ +\frac{A^3\lambda_3}{4}\left(\frac{\lambda_1}{9\omega^2-\lambda_1^2}\right)\cos(3\omega t) + \frac{A^2\lambda_2}{2} \end{array}\right]$$
$$+\left[A\left(\lambda_1 - \omega^2 + \frac{3}{4}A^2\lambda\right)\left(\frac{\lambda_1}{\lambda_1^2-\omega^2}\right) + \frac{A^2\lambda_2}{2}\left(\frac{\lambda_1}{\lambda_1^2-4\omega^2}\right) + \frac{A^3\lambda_3}{4}\left(\frac{\lambda_1}{\lambda_1^2-9\omega^2}\right) + \frac{A^2\lambda_2}{2\lambda_1}\right]\cos(\lambda_1 t) + \ldots \tag{98}$$

In order to find the natural frequency, ω, the secular term must be eliminated. In order to do this, set the coefficient of $\cos(\lambda_1 t)$ to zero.

$$A\left(\lambda_1 - \omega^2 + \frac{3}{4}A^2\lambda\right)\left(\frac{\lambda_1}{\lambda_1^2-\omega^2}\right) + \frac{A^2\lambda_2}{2}\left(\frac{\lambda_1}{\lambda_1^2-4\omega^2}\right) + \frac{A^3\lambda_3}{4}\left(\frac{\lambda_1}{\lambda_1^2-9\omega^2}\right) + \frac{A^2\lambda_2}{2\lambda_1} = 0 \tag{99}$$

After simplification of Eq. (99), we have

$$\begin{array}{l}\left(\frac{A\lambda_2}{2\lambda_1^2} - 1\right)\omega^6 + A\left[\lambda_1^2\left(13 - \frac{49A\lambda_2}{2}\right) - 36\lambda_1 + \frac{9A\lambda_2}{2} - 26\lambda_3 A\right]\omega^4 \\ A\left[\lambda_1^4 + 13\lambda_1^3 - (2A\lambda_2 - 11\lambda_3A^2)\lambda_1^2\right]\omega^2 + \lambda_1^4 A\left(\lambda_1 + \lambda_3A^2\right) = 0\end{array} \tag{100}$$

The sextic equation can be written as

$$\begin{array}{l}\left(\frac{A\lambda_2}{2\lambda_1^2} - 1\right)\omega^6 + A\left[\lambda_1^2\left(13 - \frac{49A\lambda_2}{2}\right) - 36\lambda_1 + \frac{9A\lambda_2}{2} - 26\lambda_3 A\right]\omega^4 \\ A\left[\lambda_1^4 + 13\lambda_1^3 - (2A\lambda_2 - 11\lambda_3A^2)\lambda_1^2\right]\omega^2 + \lambda_1^4 A\left(\lambda_1 + \lambda_3A^2\right) = 0\end{array} \tag{101}$$

Eq. (101) can be written as

$$\chi_1\omega^6 + \chi_2\omega^4 + \chi_3\omega^2 + \chi_4 = 0 \tag{102}$$

where

$$\chi_1 = \left(\frac{A\lambda_2}{2\lambda_1^2} - 1\right), \chi_2 = A\left[\lambda_1^2\left(13 - \frac{49A\lambda_2}{2}\right) - 36\lambda_1 + \frac{9A\lambda_2}{2} - 26\lambda_3 A\right]$$

$$\chi_3 = A\left[\lambda_1^4 + 13\lambda_1^3 - (2A\lambda_2 - 11\lambda_3A^2)\lambda_1^2\right],\ \chi_4 = \lambda_1^4 A\left(\lambda_1 + \lambda_3A^2\right) = 0$$

The roots of the sextic equation are

$$\omega_1 = \sqrt{\sqrt[3]{\left(\frac{-\chi_2^3}{27\chi_1^3}+\frac{\chi_2\chi_3}{6\chi_1^2}-\frac{\chi_4}{2\chi_1}\right)+\sqrt{\left(\frac{\chi_3}{3\chi_1}-\frac{\lambda_2^2}{9\chi_1^2}\right)^3+\left(\frac{-\chi_2^3}{27\chi_1^3}+\frac{\chi_2\chi_3}{6\chi_1^2}-\frac{\chi_4}{2\chi_1}\right)^2}} + \sqrt[3]{\left(\frac{-\chi_2^3}{27\chi_1^3}+\frac{\chi_2\chi_3}{6\chi_1^2}-\frac{\chi_4}{2\chi_1}\right)-\sqrt{\left(\frac{\chi_3}{3\chi_1}-\frac{\lambda_2^2}{9\chi_1^2}\right)^3+\left(\frac{-\chi_2^3}{27\chi_1^3}+\frac{\chi_2\chi_3}{6\chi_1^2}-\frac{\chi_4}{2\chi_1}\right)^2}} - \frac{\chi_2}{3\chi_1}} \tag{103}$$

$$\omega_2 = -\sqrt{\sqrt[3]{\left(\frac{-\chi_2^3}{27\chi_1^3}+\frac{\chi_2\chi_3}{6\chi_1^2}-\frac{\chi_4}{2\chi_1}\right)+\sqrt{\left(\frac{\chi_3}{3\chi_1}-\frac{\lambda_2^2}{9\chi_1^2}\right)^3+\left(\frac{-\chi_2^3}{27\chi_1^3}+\frac{\chi_2\chi_3}{6\chi_1^2}-\frac{\chi_4}{2\chi_1}\right)^2}} + \sqrt[3]{\left(\frac{-\chi_2^3}{27\chi_1^3}+\frac{\chi_2\chi_3}{6\chi_1^2}-\frac{\chi_4}{2\chi_1}\right)-\sqrt{\left(\frac{\chi_3}{3\chi_1}-\frac{\lambda_2^2}{9\chi_1^2}\right)^3+\left(\frac{-\chi_2^3}{27\chi_1^3}+\frac{\chi_2\chi_3}{6\chi_1^2}-\frac{\chi_4}{2\chi_1}\right)^2}} - \frac{\chi_2}{3\chi_1}} \tag{104}$$

$$\omega_3 = \sqrt{\begin{aligned}&\frac{-1}{2\chi_1}\left[\sqrt[3]{\left(\frac{-\chi_2^3}{27\chi_1^3}+\frac{\chi_2\chi_3}{6\chi_1^2}-\frac{\chi_4}{2\chi_1}\right)+\sqrt{\left(\frac{\chi_3}{3\chi_1}-\frac{\lambda_2^2}{9\chi_1^2}\right)^3+\left(\frac{-\chi_2^3}{27\chi_1^3}+\frac{\chi_2\chi_3}{6\chi_1^2}-\frac{\chi_4}{2\chi_1}\right)^2}} + \sqrt[3]{\left(\frac{-\chi_2^3}{27\chi_1^3}+\frac{\chi_2\chi_3}{6\chi_1^2}-\frac{\chi_4}{2\chi_1}\right)-\sqrt{\left(\frac{\chi_3}{3\chi_1}-\frac{\lambda_2^2}{9\chi_1^2}\right)^3+\left(\frac{-\chi_2^3}{27\chi_1^3}+\frac{\chi_2\chi_3}{6\chi_1^2}-\frac{\chi_4}{2\chi_1}\right)^2}}\right]\\ &+\frac{\sqrt{-3}}{2\chi_1}\left[\sqrt[3]{\left(\frac{-\chi_2^3}{27\chi_1^3}+\frac{\chi_2\chi_3}{6\chi_1^2}-\frac{\chi_4}{2\chi_1}\right)+\sqrt{\left(\frac{\chi_3}{3\chi_1}-\frac{\lambda_2^2}{9\chi_1^2}\right)^3+\left(\frac{-\chi_2^3}{27\chi_1^3}+\frac{\chi_2\chi_3}{6\chi_1^2}-\frac{\chi_4}{2\chi_1}\right)^2}} - \sqrt[3]{\left(\frac{-\chi_2^3}{27\chi_1^3}+\frac{\chi_2\chi_3}{6\chi_1^2}-\frac{\chi_4}{2\chi_1}\right)-\sqrt{\left(\frac{\chi_3}{3\chi_1}-\frac{\lambda_2^2}{9\chi_1^2}\right)^3+\left(\frac{-\chi_2^3}{27\chi_1^3}+\frac{\chi_2\chi_3}{6\chi_1^2}-\frac{\chi_4}{2\chi_1}\right)^2}}\right] - \frac{\chi_2}{3\chi_1}\end{aligned}} \tag{105}$$

$$\omega_4 = -\sqrt{\begin{aligned}&\frac{-1}{2\chi_1}\left[\sqrt[3]{\left(\frac{-\chi_2^3}{27\chi_1^3}+\frac{\chi_2\chi_3}{6\chi_1^2}-\frac{\chi_4}{2\chi_1}\right)+\sqrt{\left(\frac{\chi_3}{3\chi_1}-\frac{\lambda_2^2}{9\chi_1^2}\right)^3+\left(\frac{-\chi_2^3}{27\chi_1^3}+\frac{\chi_2\chi_3}{6\chi_1^2}-\frac{\chi_4}{2\chi_1}\right)^2}} + \sqrt[3]{\left(\frac{-\chi_2^3}{27\chi_1^3}+\frac{\chi_2\chi_3}{6\chi_1^2}-\frac{\chi_4}{2\chi_1}\right)-\sqrt{\left(\frac{\chi_3}{3\chi_1}-\frac{\lambda_2^2}{9\chi_1^2}\right)^3+\left(\frac{-\chi_2^3}{27\chi_1^3}+\frac{\chi_2\chi_3}{6\chi_1^2}-\frac{\chi_4}{2\chi_1}\right)^2}}\right]\\ &+\frac{\sqrt{-3}}{2\chi_1}\left[\sqrt[3]{\left(\frac{-\chi_2^3}{27\chi_1^3}+\frac{\chi_2\chi_3}{6\chi_1^2}-\frac{\chi_4}{2\chi_1}\right)+\sqrt{\left(\frac{\chi_3}{3\chi_1}-\frac{\lambda_2^2}{9\chi_1^2}\right)^3+\left(\frac{-\chi_2^3}{27\chi_1^3}+\frac{\chi_2\chi_3}{6\chi_1^2}-\frac{\chi_4}{2\chi_1}\right)^2}} - \sqrt[3]{\left(\frac{-\chi_2^3}{27\chi_1^3}+\frac{\chi_2\chi_3}{6\chi_1^2}-\frac{\chi_4}{2\chi_1}\right)-\sqrt{\left(\frac{\chi_3}{3\chi_1}-\frac{\lambda_2^2}{9\chi_1^2}\right)^3+\left(\frac{-\chi_2^3}{27\chi_1^3}+\frac{\chi_2\chi_3}{6\chi_1^2}-\frac{\chi_4}{2\chi_1}\right)^2}}\right] - \frac{\chi_2}{3\chi_1}\end{aligned}} \tag{106}$$

$$\omega_5 = \sqrt{\begin{aligned}&\frac{-1}{2\chi_1}\left[\begin{aligned}&\sqrt[3]{\left(\frac{-\chi_2^3}{27\chi_1^3}+\frac{\chi_2\chi_3}{6\chi_1^2}-\frac{\chi_4}{2\chi_1}\right)+\sqrt{\left(\frac{\chi_3}{3\chi_1}-\frac{\chi_2^2}{9\chi_1^2}\right)^3+\left(\frac{-\chi_2^3}{27\chi_1^3}+\frac{\chi_2\chi_3}{6\chi_1^2}-\frac{\chi_4}{2\chi_1}\right)^2}}\\&+\sqrt[3]{\left(\frac{-\chi_2^3}{27\chi_1^3}+\frac{\chi_2\chi_3}{6\chi_1^2}-\frac{\chi_4}{2\chi_1}\right)-\sqrt{\left(\frac{\chi_3}{3\chi_1}-\frac{\chi_2^2}{9\chi_1^2}\right)^3+\left(\frac{-\chi_2^3}{27\chi_1^3}+\frac{\chi_2\chi_3}{6\chi_1^2}-\frac{\chi_4}{2\chi_1}\right)^2}}\end{aligned}\right]\\&\frac{-\sqrt{-3}}{2\chi_1}\left[\begin{aligned}&\sqrt[3]{\left(\frac{-\chi_2^3}{27\chi_1^3}+\frac{\chi_2\chi_3}{6\chi_1^2}-\frac{\chi_4}{2\chi_1}\right)+\sqrt{\left(\frac{\chi_3}{3\chi_1}-\frac{\chi_2^2}{9\chi_1^2}\right)^3+\left(\frac{-\chi_2^3}{27\chi_1^3}+\frac{\chi_2\chi_3}{6\chi_1^2}-\frac{\chi_4}{2\chi_1}\right)^2}}\\&-\sqrt[3]{\left(\frac{-\chi_2^3}{27\chi_1^3}+\frac{\chi_2\chi_3}{6\chi_1^2}-\frac{\chi_4}{2\chi_1}\right)-\sqrt{\left(\frac{\chi_3}{3\chi_1}-\frac{\chi_2^2}{9\chi_1^2}\right)^3+\left(\frac{-\chi_2^3}{27\chi_1^3}+\frac{\chi_2\chi_3}{6\chi_1^2}-\frac{\chi_4}{2\chi_1}\right)^2}}\end{aligned}\right]-\frac{\chi_2}{3\chi_1}\end{aligned}} \tag{107}$$

$$\omega_6 = -\sqrt{\begin{aligned}&\frac{-1}{2\chi_1}\left[\begin{aligned}&\sqrt[3]{\left(\frac{-\chi_2^3}{27\chi_1^3}+\frac{\chi_2\chi_3}{6\chi_1^2}-\frac{\chi_4}{2\chi_1}\right)+\sqrt{\left(\frac{\chi_3}{3\chi_1}-\frac{\chi_2^2}{9\chi_1^2}\right)^3+\left(\frac{-\chi_2^3}{27\chi_1^3}+\frac{\chi_2\chi_3}{6\chi_1^2}\;\frac{\chi_4}{2\chi_1}\right)^2}}\\&+\sqrt[3]{\left(\frac{-\chi_2^3}{27\chi_1^3}+\frac{\chi_2\chi_3}{6\chi_1^2}-\frac{\chi_4}{2\chi_1}\right)-\sqrt{\left(\frac{\chi_3}{3\chi_1}-\frac{\chi_2^2}{9\chi_1^2}\right)^3+\left(\frac{-\chi_2^3}{27\chi_1^3}+\frac{\chi_2\chi_3}{6\chi_1^2}-\frac{\chi_4}{2\chi_1}\right)^2}}\end{aligned}\right]\\&\frac{-\sqrt{-3}}{2\chi_1}\left[\begin{aligned}&\sqrt[3]{\left(\frac{-\chi_2^3}{27\chi_1^3}+\frac{\chi_2\chi_3}{6\chi_1^2}-\frac{\chi_4}{2\chi_1}\right)+\sqrt{\left(\frac{\chi_3}{3\chi_1}-\frac{\chi_2^2}{9\chi_1^2}\right)^3+\left(\frac{-\chi_2^3}{27\chi_1^3}+\frac{\chi_2\chi_3}{6\chi_1^2}-\frac{\chi_4}{2\chi_1}\right)^2}}\\&-\sqrt[3]{\left(\frac{-\chi_2^3}{27\chi_1^3}+\frac{\chi_2\chi_3}{6\chi_1^2}-\frac{\chi_4}{2\chi_1}\right)-\sqrt{\left(\frac{\chi_3}{3\chi_1}-\frac{\chi_2^2}{9\chi_1^2}\right)^3+\left(\frac{-\chi_2^3}{27\chi_1^3}+\frac{\chi_2\chi_3}{6\chi_1^2}-\frac{\chi_4}{2\chi_1}\right)^2}}\end{aligned}\right]-\frac{\chi_2}{3\chi_1}\end{aligned}} \tag{108}$$

7. Conclusion

In this chapter, the applications of regular and homotopy perturbation methods to thermal, fluid flow and dynamic behaviors of engineering systems have been presented. Regular perturbation was used in the first example to developed approximate analytical solutions for thermal behavior of convective-radiative fin with end cooling and thermal contact resistance. In the second example, homotopy perturbation method utilized to study squeezing flow and heat transfer of Casson nanofluid between two parallel plates embedded in a porous medium under the influences of slip, Lorentz force, viscous dissipation and thermal radiation. The same method was used in the third example to analyze the dynamic behavior of piezoelectric nanobeam embedded in linear and nonlinear elastic foundations operating in a thermal-magnetic environment. It is hoped that the vivid presentation and applications of these perturbation methods in this chapter will advance better understanding of methods especially for real world applications.

Author details

Gbeminiyi M. Sobamowo
Department of Mechanical Engineering, University of Lagos, Lagos, Nigeria

*Address all correspondence to: gsobamowo@unilag.edu.ng

References

[1] Filobello-Niño U, Vazquez-Leal H, Boubaker K, Khan Y, Perez-Sesma A, Sarmiento Reyes A, Jimenez-Fernandez VM, Diaz-Sanchez A, , Herrera-May A, Sanchez-Orea J, K. Pereyra-Castro K, Perturbation Method as a Powerful Tool to Solve Highly Nonlinear Problems: The Case of Gelfand's Equation. Asian Journal of Mathematics and Statistics. 2013; 6(2):76-82. DOI: 10.3923/ajms.2013.76.82.

[2] Lewandowski R. Analysis of strongly non-linear free vibration of beams using perturbation method. Civil and Environmental Reports. 2005; 1: 153-168.

[3] Cheung YK, Chen SH, Lau, SL. A modified Lindsteadt-Poincare method for certain strongly non-linear oscillators, International Journal of Non-Linear Mechanics. 1991; 26: 367-378. doi.org/10.1016/0020-7462(91)90066-3.

[4] Lim CW, Wu BS. A modified Mickens procedure for certain non-linear oscillators, Journal of Sound and Vibration. 2002; 257: 202-206. DOI: 10.1016/j.jsv.2008.03.007

[5] Hu H. A classical perturbation technique which is valid for large parameters. Journal of Sound and Vibration. 2004; 269:409-412. DOI: 10.1016/S0022-460X(03)00318-3.

Chapter 6

SIR Model with Homotopy to Predict Corona Cases

Nahid Fatima

Abstract

In this chapter, we will discuss SIR model to study the spread of COVID-2019 pandemic of India. We will give the prediction of corona cases using homotopy method. The HM is a method for solving the ordinary differential equations. The SIR model consists of three ordinary differential equations. In this study, we have used the data of COVID-2019 Outbreak of India on 20 Jan 2021. In this data, Recovered is 102656163, Active cases are 189245 Susceptible persons are 189347782 for the experimental purpose. Data about a wide variety of infectious diseases has been analyzed with the help of SIR model. Therefore, this model has been already well tested for infectious diseases by various scientists and researchers.

Keywords: SIR model, homotopy, differential equation, corona, graph, table

1. Introduction

Novel Coronavirus, assigned as 2019-nCoV, emerged in Wuhan, China, toward the end of 2019. As of January 24, 2020, as many as 830 cases had been analyzed in nine nations: China, Thailand, Japan, South Korea, Singapore, Vietnam, Taiwan, Nepal, and the United States [1–3]. Twenty-six fatalities happened, chiefly in patients who had genuine basic sickness. Albeit numerous subtleties of the rise of this infection.

In 2019, the Centers for Disease Control and Prevention (CDC) started monitoring the outbreak of a new coronavirus, SARS-CoV-2, which causes the respiratory illness now known as COVID-19. Authorities first identified the virus in Wuhan, China. More than 74,000 people have contracted the virus in China. Health authorities have identified many other people with COVID-19 around the world, including many in the United States. On January 31, 2020, the virus passed from one person to another in the U.S. The World Health Organization (WHO) have declared a public health emergency relating to COVID-19. Since then, this strain has been diagnosed in several U.S. residents. The CDC have advised that it is likely to spread to more people. COVID-19 has started causing disruption in at least 25 other countries.

All the adjoining nations of India have revealed positive COVID-19 cases. To secure against the lethal infection, the Indian government have taken fundamental and severe measures, including setting up wellbeing check posts between the public lines to test whether individuals entering the nation have the infection. Various nations have presented salvage endeavors and reconnaissance measures for residents wishing to get back from China. The exercise gained from the SARS episode was first that the absence of lucidity and data about SARS debilitated China's worldwide standing and hampered its financial development. The episode of SARS

IntechOpen

in China was disastrous and has prompted changes in medical care and clinical frameworks. Contrasted and China, the capacity of India to counter a pandemic is by all accounts a lot of lower. A new report announced that influenced relatives had not visit the Wuhan market in China, proposing that SARS-CoV-2 may spread without showing side effects. Analysts accept that this wonder is typical for some infections. India, with a populace of more than 1.34 billion—the second biggest populace on the planet—will experience issues treating serious COVID-19 cases on the grounds that the nation has just 49,000 ventilators, which is a negligible sum. On the off chance that the quantity of COVID-19 cases increments in the country, it would be a fiasco for India.

As the characteristics of a potential vaccine become better known, mathematical models can be used to explore alternative scenarios about effectively distributing a vaccine in order to limit transmission and protect the most vulnerable population groups.

Coronaviruses can spread in the following ways:

Coughing and sneezing without covering the mouth can disperse droplets into the air. Touching or shaking hands with a person who has the virus can pass the virus between individuals. Making contact with a surface or object that has the virus and then touching the nose, eyes, or mouth.

The National Institutes of Health (NIH) suggest that several groups of people have the highest risk of developing complications due to COVID-19. These groups include:

1. Young children

2. People aged 65 years or older

3. Pregnant women.

Coronaviruses will contaminate most individuals at a few time amid their lifetime. Coronaviruses can change viably, which makes them so infectious. To anticipate transmission, individuals ought to remain at domestic and rest whereas side effects are dynamic. They ought to moreover maintain a strategic distance from near contact with other individuals. They should also avoid close contact with other people. Covering the mouth and nose with a tissue or handkerchief while coughing or sneezing.

2. Analysis of SIR model

SIR model is first introduced by W.O. Kermach and A.G Mckendrick in 1927. SIR model is a best model of an infectious disease. This model divided the population into the three groups. The groups name is.

S (t) is the Susceptible people at the time.

I (t) is the infected people at the time.

R(t) is the recovered people at the time.

This model is constructing the ordinary differential equations in this model time t is the independent variable and S, I, R is the dependent variables. These groups have taken the number of people on every day. Yet, the data is transitions with time, as human being act from one group to another group. Illustration, human being in group **S** will act to the group **I**, that is the infected. Furthermore, infected person, **I** will act to the recovered **R** group that is they are recover or die from the disease. This method has been used successfully many times before in spreading disease like yellow fever, plague, fever, influenza, avian influenza, etc. Therefore, we have made the differential equations of COVID 19 using this method. This method is very helpful in giving a mathematical model to COVID.

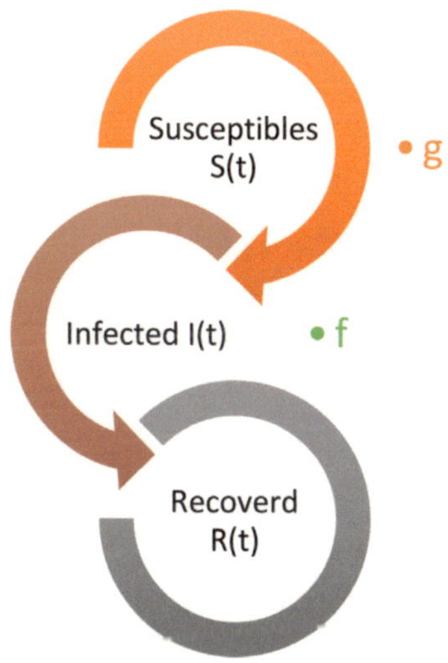

$$\frac{ds}{dt} = -gsi. \tag{1}$$

$$\frac{di}{dt} = gsi - fi \tag{2}$$

$$\frac{dr}{dt} = fi \tag{3}$$

where t is the independent variable s, i, r is dependent variables i.e.
s is denote the susceptible person at the time t
i is denote the infected person at the time t
r is denote the recovered person at the time t
g is transmission coefficient
f is recovery
If $s > 0, i > 0$ then

$$\frac{ds}{dt} = -gsi < 0, \forall s > 0, i > 0 \tag{4}$$

$$\frac{\mathrm{di}}{\mathrm{dt}} = \mathrm{i}(\mathrm{gs} - \mathrm{f}) \tag{5}$$

$$gs - f = 0$$

$$s = \frac{f}{g}$$

$$s - \frac{f}{g} = 0 \tag{6}$$

$$\text{So } \frac{di}{dt} < 0 \text{ if } s < \frac{f}{g} \tag{7}$$

$\frac{di}{dt} > 0$ if $s > \frac{f}{g}$ is defined the direction diagram of trajectories
We find the trajectories

$$\frac{di}{ds} = \frac{\frac{di}{dt}}{\frac{ds}{dt}} = \frac{gsi - fi}{-gsi}$$

$$\frac{di}{ds} = -1 + \frac{f}{gs}$$

$$di = -1ds + \frac{f}{gs}\,ds$$

$$i = -s + \frac{f}{g}\log s + c$$

Initial conditions

$$s(0) = s_0$$
$$i(0) = i_0$$

(If)

$$s \to 0, i \to -\infty$$
$$s \to \infty, i \to -\infty$$

1. It is impossible for the disease to infect all the susceptible person.

2. $s_0 > \frac{f}{g}$ for an epidemic to occur.

3. $s_0 < \frac{f}{g}$ disease dig out.

4. $\frac{gs_0}{f} > 1$ then number of infected is increase.

5. $s + i + r$ is the total population.

3. Mathematical modeling of COVID 19

3.1 Case study of India

In this research work we have discussed about the COVID 19 disease. also know about how many people got sick in India due to pandemic disease COVID19. In this article, we have given a mathematical model to COVID 19 with the help of SIR model [4–6]. We have taken data of how many people had become ill in India by COVID 19 on 20 January 2021 and using this data , we have created a mathematical model of COVID 19 with the help of SIR model. We have created three differential equations by taking the original data and solving those equations by the Homotopy Perturbation method (HPM).We got out the numerical solutions and made a table, and with the help of that table, we tried to tell what is the position of COVID 19 in India by making the graphs. There are a lot of methods which solves the differential equations [7–9], but we have used the HPM method. This method solves the biggest and most difficult equations very easily and with less of calculations.We will solve the Differential Equations of COVID 19, which is made with the help of SIR model. In these equations, we took the data of 20 January 2021 COVID 19 people of India who were caught by COVID 19 epidemics.

Total confirmed cases on 20 January 2021 in India is 10611728.

Death 152907

Recovered is 102656163

Active cases are 189245

Susceptible persons are 189347782

So, we take the

$$s(0\,) = 18.9347782$$

$$i(0) = 1.0611728$$

$$r\,(0) = 10265163 + 152907 = 1.0418070$$

$$g = \frac{active\ \ cases\ of\ \ india\ on\ 20\,january\ 2021\,for\ COVID\ 19}{susceptible\ people\ of\ \ india\ on\ 20\,january\ 2021\,for\ \ COVID\ 19}$$

$$g = \frac{189245}{18.9347782} = 0.00999457$$

$$f = \frac{1}{14} = 0.0714$$

$$\frac{ds}{dt} = -gsi \tag{8}$$

$$\frac{di}{dt} = gsi\ \ -fi \tag{9}$$

$$\frac{dr}{dt} = fi \tag{10}$$

where t is the independent variable s, i, r is dependent variables i.e.

s is denote the susceptible person at the time t

i is denote the infected person at the time t

r is denote the recovered person at the time t

g is transmission coefficient

f is recovery

Now we will solve these equations with the help of HPM method.

By the homotopy method we get,

$(\,1-p)\,\frac{dS}{dt} + p(\frac{dS}{dt} + si)$

$$\frac{dS}{dt} = p(-0.00999457si)$$

$$(1-p)\frac{di}{dt} + p\left(\frac{di}{dt} - 0.00999457si + 0.0714i\right) \tag{11}$$

$$\frac{di}{dt} = p(\,0.00999457si - 0.0714i) \tag{12}$$

$$\frac{dr}{dt} = p0.0714i$$

$$s = s_0 + p^1 s_1 + p^2 s_2 + \ldots \quad (13)$$

$$i = i_0 + p^1 i_1 + p^2 i_2 + \ldots \quad (14)$$

$$r = r_0 + p^1 r_1 + p^2 r_2 + \ldots \quad (15)$$

Putting value s, i, r we get,

$$\frac{ds}{dt} = p(-0.00999457\{ s_0 + p^1 s_1 + p^2 s_2 + \ldots\}\{i_0 + p^1 i_1 + p^2 i_2 + \ldots\}) \quad (16)$$

$$\frac{di}{dt} = p(0.00999457[\{ s_0 + p^1 s_1 + p^2 s_2 + \ldots\}\{ i_0 + p^1 i_1 + p^2 i_2 + \ldots\}] - 0.0714\{i_0 + p^1 i_1 + p^2 i_2 + \ldots\}) \quad (17)$$

$$\frac{dr}{dt} = p\ 0.0714\{i_0 + p^1 i_1 + p^2 i_2 \quad (18)$$

Both side comparing the coefficient of p we get

$$s_0 = 18.9347782$$

$$i_0 = 1.0611728$$

$$r_0 = 1.0418070$$

$$\frac{ds_1}{dt} = -0.00999457(18.9347782)(1.0611728)$$

$$s_1 = -0.2008216t$$

$$\frac{di_1}{dt} = 0.00999457(18.9347782)(1.0611728) - 0.0714(1.0611728)$$

$$i_1 = 0.1250538t$$

$$\frac{dr_1}{dt} = 0.0714\{1.0611728\}$$

$$r_1 = 0.0757677t$$

So, by the HPM we get the solution:

$$s(t) = 18.9347782 - 0.2008216t + \ldots$$

$$i(t) = 1.0611728 + 0.1250538t + \ldots$$

$$r(t) = 1.0418070 + 0.0757677t + \ldots$$

We have the table

S.NO.	DATE	S	I	R
1	20/01/2021	18.9347782	1.0611728	1.0418070
2	21 /01/2021	18.7339566	1.0662266	1.1175747
3	22/01/2021	18.533135	1.06128040	1.1933424
4	23/01/2021	18.3323134	1.06633424	1.2691101
5	24/01/2021	18.1314918	1.0613880	1.3448778

S.NO.	DATE	S	I	R
6	25/01/2021	17.9306702	1.0664418	1.4206455
7	26/01/2021	17.7298486	1.0684956	1.4964132
8	27/01/2021	17.529027	1.0705494	1.5721809
9	28/01/2021	17.3282054	1.0716032	1.6479486
10	29/01/2021	17.1273838	1.0726657	1.7237163
11	30/01/2021	16.9265622	1.0747108	1.799484
12	31/01/2021	16.7257406	1.0767646	1.8752517

From the above table we can predict that infected cases of corona on 31 January which is almost same as actual cases on 31 January. The current COVID-19 pandemic is unprecedented, but the global response draws on the lessons learned from other disease outbreaks over the past several decades.

World scientists on COVID-19 then met at the World Health Organization's Geneva headquarters on 11–12 February 2020 to assess the current level of

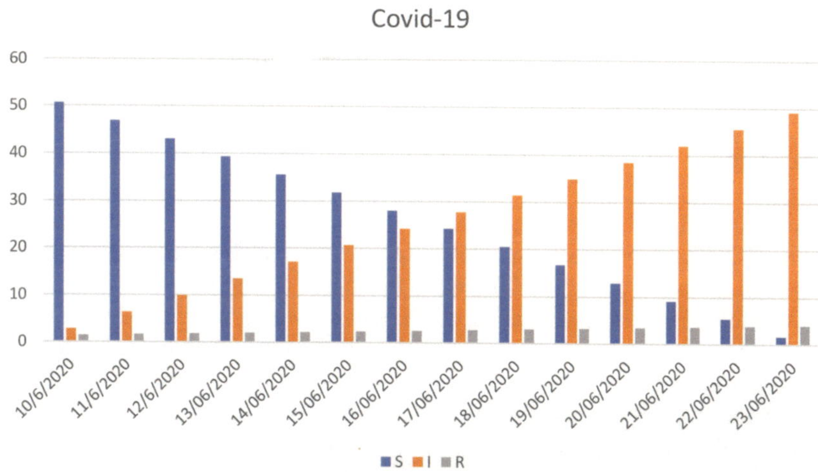

Figure 1.
SIR chart depicting no of people susceptible, infected and recovered.

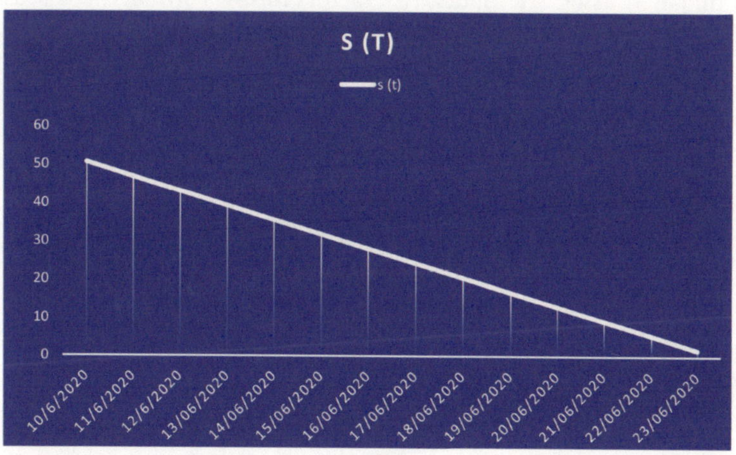

Figure 2.
SIR chart depicting no of people susceptible.

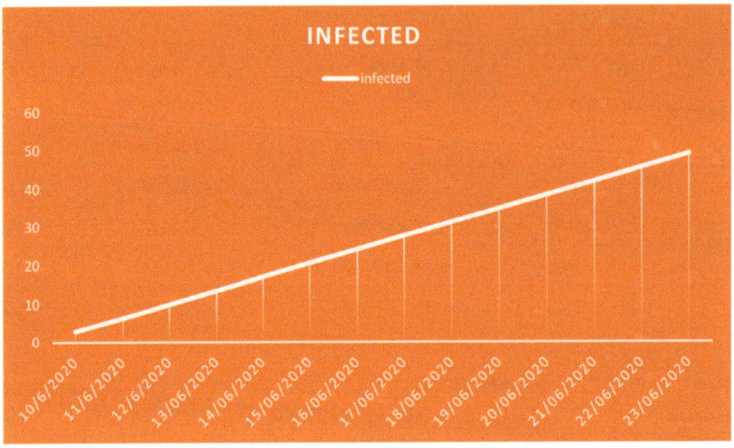

Figure 3.
SIR chart depicting no of people infected.

knowledge about the new virus, agree on critical research questions that need to be answered urgently, and to find ways to work together to accelerate and fund priority research to curtail this outbreak and prepare for those in the future see **Figures 1–3** for reference.

4. Conclusion

In this chapter, we have taken data of people affected by coronavirus in India till 20 January (1). Then we converted this data into three differential equations with the help of SIR model. We solved the equation made from SIR model with HPM. From the result of solving, we estimated the people who got infected with corona virus in the coming 5 days. We converted the result from HPM into a table and graph and from the result we saw that in the coming days, corona cases are increasing and recovering but the corona positive rate is very high, and the rate of recovery is very short. We saw that the information about Corona-positive cases being given by the Government of India was also that the rate of positive is increasing very fast, but the rate of recovery is very low. From all these, we can now say that by solving with HPM we get the result very close to the actual result. We have predicted cases of corona till Jan 31 using SIR model, risk factors for the coronavirus disease. The risk is especially high if two or three of the Cs come together.

Author details

Nahid Fatima
Prince Sultan University, Riyadh, Saudi Arabia

*Address all correspondence to: drnahidfatima@gmail.com

References

[1] Zunyou Wu and Jennifer M McGoogan. Characteristics of and important lessons from the coronavirus disease 2019 (COVID-19) outbreak in China: summary of a report of 72 314 cases from the Chinese center for disease control and prevention. Jama, 2020.

[2] Yueling Ma, Yadong Zhao, Jiangtao Liu, Xiaotao He, Bo Wang, Shihua Fu, Jun Yan, Jingping Niu, and Bin Luo. Effects of temperature variation and humidity on the mortality of covid-19 in Wuhan. MedRxiv, 2020.

[3] Miguel B. Araujo and Babak Naimi. Spread of SARS-CoV-2 Coronavirus likely to be constrained by climate. MedRxiv, 2020.

[4] Junling Ma, Jonathan Dushoff, Benjamin M Bolker, and David JD Earn. Estimating initial epidemic growth rates. Bulletin of mathematical biology, 76(1):245–260, 2014.

[5] Fred Brauer, Carlos Castillo-Chávez, Mathematical Models in Population Biology and Epidemiology, New York Springer 2001.

[6] H. W. Hethcote. The Mathematics of Infectious Diseases, SIAM Rev. 42 (2000), pp. 599-653.

[7] Samar Salman and Mohammed Labib Salem. The mystery behind childhood sparing by COVID-19. International Journal of Cancer and Biomedical Research, 2020.

[8] Wei Luo, Maimuna Majumder, Dianbo Liu, Canelle Poirier, Kenneth Mandl, Marc Lipsitch, and Mauricio Santillana. The role of absolute humidity on transmission rates of the covid-19 outbreak. 2020.

[9] J. D. Murray, Mathematical Biology, Springer-Verlag (1993).

Chapter 7

Rare Event Simulation in a Dynamical Model Describing the Spread of Traffic Congestions in Urban Network Systems

Getachew K. Befekadu

Abstract

In this chapter, we present a mathematical framework that provides a new insight for understanding the spread of traffic congestions in an urban network system. In particular, we consider a dynamical model, based on the well-known susceptible-infected-recovered (SIR) model from mathematical epidemiology, with small random perturbations, that describes the process of traffic congestion propagation and dissipation in an urban network system. Here, we provide the asymptotic probability estimate based on the Freidlin-Wentzell theory of large deviations for certain rare events that are difficult to observe in the simulation of urban traffic network dynamics. Moreover, the framework provides a computational algorithm for constructing efficient importance sampling estimators for rare event simulations of certain events associated with the spread of traffic congestions in the dynamics of the traffic network.

Keywords: diffusion processes, exit probability, HJB equations, importance sampling, large deviations, rare-event simulation, SIR model, traffic network dynamics

1. Introduction

In recent years, there have been a number of interesting studies related to modeling the spread of traffic congestion propagation and traffic dissipation in urban network systems (e.g., see [1–5] in the context of macroscopic traffic model involving traffic flux and traffic density; see [6, 7] in the context of percolation theory; see [8] for results based on machine-learning methods; and see [9, 10] for studies based on queuing theory). In this paper, without attempting to give a literature review, we consider a dynamical model, based on the well-known susceptible-infected-recovered (SIR) model from mathematical epidemiology, with small random perturbation, that describes the spread of traffic congestion propagation and dissipation in an urban network system, i.e.,

$$\begin{aligned} dc^{\varepsilon}(t) &= (-\mu + \beta k(1 - r^{\varepsilon}(t) - c^{\varepsilon}(t)))c^{\varepsilon}(t)dt \\ &\quad + \sqrt{\varepsilon}\sqrt{(\mu + \beta k(1 + r^{\varepsilon}(t) + c^{\varepsilon}(t)))c^{\varepsilon}(t)}\,dW_1(t) \end{aligned} \tag{1}$$

$$dr^{\varepsilon}(t) = \mu r^{\varepsilon}(t) + \sqrt{\varepsilon}\sqrt{\mu r^{\varepsilon}(t)}\,dW_2(t) \tag{2}$$

$$df^{\varepsilon}(t) = (-\beta k(1 - r^{\varepsilon}(t) - c^{\varepsilon}(t))c^{\varepsilon}(t)dt \\ + \sqrt{\varepsilon}\sqrt{(\beta k(1 + r^{\varepsilon}(t) + c^{\varepsilon}(t))c^{\varepsilon}(t)}dW_3(t) \tag{3}$$

where

- $c^{\varepsilon}(t)$ represents the fraction of congested links in the network
- $r^{\varepsilon}(t)$ represents the fraction of recovered links in the network
- $f^{\varepsilon}(t)$ represents the fraction of free flow links in the network
- the parameters β and μ represent respectively the propagation and recovery rates considering that a certain fraction of congested links will eventually recover as the demand for travel diminishes
- the quantity $k\beta/\mu$ represents the average number of newly congested links that, in a fully freely flowing traffic network, each already congested link can potentially create,
- $W_1(t)$, $W_2(t)$ and $W_3(t)$ are three independent standard (one-dimensional) Wiener processes, and
- ε is a small positive number that represents the level of random perturbation in the network.

Notice that Eq. (1) describes the rate at which the fraction of congested links, i.e., $c^{\varepsilon}(t)$, changes over time given the propagation rate β and recovery rate μ considering that a fraction of congested links will eventually recover as the demand for the travel volume diminishes. Moreover, Eq. (2) describes the rate at which congested links normally recover given the recovery rate μ. Finally, Eq. (3) represents how the fraction of free flow links $f^{\varepsilon}(t)$ in the network changes over time given $c^{\varepsilon}(t)$ and $r^{\varepsilon}(t)$. Note that, for a normalized SIR based traffic network dynamic model, the following mathematical condition $c^{\varepsilon}(t) + r^{\varepsilon}(t) + f^{\varepsilon}(t) = 1$ holds true for all $t > 0$, where $f^{\varepsilon}(t)$ represents links that have remained in a free flow state starting from $t = 0$ (e.g., see Saberi et al. [11] for detailed discussions related to deterministic models).

In this chapter, we provide the asymptotic probability estimate based on the Freidlin-Wentzell theory of large deviations for certain rare events that are difficult to observe in the simulation of urban traffic network dynamics. The framework considered in this study basically relies on the connection between the probability theory of large deviations and that of the values functions for a family of stochastic control problems, where such a connection also provides a desirable computational algorithm for constructing an efficient importance sampling estimator for rare event simulations of certain events associated with the spread of traffic congestions in the dynamics of the traffic network. Here, it is worth mentioning that a number of interesting studies based on various approximations techniques from the theory of large deviations have provided a framework for constructing efficient importance sampling estimators for rare event simulation problems involving the behavior of diffusion processes (e.g., [12–16] for additional discussions). The approach followed in these studies is to construct exponentially-tilted biasing distributions, which was originally introduced for proving Cramér's theorem and its extension, and later on it was found to be an efficient importance sampling distribution for certain problems with various approximations involving rare-events (e.g., see [17–19] or [13] for detailed discussions). The rationale behind our framework follows in some sense the settings of these papers. However, to our knowledge, the problem of rare event simulations involving the spread of traffic congestions in an urban network system has not been addressed in the context of large deviations and stochastic control arguments in the small noise limit; and it is important because it provides a new insight for understanding the spread of traffic congestions in an urban network system.

This chapter is organized as follows. In Section 2, we provide an asymptotic estimate on the exit probability using the Freidlin-Wentzell theory of large deviations [20] (see also [21], Chapter 4) and the stochastic control arguments from Fleming [22] (see also [23]), where such an asymptotic estimate relies on the interpretation of the exit probability function as a value function for a family of stochastic control problems that can be associated with the underlying SIR based traffic network dynamic model with small random perturbations. In Section 3, we discuss importance sampling and the necessary background upon which our main results rely. In Section 4, we provide our main results for an efficient importance sampling estimator for rare event simulations of certain events associated with the spread of traffic congestions in the dynamics of the traffic network. Finally, Section 5 provides some concluding remarks.

2. The Freidlin-Wentzell theory

In this section, we briefly review the classical Freidlin-Wentzell theory of large deviations for the stochastic differential equations (SDEs) with small noise terms. In what follows, let us denote the solution of the SDEs in Eqs. (1)–(3) by a bold face letter $\left(\mathbf{x}_t^\varepsilon\right)_{t\geq 0} = \left(x_t^{\varepsilon,1}, x_t^{\varepsilon,2}, x_t^{\varepsilon,3}\right)_{t\geq 0} \triangleq \left(c^\varepsilon(t), r^\varepsilon(t), f^\varepsilon(t)\right)_{t\geq 0}$ as an $\mathbb{R}^3$-valued diffusion process and rewrite the above equations as follows

$$d\mathbf{x}_t^\varepsilon = \mathbf{f}\left(\mathbf{x}_t^\varepsilon\right)dt + \sqrt{\varepsilon}\,\sigma\left(\mathbf{x}_t^\varepsilon\right)dW_t, \tag{4}$$

where $\mathbf{f}\left(\mathbf{x}_t^\varepsilon\right) = \left[f_1\left(\mathbf{x}_t^\varepsilon\right), f_2\left(\mathbf{x}_t^\varepsilon\right), f_3\left(\mathbf{x}_t^\varepsilon\right)\right]^T$ with

$$\begin{aligned} f_1\left(x_t^{\varepsilon,1}, x_t^{\varepsilon,2}, x_t^{\varepsilon,3}\right) &= \left(-\mu + \beta k\left(1 - x_t^{\varepsilon,2} - x_t^{\varepsilon,1}\right)\right)x_t^{\varepsilon,1} \\ f_2\left(x_t^{\varepsilon,1}, x_t^{\varepsilon,2}, x_t^{\varepsilon,3}\right) &= \mu x_t^{\varepsilon,2} \\ f_3\left(x_t^{\varepsilon,1}, x_t^{\varepsilon,2}, x_t^{\varepsilon,3}\right) &= \left(-\beta k\left(1 - x_t^{\varepsilon,2} - x_t^{\varepsilon,1}\right)\right)x_t^{\varepsilon,1} \end{aligned} \tag{5}$$

and $\sigma\left(\mathbf{x}_t^\varepsilon\right) = \left[\sigma_1\left(\mathbf{x}_t^\varepsilon\right), \sigma_2\left(\mathbf{x}_t^\varepsilon\right), \sigma_3\left(\mathbf{x}_t^\varepsilon\right)\right]^T$ with

$$\begin{aligned} \sigma_1\left(x_t^{\varepsilon,1}, x_t^{\varepsilon,2}, x_t^{\varepsilon,3}\right) &= \sqrt{\left(\mu + \beta k\left(1 + x_t^{\varepsilon,2} + x_t^{\varepsilon,1}\right)\right)x_t^{\varepsilon,1}} \\ \sigma_2\left(x_t^{\varepsilon,1}, x_t^{\varepsilon,2}, x_t^{\varepsilon,3}\right) &= \sqrt{\mu x_t^{\varepsilon,2}} \\ \sigma_3\left(x_t^{\varepsilon,1}, x_t^{\varepsilon,2}, x_t^{\varepsilon,3}\right) &= \sqrt{\left(\beta k\left(1 + x_t^{\varepsilon,2} + x_t^{\varepsilon,1}\right)\right)x_t^{\varepsilon,1}}. \end{aligned} \tag{6}$$

Moreover, W_t is a standard three-dimensional Wiener process. Note that the corresponding backward operator for the diffusion process $\mathbf{x}_t^\varepsilon$, when applied to a certain function $v^\varepsilon(t, \mathbf{x})$, is given by

$$\partial_t v^\varepsilon + \mathcal{L}^\varepsilon v \triangleq \frac{\partial v^\varepsilon(t,\mathbf{x})}{\partial t} + \frac{\varepsilon}{2}\sum_{i,j=1}^{3} a_{i,j}(\mathbf{x})\frac{\partial^2 v^\varepsilon(t,\mathbf{x})}{\partial x^i \partial x^j} + \mathbf{f}(\mathbf{x})\cdot\nabla_{\mathbf{x}} v^\varepsilon(t,\mathbf{x}), \tag{7}$$

where $a(\mathbf{x}) = \sigma(\mathbf{x})\sigma^T(\mathbf{x})$.

Let $\Omega \in \mathbb{R}^3$ be bounded open domains with smooth boundary (i.e., $\partial\Omega$ is a manifold of class C^2) and let Ω^T be an open set defined by

$$\Omega^T = (0, T) \times \Omega. \tag{8}$$

Furthermore, let us denote by $C^\infty(\Omega^T)$ the spaces of infinitely differentiable functions on Ω^T and by $C_0^\infty(\Omega^T)$ the space of the functions $\phi \in C^\infty(\Omega^T)$ with compact support in Ω^T. A locally square integrable function $v^\varepsilon(t,\mathbf{x})$ on Ω^T is said to be a distribution solution to the following equation

$$\partial_t v^\varepsilon + \mathcal{L}^\varepsilon v^\varepsilon = 0, \tag{9}$$

if, for any test function $\phi \in C_0^\infty(\Omega^T)$, the following holds true

$$\int_{\Omega^T} (-\partial_t\phi + \mathcal{L}^{\varepsilon *}\phi) v^\varepsilon d\Omega^T = 0, \tag{10}$$

where $d\Omega^T$ denotes the Lebesgue measure on $\mathbb{R}^3 \times \mathbb{R}_+$ and $\mathcal{L}^{\varepsilon *}$ is an adjoint operator corresponding to the infinitesimal generator $\mathcal{L}^\varepsilon$ of the process $\mathbf{x}_t^\varepsilon$.

Moreover, we also assume that the following statements hold for the SDE in (4).

Assumption 1

a. The function $\mathbf{f}$ is a bounded $C^\infty((0,\infty)\times\Omega)$-function, with bounded first derivatives. Moreover, σ and σ^{-1} are bounded $C^\infty((0,\infty)\times\mathbb{R}^3)$-functions, with bounded first derivatives.

b. Let $n(\mathbf{x})$ be the outer normal vector to $\partial\Omega$ and, further, let Γ^+ and Γ^0 denote the sets of points $(t,\mathbf{x})$, with $\mathbf{x} \in \partial\Omega$, such that

$$\langle \mathbf{f}(t,\mathbf{x}), n(\mathbf{x})\rangle \tag{11}$$

is positive and zero, respectively.

Remark 1 *Note that*

$$\mathbb{P}^\varepsilon_{s,\mathbf{x}_s^\varepsilon}\left\{(\tau^\varepsilon, \mathbf{x}^\varepsilon_{\tau^\varepsilon}) \in \Gamma^+ \bigcup \Gamma^0,\ \tau^\varepsilon < \infty\right\} = 1, \quad \forall (s, \mathbf{x}_s^\varepsilon) \in \Omega_0^\infty. \tag{12}$$

where $\tau^\varepsilon = \inf\left\{t > s \,|\, \mathbf{x}_t^\varepsilon \in \partial\Omega\right\}$. Moreover, if

$$\mathbb{P}^\varepsilon_{s,\mathbf{x}_s^\varepsilon}\left\{(t, \mathbf{x}_t^\varepsilon) \in \Gamma^0 \text{ for some } t \in [s, T]\right\} = 0, \quad \forall (s, \mathbf{x}_s^\varepsilon) \in \Omega_0^\infty, \tag{13}$$

and $\tau^\varepsilon \leq T$, then we have $(\tau^\varepsilon, \mathbf{x}^\varepsilon_{\tau^\varepsilon}) \in \Gamma^+$, almost surely (see [24], Section 7).

In what follows, let $\mathbf{x}_t^\varepsilon$, for $0 \leq t \leq T$, be the diffusion process associated with (4) (or Eqs. (1)–(3)) and consider the following boundary value problem

$$\left.\begin{aligned} \partial_s v^\varepsilon + \mathcal{L}^\varepsilon v^\varepsilon = 0 \quad &\text{in} \quad \Omega^T \\ v^\varepsilon(s,\mathbf{x}) = 1 \quad &\text{on} \quad \Gamma_T^+ \\ v^\varepsilon(s,\mathbf{x}) = 0 \quad &\text{on} \quad \{T\}\times\Omega \end{aligned}\right\} \tag{14}$$

where $\mathcal{L}^\varepsilon$ is the backward operator in (7) and

$$\Gamma_T^+ = \{(s, \mathbf{x}) \in \Gamma^+ \,|\, 0 < s \le T\}. \tag{15}$$

Further, let Ω^{0T} be the set consisting of $\Omega^T \cup \{T\} \times \Omega$, together with the boundary points $(s, \mathbf{x}) \in \Gamma^+$, with $0 < s < T$. Then, the following proposition, whose proof is given in [25], provides a solution to the exit probability $\mathbb{P}^\varepsilon_{s,\mathbf{x}^\varepsilon_s}\{\tau^\varepsilon \le T\}$ with which the diffusion process $\mathbf{x}^\varepsilon_t$ exits from the domain Ω.

Proposition 1 Suppose that the statements in Assumption 1 hold true. Then, the exit probability $q^\varepsilon(s, \mathbf{x}^\varepsilon) = \mathbb{P}^\varepsilon_{s,\mathbf{x}^\varepsilon_s}\{\tau^\varepsilon \le T\}$ is a smooth solution to the boundary value problem in (14) and, moreover, it is a continuous function on Ω^{0T}.

Note that, from Proposition 1, the exit probability $q^\varepsilon(s, \mathbf{x}^\varepsilon)$ is a smooth solution to the boundary value problem in (14). Further, if we introduce the following logarithmic transformation (e.g., see [22, 26] or [23])

$$I^\varepsilon(s, \mathbf{x}^\varepsilon) = -\varepsilon \log q^\varepsilon(s, \mathbf{x}^\varepsilon). \tag{16}$$

Then, using ideas from stochastic control theory (see [22] for similar arguments), we present results useful for proving the following asymptotic property

$$I^\varepsilon(s, \mathbf{x}^\varepsilon) \to I^0(s, \mathbf{x}^\varepsilon) \quad \text{as} \quad \varepsilon \to 0. \tag{17}$$

The starting point for such an analysis is to introduce a family of related stochastic control problems whose dynamic programming equation, for $\varepsilon > 0$, is given below by (21). Then, this also allows us to reinterpret the exit probability function as a value function for a family of stochastic control problems associated with the underlying urban traffic network dynamics with small random perturbation. Moreover, as discussed later in Section 5, such a connection provides a computational paradigm – based on an exponentially-tilted biasing distribution – for constructing an efficient importance sampling estimators for rare-event simulations that further improves the efficiency of Monte Carlo simulations.

Then, we consider the following boundary value problem

$$\left.\begin{aligned} \partial_s g^\varepsilon + \frac{\varepsilon}{2}\mathcal{L}^\varepsilon = 0 \quad &\text{in} \quad \Omega^T \\ g^\varepsilon = \mathbb{E}^\varepsilon_{s,\mathbf{x}}\left\{ \exp\left(-\frac{1}{\varepsilon}\Phi^\varepsilon\right)\right\} \quad &\text{on} \quad \partial^*\Omega^T \end{aligned}\right\} \tag{18}$$

where $\Phi^\varepsilon(s, \mathbf{x}^\varepsilon)$ is a bounded, nonnegative Lipschitz function such that

$$\Phi^\varepsilon(s, \mathbf{x}^\varepsilon) = 0, \quad \forall (s, \mathbf{x}^\varepsilon) \in \Gamma_T^+. \tag{19}$$

Observe that the function $g^\varepsilon(s, \mathbf{x}^\varepsilon)$ is a smooth solution in Ω^T to the backward operator in (9); and it is also continuous on $\partial^*\Omega^T$. Moreover, if we introduce the following logarithm transformation

$$J^\varepsilon(s, \mathbf{x}^\varepsilon) = -\varepsilon \log g^\varepsilon(s, \mathbf{x}^\varepsilon). \tag{20}$$

Then, $J^\varepsilon(s, \mathbf{x}^\varepsilon)$ satisfies the following dynamic programming equation (i.e., the Hamilton-Jacobi-Bellman equation)

$$\partial_s J^\varepsilon + \frac{\varepsilon}{2}\sum_{i,j=1}^{3} a_{i,j}\frac{\partial^2 J^\varepsilon}{\partial x^i \partial x^j} + H^\varepsilon = 0, \quad \text{in} \quad \Omega^T, \tag{21}$$

where $H^\varepsilon = H^\varepsilon(s, \mathbf{x}^\varepsilon, \nabla_{\mathbf{x}} J^\varepsilon)$ is given by

$$H^\varepsilon(s, \mathbf{x}^\varepsilon, \nabla_{\mathbf{x}} J^\varepsilon) = \mathbf{f}(\mathbf{x}^\varepsilon) \cdot \nabla_{\mathbf{x}} J^\varepsilon(s, \mathbf{x}^\varepsilon) - \frac{1}{2}(\nabla_{\mathbf{x}} J^\varepsilon(s, \mathbf{x}^\varepsilon))^T a(\mathbf{x}^\varepsilon) \nabla_{\mathbf{x}} J^\varepsilon(s, \mathbf{x}^\varepsilon). \tag{22}$$

Note that the duality relation between $H^\varepsilon(s, \mathbf{x}^\varepsilon, \cdot)$ and $L^\varepsilon(s, \mathbf{x}^\varepsilon, \cdot)$, i.e.,

$$H^\varepsilon(s, \mathbf{x}^\varepsilon, \nabla_{\mathbf{x}} J^\varepsilon) = \inf_{\hat{u}} \{L^\varepsilon(s, \mathbf{x}^\varepsilon, \hat{u}) + \nabla_{\mathbf{x}} J^\varepsilon \cdot \hat{u}\rangle\}, \tag{23}$$

with

$$L^\varepsilon(s, \mathbf{x}^\varepsilon, \hat{u}) = \frac{1}{2}\|\mathbf{f}(\mathbf{x}^\varepsilon) - \hat{u}\|^2_{[a(\mathbf{x}^\varepsilon)]^{-1}}, \tag{24}$$

where $\| \cdot \|^2_{[a(\mathbf{x}^\varepsilon)]^{-1}}$ denotes the Riemannian norm of a tangent vector.

Then, it is easy to see that $J^\varepsilon(s, \mathbf{x}^\varepsilon)$ is a solution in Ω^T, with $J^\varepsilon = \Phi^\varepsilon$ on $\partial^* \Omega^T$, to the dynamic programming in (21), where the latter is associated with the following stochastic control problem

$$J^\varepsilon(s, \mathbf{x}^\varepsilon) = \inf_{\hat{u} \in \hat{U}(s, \mathbf{x}_s^\varepsilon)} \mathbb{E}_{s, \mathbf{x}_s^\varepsilon} \left\{ \int_s^\theta L^\varepsilon(s, \mathbf{x}^\varepsilon, \hat{u}) dt + \Phi^\varepsilon(\theta, \mathbf{x}^\varepsilon) \right\} \tag{25}$$

that corresponds to the following system of SDEs

$$d\mathbf{x}_t^\varepsilon = \hat{u}(t)dt + \sqrt{\varepsilon}\sigma(\mathbf{x}_t^\varepsilon)dW_t, \tag{26}$$

with an initial condition $\mathbf{x}_s^\varepsilon = \mathbf{x}^\varepsilon$ and $\hat{U}(s, \mathbf{x}^\varepsilon)$ is a class of continuous functions for which $\theta \le T$ and $(\theta, x_\theta^\varepsilon) \in \Gamma_T^+$.

Next, we provide bounds, i.e., the asymptotic lower and upper bounds, on the exit probability $q^\varepsilon(s, \mathbf{x}^\varepsilon)$.

Define

$$\begin{aligned} I_\Omega^\varepsilon((s, \mathbf{x}^\varepsilon); \partial\Omega) &= -\lim_{\varepsilon \to 0} \varepsilon \log \mathbb{P}_{s, \mathbf{x}_s^\varepsilon}^\varepsilon \{\mathbf{x}_\theta^\varepsilon \in \partial\Omega\}, \\ &\triangleq -\lim_{\varepsilon \to 0} \varepsilon \log q^\varepsilon(s, \mathbf{x}^\varepsilon), \end{aligned} \tag{27}$$

where θ (or $\theta = \tau^\varepsilon \wedge T$) is the first exit-time of $\mathbf{x}_t^\varepsilon$ from the domain Ω. Furthermore, let us introduce the following supplementary minimization problem

$$\tilde{I}_\Omega^\varepsilon(s, \varphi, \theta) = \inf_{\varphi \in C_{sT}([s, T], \mathbb{R}^3), \theta \ge s} \int_s^\theta L^\varepsilon(t, \varphi(t), \dot{\varphi}(t)) dt, \tag{28}$$

where the infimum is taken among all $\varphi(\cdot) \in C_{sT}([s, T], \mathbb{R}^3)$ (i.e., from the space of $\mathbb{R}^d$-valued locally absolutely continuous functions, with $\int_s^T |\dot{\varphi}(t)|^2 dt < \infty$ for each $T > s$) and $\theta \ge s > 0$ such that $\varphi(s) \in \Omega^T$, for all $t \in [s, \theta)$, and $(\theta, \varphi(\theta)) \in \Gamma_T^+$. Then, it is easy to see that

$$\tilde{I}_\Omega^\varepsilon(s, \varphi, \theta) = I_\Omega^\varepsilon((s, \mathbf{x}^\varepsilon); \partial\Omega). \tag{29}$$

Next, we state the following lemma that will be useful for proving Proposition 2 (cf. [22], Lemma 3.1).

Lemma 1 If $\varphi \in C_{sT}([s,T],\mathbb{R}^3)$, for $s > 0$, and $\varphi(s) = \mathbf{x}_s^\varepsilon$, $(t,\varphi(t)) \in \Omega^T$, for all $t \in [s,T)$, then $\lim_{T\to\infty} \int_s^T L^\varepsilon(t,\varphi(t),\dot{\varphi}(t))dt = +\infty$.

Consider again the stochastic control problem in (25) together with (26). Suppose that Φ_M^ε (with $\Phi_M^\varepsilon \geq 0$) is class C^2 such that $\Phi_M^\varepsilon \to +\infty$ as $M \to \infty$ uniformly on any compact subset of $\Omega^T \backslash \overline{\Gamma}_T^+$ and Φ_M^ε on Γ_T^+. Further, if we let $J^\varepsilon = J_{\Phi_M}^\varepsilon$, when $\Phi^\varepsilon = \Phi_M^\varepsilon$, then we have the following lemma.

Lemma 2 Suppose that Lemma 1 holds, then we have

$$\liminf_{\substack{M\to\infty \\ (t,\mathbf{x}_t^\varepsilon)\to(s,\mathbf{x}_s^\varepsilon)}} J_{\Phi_M}^\varepsilon((s,\mathbf{x}^\varepsilon)) \geq I^\varepsilon(s,\mathbf{x}^\varepsilon). \tag{30}$$

Then, we have the following result.

Proposition 2 [25, Proposition 2.8] Suppose that Lemma 1 holds, then we have

$$I^\varepsilon(s,\mathbf{x}^\varepsilon) \to I^0(s,\mathbf{x}^\varepsilon) \quad \text{as} \quad \varepsilon \to 0, \tag{31}$$

uniformly for all $(s,\mathbf{x}_s^\varepsilon)$ in any compact subset $\overline{\Omega}^T$.

Proof: It is suffices to show the following conditions

$$\limsup_{\varepsilon\to 0} \varepsilon \log \mathbb{P}_{s,\mathbf{x}_s^\varepsilon}^\varepsilon\{\mathbf{x}_\theta^\varepsilon \in \partial\Omega\} \leq -I_\Omega^\varepsilon((s,\mathbf{x}^\varepsilon);\partial\Omega) \tag{32}$$

and

$$\liminf_{\varepsilon\to 0} \varepsilon \log \mathbb{P}_{s,\mathbf{x}_s^\varepsilon}^\varepsilon\{\mathbf{x}_\theta^\varepsilon \in \partial\Omega\} \geq -I_\Omega^\varepsilon((s,\mathbf{x}^\varepsilon);\partial\Omega), \tag{33}$$

uniformly for all $(s,\mathbf{x}_s^\varepsilon)$ in any compact subset $\overline{\Omega}^T$. Note that $I_\Omega^\varepsilon((s,\mathbf{x}^\varepsilon);\partial\Omega) = I^\varepsilon(s,\mathbf{x}^\varepsilon)$ (cf. Eq. (29)), then the upper bound in (32) can be verified using the Freidlin-Wentzell asymptotic estimates (e.g., see [27], pp. 332–334, [20] or [28]).

On the other hand, to prove the lower bound in (33), we introduce a penalty function Φ_M^ε (with $\Phi_M^\varepsilon(t,\mathbf{y}) = 0$ for $(t,\mathbf{y}) \in \Gamma_T^+$); and write $g^\varepsilon = g_M^\varepsilon$ $(\equiv \mathbb{E}_{s,\mathbf{x}_s^\varepsilon}^\varepsilon\{\exp(-\frac{1}{\varepsilon}\Phi_M^\varepsilon)\})$ and $J^\varepsilon = J_{\Phi_M}^\varepsilon$, with $\Phi^\varepsilon = \Phi_M^\varepsilon$. From the boundary condition in (18), then, for each M, we have

$$g^\varepsilon(s,\mathbf{x}^\varepsilon) \leq g_M^\varepsilon(s,\mathbf{x}^\varepsilon). \tag{34}$$

Using Lemma 2 and noting further the following

$$J_{\Phi_M}^\varepsilon(s,\mathbf{x}^\varepsilon) \geq I_\Omega^\varepsilon((s,\mathbf{x}^\varepsilon);\partial\Omega). \tag{35}$$

Then, the lower bound in (33) holds uniformly for all $(s,\mathbf{x}_s^\varepsilon)$ in any compact subset $\overline{\Omega}^T$. This completes the proof of Proposition 2. □

3. Importance sampling

In this paper, we are mainly concerned with estimating the following quantity

$$\mathbb{E}_{s,\mathbf{x}_s^\varepsilon}^\varepsilon\left[\exp\left(-\frac{1}{\varepsilon}\Phi^\varepsilon(\mathbf{x}^\varepsilon)\right)\right], \tag{36}$$

where Φ^{ε} is an appropriate functional on $C([0,T];\mathbb{R}^3)$ and $\mathbf{x}^{\varepsilon}$ is a solution of the SDE in (4) and our analysis is in the situation where the level of the random perturbation is small, i.e., $\varepsilon \ll 1$, and the functional $\mathbb{E}^{\varepsilon}_{s,\mathbf{x}^{\varepsilon}_s}\left[\exp\left(-\frac{1}{\varepsilon}\Phi^{\varepsilon}(\mathbf{x}^{\varepsilon})\right)\right]$ is rapidly varying in $\mathbf{x}^{\varepsilon}$. Note that the challenge presented by such an analysis of rare event probabilities is well documented (see [12, 18, 29] for additional discussions). In the following (and see also Section 4), we specifically consider the case when the functional Φ^{ε} is bounded and nonnegative Lipschitz, with $\Phi^{\varepsilon} = 0$, if $\mathbf{x}^{\varepsilon}_t \in \Omega^T \subset C([0,T]:\mathbb{R}^3)$ and $\Phi^{\varepsilon} = \infty$ otherwise; and we further consider analysis on the asymptotic estimates for exit probabilities from a given bounded open domain in the small noise limit case.

Consider the following simple estimator for the quantity of interest in (36)

$$\rho(\varepsilon) = \frac{1}{N}\sum_{j=1}^{N} \exp\left(-\frac{1}{\varepsilon}\Phi^{\varepsilon}\left(\mathbf{x}^{\varepsilon(j)}\right)\right), \tag{37}$$

where $\left\{\mathbf{x}^{\varepsilon(j)}\right\}_{j=1}^{N}$ are N-copies of independent samples of $\mathbf{x}^{\varepsilon}$. Here we remark that such an estimator is unbiased in the sense that

$$\mathbb{E}^{\varepsilon}_{s,\mathbf{x}^{\varepsilon}_s}[\rho(\varepsilon)] = \mathbb{E}^{\varepsilon}_{s,\mathbf{x}^{\varepsilon}_s}\left[\exp\left(-\frac{1}{\varepsilon}\Phi^{\varepsilon}(\mathbf{x}^{\varepsilon})\right)\right], \tag{38}$$

Moreover, its variance is given by

$$\mathrm{Var}(\rho(\varepsilon)) = \frac{1}{N}\left(\mathbb{E}^{\varepsilon}_{s,\mathbf{x}^{\varepsilon}_s}\left[\exp\left(-\frac{2}{\varepsilon}\Phi^{\varepsilon}(\mathbf{x}^{\varepsilon})\right)\right] - \mathbb{E}^{\varepsilon}_{s,\mathbf{x}^{\varepsilon}_s}\left[\exp\left(-\frac{1}{\varepsilon}\Phi^{\varepsilon}(\mathbf{x}^{\varepsilon})\right)\right]^2\right). \tag{39}$$

Then, we have the following for the relative estimation error

$$R_{\mathrm{err}}(\rho(\varepsilon)) = \frac{\sqrt{\mathrm{Var}(\rho(\varepsilon))}}{\mathbb{E}^{\varepsilon}_{s,\mathbf{x}^{\varepsilon}_s}[\rho(\varepsilon)]} \tag{40}$$

which can be further rewritten as follows

$$R_{\mathrm{err}}(\rho(\varepsilon)) = \left(1/\sqrt{N}\right)\sqrt{\Delta(\rho(\varepsilon)) - 1}, \tag{41}$$

where

$$\Delta(\rho(\varepsilon)) = \frac{\mathbb{E}^{\varepsilon}_{s,\mathbf{x}^{\varepsilon}_s}\left[\exp\left(-\frac{2}{\varepsilon}\Phi^{\varepsilon}(\mathbf{x}^{\varepsilon})\right)\right]}{\mathbb{E}^{\varepsilon}_{s,\mathbf{x}^{\varepsilon}_s}\left[\exp\left(-\frac{1}{\varepsilon}\Phi^{\varepsilon}(\mathbf{x}^{\varepsilon})\right)\right]^2}. \tag{42}$$

Note that, as we might expect, the relative estimation error may decrease with increasing the number of the sample size N. However, from Varahhan's lemma (e.g., see [30]; see also [20, 28]), under suitable assumptions, we also have the following conditions

$$\limsup_{\varepsilon\to 0} \varepsilon \log \mathbb{E}^{\varepsilon}_{s,\mathbf{x}^{\varepsilon}_s}\left[\exp\left(-\frac{1}{\varepsilon}\Phi^{\varepsilon}(\mathbf{x}^{\varepsilon})\right)\right] = -\inf_{\substack{\varphi\in C_{sT}([s,T],\mathbb{R}^{nd})\\ \varphi(s)=\mathbf{x}_s}} \{I(\varphi) + \Phi^{\varepsilon}(\varphi)\} \tag{43}$$

and

$$\limsup_{\varepsilon\to 0} \varepsilon \log \mathbb{E}^{\varepsilon}_{s,\mathbf{x}^{\varepsilon}_s}\left[\exp\left(-\frac{2}{\varepsilon}\Phi^{\varepsilon}(\mathbf{x}^{\varepsilon})\right)\right] = - \inf_{\substack{\varphi\in C_{sT}\left([s,T],\mathbb{R}^{nd}\right)\\ \varphi(s)=\mathbf{x}_s}} \{I(\varphi) + 2\Phi^{\varepsilon}(\varphi)\} \quad (44)$$

where $C_{sT}\left([s,T],\mathbb{R}^3\right)$ is the set of absolutely continuous functions from $[s,T]$ into $\mathbb{R}^3$, with $0 \le s \le t \le T$, and $I(\varphi)$ is the rate functional for the diffusion process $\mathbf{x}^{\varepsilon}_t$. From Jensen's inequality, the above equations in (43) and (44) also imply the following condition $\Delta(\rho(\varepsilon)) \ge 1$.

4. Main results

In this section, we present our main result that asserts the relative error decreases to zero as the small random perturbation tends to zero, which in turn implies the uniform log-efficiency for the estimation problem in (36).

In what follows, let $\hat{\mathbf{x}}^{\varepsilon}_t$ be the solution to the following SDE

$$\begin{aligned} d\hat{\mathbf{x}}^{\varepsilon}_t &= \mathbf{f}\left(t,\hat{\mathbf{x}}^{\varepsilon}_t\right)dt + \mathbf{b}\sigma\left(t,\hat{\mathbf{x}}^{\varepsilon}_t\right)v^{\varepsilon}\left(t,\hat{\mathbf{x}}^{\varepsilon}_t\right)dt + \sqrt{\varepsilon}\mathbf{b}\sigma\left(t,\hat{\mathbf{x}}^{\varepsilon}_t\right)dW_t, \\ &\text{with an initial condition} \quad \hat{\mathbf{x}}^{\varepsilon}_s = \mathbf{x}^{\varepsilon}_s, \end{aligned} \quad (45)$$

where v^{ε} is an appropriate control function (which also depends on ε) to be chosen so as to reduce the variance of the importance sampling estimator.

Let

$$z^{\varepsilon} = \exp\left(-\frac{1}{\sqrt{\varepsilon}}\int_s^T \langle v^{\varepsilon}\left(t,\hat{\mathbf{x}}^{\varepsilon}_t\right), dW_t\rangle - \frac{1}{2\varepsilon}\int_s^T \left|v^{\varepsilon}\left(t,\hat{\mathbf{x}}^{\varepsilon}_t\right)\right|^2 dt\right). \quad (46)$$

Then, the corresponding importance sampling estimator is given by

$$\hat{\rho}(\varepsilon) = \frac{1}{N}\sum_{j=1}^{N} \exp\left(-\frac{1}{\varepsilon}\Phi^{\varepsilon}\left(\hat{\mathbf{x}}^{\varepsilon(j)}\right)\right)z^{\varepsilon(j)}, \quad (47)$$

where $\left\{\left(\hat{\mathbf{x}}^{\varepsilon(j)}, z^{\varepsilon(j)}\right)\right\}_{j=1}^{N}$ are N-copies of independent samples of $(\hat{\mathbf{x}}^{\varepsilon}, z^{\varepsilon})$. Note that, for an appropriately chosen control function v^{ε}, the above importance sampling estimator in (47) is an unbiased estimator for (37), i.e.,

$$\begin{aligned} \mathbb{E}^{\varepsilon}_{s,\mathbf{x}^{\varepsilon}_s}[\hat{\rho}(\varepsilon)] &= \mathbb{E}^{\varepsilon}_{s,\mathbf{x}^{\varepsilon}_s}\left[\exp\left(-\frac{1}{\varepsilon}\Phi^{\varepsilon}(\mathbf{x}^{\varepsilon})\right)\right] \\ &\equiv \mathbb{E}^{\varepsilon}_{s,\mathbf{x}^{\varepsilon}_s}[\rho(\varepsilon)]. \end{aligned} \quad (48)$$

Moreover, the relative estimation error is given by

$$R_{\text{err}}(\hat{\rho}(\varepsilon)) = \frac{\sqrt{\text{Var}(\hat{\rho}(\varepsilon))}}{\mathbb{E}^{\varepsilon}_{s,\mathbf{x}^{\varepsilon}_s}[\hat{\rho}(\varepsilon)]} \quad (49)$$

which can be rewritten as follows

$$R_{\text{err}}(\hat{\rho}(\varepsilon)) = \left(1/\sqrt{N}\right)\sqrt{\Delta(\hat{\rho}(\varepsilon)) - 1}, \quad (50)$$

where

$$\Delta(\hat{\rho}(\varepsilon)) = \frac{\mathbb{E}^{\varepsilon}_{s,\mathbf{x}^{\varepsilon}_s}\left[\exp\left(-\frac{2}{\varepsilon}\Phi^{\varepsilon}(\hat{\mathbf{x}}^{\varepsilon})\right)\right](z^{\varepsilon})^2}{\mathbb{E}^{\varepsilon}_{s,\mathbf{x}^{\varepsilon}_s}\left[\exp\left(-\frac{1}{\varepsilon}\Phi^{\varepsilon}(\mathbf{x}^{\varepsilon})\right)\right]^2}. \tag{51}$$

Hence, in order to reduce the relative estimation error $R_{err}(\hat{\rho}(\varepsilon))$, we need to control the term $\Delta(\hat{\rho}(\varepsilon))$ in (50). Note that, from Jensen's inequality, we have the following condition

$$\limsup_{\varepsilon\to 0} -\varepsilon\log\mathbb{E}^{\varepsilon}_{s,\mathbf{x}^{\varepsilon}_s}\left[\exp\left(-\frac{2}{\varepsilon}\Phi^{\varepsilon}(\hat{\mathbf{x}}^{\varepsilon})\right)\right] \leq 2\lim_{\varepsilon\to 0} -\varepsilon\log\mathbb{E}^{\varepsilon}_{s,\mathbf{x}^{\varepsilon}_s}\left[\exp\left(-\frac{1}{\varepsilon}\Phi^{\varepsilon}(\hat{\mathbf{x}}^{\varepsilon})\right)\right] \tag{52}$$

which also implies $\Delta(\hat{\rho}(\varepsilon)) \geq 1$ with $\lim_{\varepsilon\to 0}\Delta(\hat{\rho}(\varepsilon)) = 1$. Moreover, the statement in (49) further implies the following

$$R_{err}(\hat{\rho}(\varepsilon)) = \frac{1}{\sqrt{N}}\exp\left(o(1)/\varepsilon\right) \quad \text{as} \quad \varepsilon\to 0, \tag{53}$$

which is generally referred as asymptotic efficiency or optimality. In this paper, our main objective is to choose appropriately the control function v^{ε} in (45), so that the resulting importance sampling estimator achieves a minimum rate of error growth. For this reason, we introduce the following standard definition from simulation theory (e.g., see [29] or [12]) which is useful for interpreting our main result.

Definition 1 An importance sampling estimator of the form (47) is log-efficient (i.e., asymptotic efficiency or optimal) if

$$\lim_{\varepsilon\to 0} -\varepsilon\log\Delta(\hat{\rho}(\varepsilon)) = 0. \tag{54}$$

Then, we state the following result as follows.

Proposition 3 Suppose that the importance sampling estimator $\hat{\rho}(\varepsilon)$ in (47), with $v^{\varepsilon}(t,\mathbf{x}) = -\sigma^T(\mathbf{x})\nabla_{\mathbf{x}}J^{\varepsilon}(t,\mathbf{x})$, is uniformly log-efficient (i.e., asymptotic efficient), where $J^{\varepsilon}(t,\mathbf{x})$ satisfies the corresponding dynamic programming equation in Ω^T with respect to the system in (45), with $J^{\varepsilon} = \Phi^{\varepsilon}$ on $\partial^*\Omega^T$. Then, there exits a set $\mathbb{A}\subset\mathbb{R}^3$ such that the Hausedorf dimension of $\mathbb{A}^c$ is zero and

$$\lim_{\varepsilon\to 0} R_{err}(\hat{\rho}(\varepsilon)) = 0, \tag{55}$$

for all $x\in\mathbb{A}$.

Proof: The above proposition basically asserts that the relative error $R_{err}(\hat{\rho}(\varepsilon))$ decreases to zero as the small random perturbation level ε tends to zero. Note that, if $J^{\varepsilon}(s,\mathbf{x}^{\varepsilon})$ satisfies the dynamic programming equation in (21), then, with $v^{\varepsilon}(t,\mathbf{x}) = -\sigma^T(\mathbf{x})\nabla_{\mathbf{x}}J^{\varepsilon}(t,\mathbf{x})$, the importance sampling for the estimation problem in (36), i.e., $\mathbb{E}^{\varepsilon}_{s,\mathbf{x}^{\varepsilon}_s}\left[\exp\left(-\frac{1}{\varepsilon}\Phi^{\varepsilon}(\mathbf{x}^{\varepsilon})\right)\right]$, is uniformly log-efficient if the point $(s,\mathbf{x}^{\varepsilon}_s)$ is contained in a region of sufficient regularity that encompasses almost all $\mathbb{R}^3$. As a result of this, it only suffices to show that

$$\lim_{\varepsilon\to 0}\frac{\mathbb{E}^{\varepsilon}_{s,\mathbf{x}^{\varepsilon}_s}\left[\exp\left(-\frac{2}{\varepsilon}\Phi^{\varepsilon}(\hat{\mathbf{x}}^{\varepsilon})\right)(z^{\varepsilon})^2\right]}{\mathbb{E}^{\varepsilon}_{s,\mathbf{x}^{\varepsilon}_s}\left[\exp\left(-\frac{1}{\varepsilon}\Phi^{\varepsilon}(\mathbf{x}^{\varepsilon})\right)\right]^2} = 1 \tag{56}$$

holds uniformly for all $(s,\mathbf{x}^{\varepsilon}_s)$ in any compact subset $\overline{\Omega}^T$.

Let us define following two functions

$$\psi_1^\varepsilon(s, \mathbf{x}_s^\varepsilon) = -\varepsilon \log \mathbb{E}_{s,\mathbf{x}_s^\varepsilon}^\varepsilon \left[\exp\left(-\frac{2}{\varepsilon} \Phi^\varepsilon(\hat{\mathbf{x}}^\varepsilon) \right) \right] \tag{57}$$

and

$$\begin{aligned} \psi_2^\varepsilon(s, \mathbf{x}_s^\varepsilon) &= -\varepsilon \log \mathbb{E}_{s,\mathbf{x}_s^\varepsilon}^\varepsilon \left[\exp\left(-\frac{2}{\varepsilon} \Phi^\varepsilon(\hat{\mathbf{x}}^\varepsilon) \right) (z^\varepsilon)^2 \right] \\ &= -\varepsilon \log \mathbb{E}_{s,\mathbf{x}_s^\varepsilon}^\varepsilon \left[\exp\left(-\frac{2}{\varepsilon} \Phi^\varepsilon(\hat{\mathbf{x}}^\varepsilon) - \frac{2}{\sqrt{\varepsilon}} \int_s^T \langle v^\varepsilon(t, \hat{\mathbf{x}}_t^\varepsilon), dW_t \rangle \right.\right. \\ &\qquad \left.\left. - \frac{1}{\varepsilon} \int_s^T \left| v^\varepsilon(t, \hat{\mathbf{x}}_t^\varepsilon) \right|^2 dt \right) \right]. \end{aligned} \tag{58}$$

Note that, from the large deviations results for the diffusion process $\hat{\mathbf{x}}_t^\varepsilon$ (e.g., see [21], Chapter 4, [30] or [27], pp.332–334; and see also the asymptotic estimates in Proposition 2 of Section 3), then there exists a constant C, $\gamma > 0$ and ε_0, with $\varepsilon \in (0, \varepsilon_0)$, such that

$$\begin{aligned} \mathbb{E}_{0,\mathbf{x}_0^\varepsilon}^\varepsilon \left[\exp\left(-\frac{1}{\varepsilon} (\psi_2^\varepsilon(\hat{\tau}^\varepsilon, \hat{\mathbf{x}}_{\hat{\tau}^\varepsilon}^\varepsilon) - 2\psi_1^\varepsilon(\hat{\tau}^\varepsilon, \hat{\mathbf{x}}_{\hat{\tau}^\varepsilon}^\varepsilon)) - \int_0^{\hat{\tau}^\varepsilon} \sum_{i,j=1}^3 a_{i,j}(\mathbf{x}_s^\varepsilon) \frac{\partial^2 {\psi_1}^\varepsilon(s, \hat{\mathbf{x}}_s^\varepsilon)}{\partial x^i \partial x^j} ds \right) \right] \\ \le C \exp(-\gamma/2\varepsilon), \end{aligned} \tag{59}$$

where $\hat{\tau}^\varepsilon = \inf\{t > s \,|\, \hat{\mathbf{x}}_t^\varepsilon \in \partial\Omega\} \wedge T$. Note that the above relation further implies that

$$\lim_{\varepsilon \to 0} \exp\left(-\frac{1}{\varepsilon} \left(\psi_2^\varepsilon(0, \mathbf{x}_0) - 2\psi_1^0(0, \mathbf{x}_0)) \right) \right) = \exp\left(\left(\int_0^T \sum_{i,j=1}^3 a_{i,j}(\mathbf{x}_s^\varepsilon) \frac{\partial^2 {\psi_1}^\varepsilon(s, \hat{\mathbf{x}}_s^\varepsilon)}{\partial x^i \partial x^j} ds \right) \right). \tag{60}$$

Moreover, in the same way, we can also show the following relation

$$\lim_{\varepsilon \to 0} \exp\left(-\frac{1}{\varepsilon} \left(\psi_1^\varepsilon(0, \mathbf{x}_0) - \psi_1^0(0, \mathbf{x}_0)) \right) \right) = \exp\left(\left(\int_0^T \sum_{i,j=1}^3 a_{i,j}(\mathbf{x}_s^\varepsilon) \frac{\partial^2 {\psi_1}^\varepsilon(s, \hat{\mathbf{x}}_s^\varepsilon)}{\partial x^i \partial x^j} ds \right) \right). \tag{61}$$

Finally, if we combine the above two equations, then we have the condition following

$$\lim_{\varepsilon \to 0} \exp\left(-\frac{1}{\varepsilon} (\psi_2^\varepsilon(0, \mathbf{x}_0) - \psi_1^\varepsilon(0, \mathbf{x}_0)) \right) = 1, \tag{62}$$

which implies the uniform log-efficiency for the estimation problem in (36). This completes the proof of Proposition 3.

Remark 2 The above proposition basically ensures a minimum relative estimation error in the small noise limit case for the estimation problem in (36). Note that, if $J^\varepsilon(t, \mathbf{x}^\varepsilon)$ satisfies the dynamic programming equation in (21) (i.e., if it is the solution for the family of stochastic control problems that are associated with the underlying distributed system with small random perturbation). Then, with $v^{\varepsilon}(t, \mathbf{x}) = -\sigma^{T}(\mathbf{x})\nabla_{\mathbf{x}} J^{\varepsilon}(t, \mathbf{x})$, one can provide a numerical computational framework for constructing efficient importance sampling estimators, with an exponential variance decay rate – based on an exponentially-tilted biasing distribution – for rare-event simulations involving the behavior of the diffusion process $\mathbf{x}^{\varepsilon}$.

Remark 3 Here, our primary intent is to provide a theoretical framework, rather than considering some specific numerical simulation results with respect to system parameters (such as the propagation rate β and recovery rate μ of the network), which is an ongoing research area.

5. Concluding remarks

In this chapter, we presented a mathematical framework that provides a new insight for understanding the spread of traffic congestions in an urban network system. In particular, we considered a dynamical model, based on the well-known susceptible-infected-recovered (SIR) model from mathematical epidemiology, with small random perturbations, that describes the process of traffic congestion propagation and dissipation in an urban network system. Moreover, we also provided the asymptotic probability estimate based on the Freidlin-Wentzell theory of large deviations for certain rare events that are difficult to observe in the simulation of an urban traffic network dynamic, where such a framework provides a computational algorithm for constructing efficient importance sampling estimators for rare event simulations of certain events associated with the spread of traffic congestions in the traffic network.

Author details

Getachew K. Befekadu
Department of Electrical and Computer Engineering, Morgan State University, Baltimore, USA

*Address all correspondence to: getachew.befekadu@morgan.edu

References

[1] Meead Saberi, Mahmassani HS. Exploring properties of network-wide flow-density relations in a freeway network. Transp. Res. Rec. 2012;**2315**: 153-163. https://doi.org/10.3141/2315-16 Accessed: 27 December 2020

[2] Meead Saberi, Hani S. Mahmassani: Hysteresis and capacity drop phenomena in freeway networks: empirical characterization and interpretation. Transp. Res. Rec. 2013; 2391:44–55. https://doi.org/10.3141/2391-05 [Accessed: 27 December 2020]

[3] Nikolas Geroliminis, Carlos F. Daganzo: Existence of urban-scale macroscopic fundamental diagrams: some experimental findings. Transp. Res. B. 2008; 42:759–770. https://doi.org/10.1016/j.trb.2008.02.002 [Accessed: 27 December 2020]

[4] Yuxuan Ji, Nikolas Geroliminis: On the spatial partitioning of urban transportation networks. Transp. Res. B. 2012; 46:1639–1656. https://doi.org/10.1016/j.trb.2012.08.005 [Accessed: 27 December 2020]

[5] Mohammadreza Saeedmanesh, Nikolas Geroliminis: Dynamic clustering and propagation of congestion in heterogeneously congested urban traffic networks. Transp. Res. Procedia. 2017; 23:962–979. https://doi.org/10.1016/j.trb.2017.08.021 [Accessed: 27 December 2020]

[6] Guanwen Zeng, Daqing Li, Shengmin Guo, Liang Gao, Ziyou Gao, H. Eugene Stanley, Shlomo Havlin: Switch between critical percolation modes in city traffic dynamics. Proc. Natl Acad Sci. USA. 2019; 116:23–28. https://doi.org/10.1073/pnas.1801545116 [Accessed: 27 December 2020]

[7] Daqing Li, Bowen Fu, Yunpeng Wang, Guangquan Lu, Yehiel Berezin, H. Eugene Stanley, Shlomo Havlin: Percolation transition in dynamical traffic network with evolving critical bottlenecks. Proc. Natl Acad. Sci. USA. 2015; 112:669–672. https://doi.org/10.1073/pnas.1419185112 [Accessed: 27 December 2020]

[8] M.T. Asif, J. Dauwels, C.Y. Goh, A. Oran, E. Fathi, M. Xu, M.M. Dhanya, N. Mitrovic and P. Jaillet: Spatio-temporal patterns in large-scale traffic speed prediction. IEEE Trans. Intell. Transp. Syst. 2014; 15:794–804. https://doi.org/10.1109/TITS.2013.2290285 [Accessed: 27 December 2020]

[9] Richard Arnott: A bathtub model of downtown traffic congestion. J. Urban Econ. 2013; 76:110–121. https://doi.org/10.1016/j.jue.2013.01.001 [Accessed: 27 December 2020]

[10] William S. Vickrey: Congestion theory and transport investment. Am. Econ. Rev. 1969; 59:251–260 https://www.jstor.org/stable/1823678 [Accessed: 27 December 2020]

[11] Meead Saberi, H. Hamedmoghadam, M. Ashfaq, et al. : A simple contagion process describes spreading of traffic jams in urban networks. Nat Commun. 2020; 11:1616. https://doi.org/10.1038/s41467-020-15353-2 [Accessed: 27 December 2020]

[12] Amarjit Budhiraja, Paul Dupuis: Analysis and approximation of rare events: representations and weak convergence methods. Prob. Theory and Stoch. Modelling Series, 94, Springer, 2019. https://doi.org/10.1007/978-1-4939-9579-0 [Accessed: 27 December 2020]

[13] Paul Dupuis Richard S. Ellis: A weak convergence approach to the theory of large deviations. Wiley, New York, 1997.

[14] Paul Dupuis, Hui Wang: Importance sampling, large deviations, and

differential games. Stoch. Stoch. Rep. 2004; 76:481–508. https://doi.org/10.1080/10451120410001733845 [Accessed: 27 December 2020]

[15] Weinan E, Weiqing Ren, Eric Vanden-Eijnden: Minimum action method for the study of rare events. Comm. Pure Appl. Math. 2004; 57:637–656. https://doi.org/10.1002/cpa.20005 [Accessed: 27 December 2020]

[16] Eric Vanden-Eijnden, Jonathan Weare: Rare event simulation of small noise diffusions. Comm. Pure Appl. Math. 2012; 65:1770–1803. https://doi.org/10.1002/cpa.21428 [Accessed: 27 December 2020]

[17] D. Siegmund: Importance sampling in the monte carlo study of sequential tests. The Annals of Statistics. 1976; 673–684. https://doi.org/10.1214/aos/1176343541 [Accessed: 27 December 2020]

[18] Soren Asmussen, Peter W. Glynn: Stochastic simulation: algorithms and analysis. Springer, New York, 2010. https://doi.org/10.1007/978-0-387-69033-9 [Accessed: 27 December 2020]

[19] Jügen Gärtner: On large deviations from the invariant measure. Theory Probab. Appl. 1977; 22(1), 24–39. https://doi.org/10.1137/1122003 [Accessed: 27 December 2020]

[20] A. D. Ventsel, M. I. Freidlin: On small random perturbations of dynamical systems. Russian Math. Surv. 1970; 25:1–55. https://doi.org/10.1070/RM1970v025n01ABEH001254 [Accessed: 27 December 2020]

[21] Mark I. Freidlin, and Alexander D. Wentzell: Random perturbations of dynamical systems. Springer, Berlin, 1984. https://doi.org/10.1007/978-3-642-25847-3 [Accessed: 27 December 2020]

[22] Wendell H. Fleming: Exit probabilities and optimal stochastic control. Appl. Math. Optim. 1978; 4: 329–346. https://doi.org/10.1007/BF01442148 [Accessed: 27 December 2020]

[23] Wendell Fleming, Raymond Rishel: Deterministic and stochastic optimal control. Springer-Verlag, New York, 1975. https://doi.org/10.1007/978-1-4612-6380-7 [Accessed: 27 December 2020]

[24] D. Stroock, S. R. S. Varadhan: On degenerate elliptic-parabolic operators of second order and their associated diffusions. Comm. Pure Appl. Math., 1972; 25:651–713. https://doi.org/10.1002/cpa.3160250603 [Accessed: 27 December 2020]

[25] Getachew K. Befekadu, Panos J. Antsaklis: On the asymptotic estimates for exit probabilities and minimum exit rates of diffusion processes pertaining to a chain of distributed control systems. SIAM J. Control Optim. 2015; 53:2297–2318. https://doi.org/10.1137/140990322 [Accessed: 27 December 2020]

[26] L. C. Evans, H. Ishii: A PDE approach to some asymptotic problems concerning random differential equations with small noise intensities. Ann. Inst. H. Poincaré Anal. Non Linearé. 1985; 2:1–20. https://doi.org/10.1016/S0294-1449(16)30409-7 [Accessed: 27 December 2020]

[27] Avner Friedman: Stochastic differential equations and applications. Dover Publisher, Inc. Mineola, New York, 2006.

[28] A. D. Ventcel: Limit theorems on large deviations for stochastic processes. Theo. Prob. Appl. 1973; 18:817–821. https://doi.org/10.1137/1121030 [Accessed: 27 December 2020]

[29] James A. Bucklew: Introduction to rare-event simulation. Springer Series in

Statistics. Springer, New York, 2004. https://doi.org/10.1007/978-1-4757-4078-3 [Accessed: 27 December 2020]

[30] S. R. S. Varadhan: Large deviations and applications, CBMS-NSF Regional Conference Series in Applied Mathematics, 46. SIAM, Philadelphia, 1985. https://doi.org/10.1137/1.9781611970241 [Accessed: 27 December 2020]

Chapter 8

Perturbation Expansion to the Solution of Differential Equations

Jugal Mohapatra

Abstract

The main purpose of this chapter is to describe the application of perturbation expansion techniques to the solution of differential equations. Approximate expressions are generated in the form of asymptotic series. These may not and often do not converge but in a truncated form of only two or three terms, provide a useful approximation to the original problem. These analytical techniques provide an alternative to the direct computer solution. Before attempting to solve these problems numerically, one should have an awareness of the perturbation approach.

Keywords: perturbation methods, asymptotic expansion, boundary layer, principle of least degeneracy, inner and outer expansion

1. Introduction

The governing equations of physical, biological and economical models often involve features which make it impossible to obtain their exact solution. For instance, problems where we observe "a complicated algebraic equations", "the occurrence of a complicated integral", in case of differential equations (DE), "a varying coefficients or nonlinear term" sometimes problems with an awkwardly shaped boundary are tough to solve with the limited methods for finding analytical solutions. The main purpose of this chapter is to describe the application of perturbation expansion techniques to the solution of DE. Approximate expressions are generated in the form of asymptotic series. These may not and often do not converge but in a truncated form of only two or three terms, provide a useful approximation to the original problem. These analytical techniques provide an alternative to the direct computer solution. Before attempting to solve these DE numerically, one should have an awareness of the perturbation approach. An example of this occurs in boundary layer problems where there are regions of rapid change of quantities such as fluid velocity, temperature or concentration. Appropriate scaling of the boundary layer dimension is required before a numerical solution can be generated which will capture the behavior in the rapidly changing region.

When a large or small parameter occurs in a mathematical model of a process there are various methods of constructing perturbation expansions for the solution of the governing equations. Often the terms in the perturbation expansions are governed by simpler equations for which the exact solution techniques are available. Even if exact solutions cannot be obtained, the numerical methods used to solve the perturbation equations approximately are often easier to construct than the numerical approximation for the original governing equation.

First, we consider a model problem for which an exact solution is available against which the perturbation expansion can be compared. A feature of the perturbation expansions is that they often form divergence series. The concept of an asymptotic expansion will be introduced and the value of a truncated divergent series will be demonstrated.

2. Projectile motion

This example studies the effect of small damping on the motion of a particle. Consider a particle of mass M which is projected vertically upward with an initial speed U_0. Let U denote the speed at some general time T. If air resistance is neglected then the only force acting on the particle is gravity, $-Mg$, where g is the acceleration due to gravity and the minus sign occurs because the upward direction is chosen to be the positive direction. Newton's second law governs the motion of the projectile, i.e.,

$$M\frac{dU}{dT} = -Mg. \tag{1}$$

Integrating (1), we obtain the solution $U = C - gT$. The constant of integration is determined from the initial condition $U(0) = U_0$, so that

$$U = U_0 - gT. \tag{2}$$

On defining the non-dimensional velocity v, and time t, by $v = U/U_0$ and $t = gT/U_0$, the governing equation becomes

$$\frac{dv}{dt} = -1, \quad v(0) = 1, \tag{3}$$

with the solution $v(t) = 1 - t$.

Taking account of the air resistance, and is included in the Newton's second law as a force dependent on the velocity in a linear way, we obtain the following linear equation

$$M\frac{dU}{dT} = -Mg - KU, \tag{4}$$

where the drag constant K is the dimensions of masa/time. In the non-dimensional variables, it becomes

$$\frac{dv}{dt} = -1 - \left(\frac{KU_0}{Mg}\right)v. \tag{5}$$

Let us denote the dimensionless drag constant by ε, then the governing equation is

$$\frac{dv}{dt} = -1 - \varepsilon v, \qquad v(0) = 1, \tag{6}$$

where $\varepsilon > 0$ is a "small" parameter and the disturbances are very "small". The damping constant K in (4) is small, since K has the dimensions of mass/time and a small quantity in units of kilograms per second.

2.1 Perturbation expansion

It is possible to solve (6) exactly since it is of variables separable form. Here, we solve by an iterative process, known as perturbation expansion for the solution.

Let $v^{(i)}$ denotes the ith iterate, which is obtained from the equation

$$\frac{dv^{(i)}}{dt} = -1 - \varepsilon v^{(i-1)}, \qquad v^{(i)}(0) = 1, \tag{7}$$

The justification for this iterative scheme is that the term εv involves the small multiplying coefficient ε, and so the term itself may be expected to be small. Thus, the term $\varepsilon v^{(i)}$ which should appear on the RHS of (7) to make it exact, may be replaced by $\varepsilon v^{(i-1)}$ with an error which is expected to be small.

The first iterate is obtained by neglecting the perturbation, thus

$$\frac{dv^{(0)}}{dt} = -1, \quad v^{(0)} = 1.$$

This is known as the unperturbed problem, and direct integration yields

$$v^{(0)} = 1 - t.$$

The next iterate $v^{(1)}$, satisfies

$$\frac{dv^{(1)}}{dt} = -1 - \varepsilon(1 - t), \quad v^{(1)} = 1.$$

and integration yields

$$v^{(1)} = 1 - t(1 + \varepsilon) + \frac{1}{2}\varepsilon t^2$$

Similarly, $v^{(2)}$ satisfies

$$\frac{dv^{(2)}}{dt} = -1 - \varepsilon\left[1 - t(1 + \varepsilon) + \frac{1}{2}\varepsilon t^2\right], \quad v^{(2)} = 1.$$

Direct integration yields the solution

$$v^{(2)} = 1 - t(1 + \varepsilon) + \varepsilon(1 + \varepsilon)\frac{t^2}{2} - \frac{1}{6}\varepsilon^2 t^3.$$

Rearranging the terms in these iterates in ascending powers of ε, we obtain

$$\begin{aligned} v^{(0)} &= 1 - t, \\ v^{(1)} &= 1 - t + \varepsilon\left(\frac{t^2}{2} - t\right), \\ v^{(2)} &= 1 - t + \varepsilon\left(\frac{t^2}{2} - t\right) + \varepsilon^2\left(\frac{t^2}{2} - \frac{t^3}{6}\right). \end{aligned} \tag{8}$$

Clearly as the iteration proceeds the expressions are refined by terms which involve increasing powers of ε. These terms become progressively smaller since ε is a small parameter. This is an example of a perturbation expansion. It will often be the case that perturbation expansions involve ascending integer powers of the small parameter, i.e., $\{\varepsilon^0, \varepsilon^1, \varepsilon^2, \cdots\}$. Such a sequence is called an *asymptotic sequence*. Although this is the most common sequence which we shall meet, it is by no means unique. Examples of other asymptotic sequences are $\{\varepsilon^{1/2}, \varepsilon, \varepsilon^{3/2}, \varepsilon^2, \cdots\}$ and $\{\varepsilon^0, \varepsilon^2, \varepsilon^4, \cdots\}$. In each case the essential feature is that subsequent terms tend to zero faster than previous terms as $\varepsilon \to 0$.

An alternative procedure to that of developing the expansion by iteration is to assume the form of the expansion at the outset. Thus, if we assume that the perturbation expansion involves the standard asymptotic sequence $\{\varepsilon^0, \varepsilon^1, \varepsilon^2, \cdots\}$, then the solution v, which depends on the variable t, and the parameter ε, is expressed in the form

$$v(t;\varepsilon) = \varepsilon^0 v_0(t) + \varepsilon^1 v_1(t) + \varepsilon^2 v_2(t) + \cdots. \tag{9}$$

The coefficients $v_0(t), v_1(t), \cdots$ of powers of ε are functions of t only. Substituting expansion (9) in the governing Eq. (6) yields the following

$$\begin{cases} \dfrac{dv_0}{dt} + \varepsilon\dfrac{dv_1}{dt} + \varepsilon^2\dfrac{dv_2}{dt} + \cdots = -1 - \varepsilon v_0 - \varepsilon^2 v_1 - \cdots \\ v_0(0) + \varepsilon v_1(0) + \varepsilon^2 v_2(0) + \cdots = 1. \end{cases} \tag{10}$$

Thus, the coefficients of powers of ε can be equated on the left– and right–hand sides of (10):

$$\begin{cases} \varepsilon^0: & \dfrac{dv_0}{dt} = -1, \quad v_0(0) = 1, \\ \varepsilon^1: & \dfrac{dv_1}{dt} = -v_0, \quad v_1(0) = 0 \\ \varepsilon^2: & \dfrac{dv_2}{dt} = -v_1, \quad v_2(0) = 0, \quad \text{etc.} \end{cases} \tag{11}$$

The proof of validity of this fundamental procedure can be developed by first setting $\varepsilon = 0$ in (10) which yields the first equation of (11). This result allows the first member of the left– and right–hand side of Eq. (10) to be removed. Then, after dividing the remaining terms by ε we obtain the equation

$$\frac{dv_1}{dt} + \varepsilon\frac{dv_2}{dt} + \cdots = -v_0 - \varepsilon v_1 - \cdots$$

This is valid for all nonzero values of ε so that on taking the limit as $\varepsilon \to 0$ we obtain the second equation of (11). Repeating the procedure, we obtain the other equations.

Integrating the equations in (11), we obtain

$$v_0 = 1 - t, \quad v_1 = t^2/2 - t, \quad v_2 = t^2/2/ - t^3/6.$$

Using these values in (9), we obtain that

$$v(t;\varepsilon) = 1 - t + \varepsilon(t^2 - t) + \varepsilon^2(t^2/2 - t^3/6) + \cdots \tag{12}$$

This is the same as the expansion (8) which is generated by iteration.

The IVP (6) can be solved exactly as

$$v(t) = [(1+\varepsilon)e^{-\varepsilon t} - 1]\varepsilon^{-1}.$$

The perturbation expansion can be obtained from (12) by replacing the exponential function by its Maclaurin expansion, i.e.,

$$v(t) = \frac{1}{\varepsilon}\left[1 - \varepsilon t + \frac{\varepsilon^2 t^2}{2} - \frac{\varepsilon^3 t^3}{6} + \cdots + \varepsilon - \varepsilon^2 t + \frac{\varepsilon^3 t^2}{2} + \cdots - 1\right] \tag{13}$$

$$= (1-t) + \varepsilon\left(\frac{t^2}{2} - t\right) + \varepsilon^2\left(\frac{t^2}{2} - \frac{t^3}{6}\right) + \cdots \tag{14}$$

This is the same as the expansion (12). Thus, the perturbation expansion approach is justified in this case. One can refer the books [1, 2].

3. Asymptotics

The letters O and o are order symbols. They are used to describe the rate at which functions approach limit values. We will consider the types of limit values, namely zero, a finite number but nonzero and infinite.

If a function $f(x)$ approaches a limiting value at the same rate of another function $g(x)$ as $x \to x_0$, then we write

$$f(x) = O(g(x)), \quad \text{as } x \to x_0 \tag{15}$$

The functions are said to be of the same order as $x \to x_0$. The test for this is the limit of the ratio. Thus, if $\lim_{x \to x_0} \frac{f(x)}{g(x)} = C$, where C is finite, then we say (15) holds.

For example, we have the following functions:

$$\begin{aligned} x^2 &= O(x), & |x| < 2, \\ \sin(x) &= O(\sqrt{x}), & x \to 0, \\ \sin(x) &= O(x), & -\infty < x < \infty. \end{aligned}$$

The expression

$$f(x) = o(g(x)), \quad \text{as } x \to x_0 \tag{16}$$

means that $\lim_{x \to x_0} \frac{f(x)}{g(x)} = 0$. This is a stronger assertion that the corresponding O–formula. The relation (16) implies the relation (15), as convergence implies boundedness from a certain point onwards.

We have the following functions satisfy the o–relation:

$$\begin{aligned} \cos(x) &= 1 + o(x), & |x| < 2, \\ e^x &= 1 + o(x), & x \to 0 \\ n! &= e^{-n} \cdot n^n \sqrt{2\pi n}(1 + o(1)), & n \to \infty. \end{aligned}$$

3.1 Asymptotic expansions

Consider the expansion

$$f(x) = a_0 + \frac{a_1}{x} + \frac{a_2}{x^2} + \cdots + \frac{a_N}{x^N} + R_N, \tag{17}$$

is an asymptotic expansion as $x \to \infty$, if, for any N,

$$R_N = O\left(\frac{1}{x^{N+1}}\right), \quad \text{as } x \to \infty \tag{18}$$

The following expansion is used when (17) and (18) hold,

$$f(x) \sim \sum_{n=0}^{\infty} \frac{a_n}{x^n}, \quad \text{as } x \to \infty \tag{19}$$

Here, $\lim_{n\to\infty} R_N = 0$, for any value of N.

The sequence $\{1, 1/x, 1/x^2, \cdots\}$ is an *asymptotic sequence* as $x \to \infty$. The characteristic feature of such sequences is that each member is dominated by the previous member. In constructing examples it is easier to deal with the limit zero than any other. Thus, for the case $x \to \infty$, we let $\varepsilon = 1/x$, which for $x \to x_0$, we let $\varepsilon = x - x_0$ so that without loss of generality we may confirm our attention to the limit $\varepsilon \to 0$. The standard asymptotic sequence is $\{1, \varepsilon, \varepsilon^2, \cdots\}$ as $\varepsilon \to 0$. If we let $\delta_n(\varepsilon)$ represent members of an asymptotic sequence $\{\delta_0(\varepsilon), \delta_1(\varepsilon), \cdots\}$ as $\varepsilon \to 0$, then the following condition must hold

$$\delta_{n+1}(\varepsilon) = o(\delta_n(\varepsilon)), \quad \text{as } \varepsilon \to 0.$$

Some examples of asymptotic sequences are

i. $\left\{1, \sin(\varepsilon), (\sin(\varepsilon))^2, (\sin(\varepsilon))^3, \cdots\right\}$, here we have

$$\lim_{\varepsilon\to 0} \frac{\delta_{n+1}}{\delta_n} = \lim_{\varepsilon\to 0} \sin(\varepsilon) = 0.$$

ii. $\{1, \ln(1+\varepsilon), \ln(1+\varepsilon^2), \ln(1+\varepsilon^3), \cdots\}$, with $\delta_0 = 1, \delta_n = \ln(1+\varepsilon^n) n \geq 1$, we have

$$\lim_{\varepsilon\to 0} \frac{\delta_1}{\delta_0} = \lim_{\varepsilon\to 0} \ln(1+\varepsilon) = 0,$$

$$\lim_{\varepsilon\to 0} \frac{\delta_{n+1}}{\delta_n} = \lim_{\varepsilon\to 0} \frac{\ln(1+\varepsilon^{n+1})}{\ln(1+\varepsilon^n)} = \lim_{\varepsilon\to 0} \frac{\varepsilon^{n+1} + O(\varepsilon^{2n+2})}{\varepsilon^n + O(\varepsilon^{2n})} = 0.$$

The general expression for an asymptotic expansion of a function $f(\varepsilon)$, in terms of an asymptotic sequence $\delta_n(\varepsilon)$ is

$$f(x) \sim \sum_{n=0}^{\infty} a_n \delta_n(\varepsilon), \quad \text{as } \varepsilon \to 0, \tag{20}$$

where the coefficients a_n are independent of ε. The expression (20) involving the symbol $\sim$, means that for all N,

$$f(x) = \sum_{n=0}^{N} a_n \delta_n(\varepsilon) + R_N, \tag{21}$$

where

$$R_N = O[\delta_{N+1}(\varepsilon)], \quad \text{as } \varepsilon \to 0, \tag{22}$$

$$a_n = \lim_{\varepsilon \to 0} \left(\frac{f(\varepsilon) - \sum_{n=0}^{N-1} a_n \delta_n(\varepsilon)}{\delta_N(\varepsilon)} \right). \tag{23}$$

If a function possesses an asymptotic expansion involving the sequence $\{\delta_0(\varepsilon), \delta_1(\varepsilon), \cdots\}$ then the coefficients a_n of the expansion (21) given by the expression (24) are unique. However, another function may share the same set of coefficients. Thus, while functions have unique expansions, an expansion does not correspond to a unique function.

Consider a function $f(x;\varepsilon)$, which depends on both an independent variable x, and a small parameter ε. Suppose that $f(x;\varepsilon)$ is expanded using an asymptotic sequence $\{\delta_n(\varepsilon)\}$,

$$f(x;\varepsilon) = \sum_{n=0}^{N} a_n(x)\delta_n(\varepsilon) + R_N(x;\varepsilon). \tag{24}$$

The coefficients of the gauge functions $\delta_n(\varepsilon)$ are functions of x, and the remainder after N terms is a function of both x and ε. For this to be an asymptotic expansion, we require

$$R_N(x;\varepsilon) = O[\delta_{N+1}(\varepsilon)], \quad \text{as } \varepsilon \to 0. \tag{25}$$

Refer [3, 4] for more details. For (24) to be a uniform asymptotic expansion the ultimate proportionality between R_N and δ_{N+1} must be bounded by a number independent of x, i.e.,

$$|R_N(x;\varepsilon)| \le K|\delta_{N+1}(\varepsilon)|, \tag{26}$$

for ε in the neighborhood near zero, where K is a fixed constant.

An example of a uniform asymptotic expansion is $f(x;\varepsilon) = \frac{1}{1-\varepsilon \sin(x)}$.

An example of a nonuniform expansion is

$$f(x;\varepsilon) \sim \sum_{n=0}^{N} x^n \varepsilon^n + R_N(x;\varepsilon), \quad \text{as } \varepsilon \to 0. \tag{27}$$

Here, one cannot find a fixed K which satisfy $|R_N| \le K|\varepsilon^{N+1}|$, because for any choice of K, x can be chosen so that x^{N+1} exceeds this value.

3.2 Nonuniformity

The expansion (27) becomes nonuniform when subsequent terms are no longer small corrections to previous terms. This occurs when subsequent terms are of the same order or of dominant order than previous terms. Subsequent terms dominate previous terms for larger x, for example, when $x = O(1/\varepsilon^2)$. The expansion is valid for $x = O(1)$ since then subsequent terms decrease by a factor of ε. The expansion remains valid for large x, provided x is not as large as $1/\varepsilon$. For instance, the expansion is valid for $x = O(1/\sqrt{\varepsilon})$, as $\varepsilon \to 0$.

The critical case is such that subsequent terms are of the same order. This determines the region of nonuniformity. In (27), the region of nonuniformity occurs when $\varepsilon x = O(1)$, i.e., $x = O(\varepsilon^{-1})$, as $\varepsilon \to 0$.

3.2.1 Sources of nonuniformity

There are two common reasons for nonuniformities in asymptotic expansions, they are

1. Infinite domains which allow long-term effects of small perturbations to accumulate.

2. Singularities in governing equations which lead to localized regions of rapid change.

Consider the nonlinear Duffing equation

$$\begin{cases} \dfrac{d^2u}{dt^2} + u + \varepsilon u^3 = 0, \quad t \in [0, \infty) \\ u(0) = a, \quad \dfrac{du}{dt}(0) = 0. \end{cases} \tag{28}$$

Suppose the solution may be expanded using the standard asymptotic sequence

$$u(t;\varepsilon) \sim u_0(t) + \varepsilon u_1(t) + \varepsilon^2 u_2(t) + \cdots. \tag{29}$$

On substituting this in (28) and in the initial conditions, we get

$$\begin{cases} \dfrac{d^2u_0}{dt^2} + \varepsilon\dfrac{d^2u_1}{dt^2} + \cdots + u_0 + \varepsilon u_1 + \cdots + \varepsilon u_0^3 + \cdots \sim 0, \\ u_0(0) + \varepsilon u_1(0) + \cdots = a + 0 \cdot \varepsilon + \cdots, \\ \dfrac{du_0}{dt}(0) + \varepsilon\dfrac{du_1}{dt}(0) + \cdots = 0 + 0 \cdot \varepsilon + \cdots. \end{cases}$$

Equating like of powers of ε on both sides, we get

$$\left. \begin{aligned} O(1): \quad & \frac{d^2u_0}{dt^2} + u_0 = 0, \\ & u_0(0) = a, \quad \frac{du_0}{dt}(0) = 0, \end{aligned} \right\} \tag{30}$$

and

$$\left. \begin{aligned} O(\varepsilon): \quad & \frac{d^2u_1}{dt^2} + u_1 = -u_0^3, \\ & u_1(0) = 0, \quad \frac{du_1}{dt}(0) = 0. \end{aligned} \right\} \tag{31}$$

Solving Eqs. (30) and (31), we obtain

$$u \sim a\cos(t) + \varepsilon\left[\frac{a^3}{32}(\cos(3t) - \cos(t)) - \frac{3a^3}{8}t\sin(t)\right] + \cdots. \tag{32}$$

The term $t\sin(t)$ in the expansion (32) is called a *secular term*. It is an oscillating term of growing amplitude. All other terms are oscillating of fixed amplitude. The secular term leads to a nonuniformity for large t. The region of nonuniformity is obtained by equating the order of the first and second terms,

$$\cos(t) = O(\varepsilon t \sin(t)), \quad \text{as } \varepsilon \to 0.$$

The trigonometric functions are treated as $O(1)$ terms. Thus, the region of nonuniformity is $t = O(1/\varepsilon)$, as $\varepsilon \to 0$.

The second common source of nonuniformities is associated with the presence of singularities. Consider, the following initial-value problem:

$$\begin{cases} \varepsilon \dfrac{dy}{dx} + y = e^{-x}, & x > 0 \\ y(0) = 2, \end{cases} \tag{33}$$

where $\varepsilon > 0$ is a small parameter. Suppose y has the expansion

$$y \sim y_0(x) + \varepsilon y_1(x) + \varepsilon^2 y_2(x) + \cdots. \tag{34}$$

Substituting (34) in (33), we have

$$\begin{cases} \varepsilon\left(\dfrac{dy_0}{dx} + \varepsilon\dfrac{dy_1}{dx} + \cdots\right) + (y_0 + \varepsilon y_1 + \cdots) = e^{-x}, \\ y_0(0) + \varepsilon y_1(0) + \cdots = 2. \end{cases} \tag{35}$$

Equating coefficients of like powers of ε on both sides, we get

$$O(1): \quad y_0 = e^{-x}, \quad y_0(0) = 2,$$

$$O(\varepsilon): \quad y_1 = -\frac{dy_0}{dx} = e^{-x}, \quad y_1(0) = 0,$$

$$O(\varepsilon^2): \quad y_2 = -\frac{dy_1}{dx} = e^{-x}, \quad y_2(0) = 0.$$

Clearly, y_0 cannot satisfy the boundary condition $y_0(0) = 2$ as no constant of integration is available because the equation determining y_0 is an algebraic equation not a differential equation, and no additional conditions are required. Thus, we have obtain the expression

$$y \sim e^{-x} + \varepsilon e^{-x} + \varepsilon^2 e^{-x} + \cdots, \tag{36}$$

but the initial condition $y(0) = 2$ has not been satisfied.

The unperturbed problem, obtained by setting $\varepsilon = 0$ is not a DE, but an algebraic equation $y = e^{-x}$. This cannot satisfy an arbitrarily imposed condition at $x = 0$. For any nonzero value of ε, (33) becomes a first-order DE which can satisfy an initial condition. This is an example of a singular perturbation problem (SPP), where the behavior of the perturbed problem is very different from that of the unperturbed problem.

Thus, the perturbation expansion (36) is a good approximation of the exact solution away from the region $x = 0$. To see this, let us compare (36) with the following exact solution:

$$y_{ex}=\frac{1-2\varepsilon}{1-\varepsilon}e^{-x/\varepsilon}+\frac{e^{-x}}{1-\varepsilon}=\left[(1-\varepsilon-\varepsilon^2-\cdots)e^{-x/\varepsilon}\right]+\left[(1+\varepsilon+\varepsilon^2+\cdots)e^{-x}\right]. \tag{37}$$

The perturbation expansion (36) generates the second member of (37), but not the first member. The coefficient $e^{-x/\varepsilon}$) is a rapidly varying function which takes the value of unity at $x=0$, and rapidly decays to zero for $x>0$. Clearly, y_0 provides a good approximation away from the region $x=0$. The region near $x=0$ is called the *boundary layer*. These regions usually occur when the highest order derivative of a DE is multiplied by a small parameter. The unperturbed problem, obtained by setting $\varepsilon=0$ is of lower order and consequently cannot satisfy all the boundary conditions. This leads to boundary layer regions where the solution varies rapidly in order to satisfy the boundary condition.

Boundary layers are regions of nonuniformity in perturbation expansions of the form (36).

4. Boundary layer

Boundary layers are regions in which a rapid change occurs in the value of a variable. Some physical examples include "the fluid velocity near a solid wall", "the velocity at the edge of a jet of fluid", "the temperature of a fluid near a solid wall." Ludwig Prandtl pioneered the subject of boundary layer theory in his explanation of how a quantity as small as the viscosity of common fluids such as water and air could nevertheless play a crucial role in determining their flow. The viscosity of many fluids is very small and yet taking account of this small quantity is vital. The essential point is that the viscous term involves higher order derivatives so that its omission necessitates the loss of a boundary condition. The ideal flow solution allow slip to occur between a solid and fluid. In reality the tangential velocity of a fluid relative to a solid is zero. The fluid is brought to rest by the action of a tangential stress resulting from the viscous force.

Mathematically the occurrence of boundary layers is associated with the presence of a small parameter multiplying the highest derivative in the governing equation of a process. A straightforward perturbation expansion using an asymptotic sequence in the small parameter leads to differential equations of lower order than the original governing equation. In consequence not all of the boundary and initial conditions can be satisfied by the perturbation expansion. This is an example of what is commonly referred to as a *singular perturbation problem*. The technique for overcoming the difficulty is to combine the straightforward expansion, which is valid away from the layer adjacent to the boundary. The straightforward expansion is referred to as the *outer expansion*. The *inner expansion* associated with the boundary layer region is expressed in terms of a stretched variable, rather than the original independent variable, which takes due account of the scale of certain derivative terms. The inner and outer expansions are matched over a region located at the edge of the boundary layer. The technique is called the method of *matched asymptotic expansions*.

Consider the following two-point boundary value problem:

$$\begin{cases}\varepsilon\dfrac{d^2u}{dx^2}+\dfrac{du}{dx}=2x+1, \quad x\in(0,1)\\[2ex] u(0)=1, \quad u(1)=4,\end{cases} \tag{38}$$

where $\varepsilon > 0$ is a small parameter. If we assume that u possesses a straightforward expansion in powers of ε,

$$u(x;\varepsilon) \sim u_0(x) + \varepsilon u_1(x) + \varepsilon^2 u_2(x) + \cdots, \tag{39}$$

then the equations associated with powers of ε leads to

$$O(1): \qquad \frac{du_0}{dx} = 2x + 1, \tag{40}$$

$$O(\varepsilon^n): \qquad \frac{du_n}{dx} = -\frac{d^2 u_{n-1}}{dx^2}, \quad \text{for } n = 1, 2, 3, \cdots. \tag{41}$$

and the boundary conditions require

$$u_0(0) + \varepsilon u_1(0) + \cdots \sim 1 + \varepsilon \cdot 0 + \cdots,$$
$$u_0(1) + \varepsilon u_1(1) + \cdots \sim 4 + \varepsilon \cdot 0 + \cdots,$$

which leads to

$$\begin{aligned} &u_0(0) = 1, \quad u_0(1) = 4, \\ &u_n(0) = 0, \quad u_n(1) = 0, \quad \text{for } n = 1, 2, \cdots. \end{aligned} \tag{42}$$

Equation (42) require that each $u_n(x)$ satisfy two boundary conditions. This is in general impossible since Eqs. (41) and (42) governing each u_n are of first-order. Now the question is which one of the boundary condition has to be taken into account. We will find out that the boundary condition at $x = 0$ must be abandoned and consequently the expansion (39) is invalid near $x = 0$.

The general solution of (42) is $u_0(x) = x^2 + x + C$, using the boundary condition $u_0(1) = 4$, we obtain

$$u_0(x) = x^2 + x + 2.$$

From (42), we obtain the equations

$$\frac{du_1}{dx} = -2, \quad u_1(1) = 0,$$

$$\frac{du_2}{dx} = 0, \quad u_2(1) = 0,$$

and its solutions are

$$u_1(x) = -2(x-1), \quad u_n(x) = 0, \quad n \geq 2.$$

Therefore, the outer expansion is

$$u^{\text{out}}(x;\varepsilon) = \left(x^2 + x + 2\right) + \varepsilon 2(1 - x), \tag{43}$$

where 'out' label is used to indicate that the solution is valid away from the region near $x = 0$. Clearly u^{out} fails to satisfy the boundary condition at $x = 0$. The reason why the outer solution is of use is that it closely follows the exact solution of the problem except in a narrow region near $x = 0$, where the exact solution changes rapidly in order to satisfy the boundary condition.

The exact solution of the BVP (38) can be obtained as

$$u(x) = A + Be^{-x/\varepsilon} + x^2 + x(1 - 2\varepsilon). \tag{44}$$

The constants A and B are determined from the boundary conditions:

$$\begin{cases} A + B = 1, \\ A + Be^{-1/\varepsilon} + 2 - 2\varepsilon = 4 \end{cases} \tag{45}$$

We know that $e^{-1/\varepsilon} = o(\varepsilon^N)$, as $\varepsilon \to 0$, for all N. This means that the exponential term tends to zero faster than any power of ε, as $\varepsilon \to 0$. It is called a *transcendentally small term* (T.S.T.) and can always be neglected since its contribution is asymptotically always less than any power of ε. Thus, (45) gives

$$A = 2(1 + \varepsilon), \quad B = -(1 + 2\varepsilon),$$

and the exact solution is

$$u^{\text{ex}}(x) = 2(1 + \varepsilon) - (1 + 2\varepsilon)e^{-x/\varepsilon} + x^2 + x(1 - 2\varepsilon), \tag{46}$$

after rearranging the terms in asymptotic order, we obtain

$$u^{\text{ex}}(x) = \left(x^2 + x + 2\right) - e^{-x/\varepsilon} + \varepsilon\left[2(1 - x) - 2e^{-x/\varepsilon}\right]. \tag{47}$$

Comparing the exact solution with the outer expansion shows that the terms involving $e^{-x/\varepsilon}$ are absent. The effect of these terms is negligible when $x = O(1)$. But, when $x = O(\varepsilon)$, then $e^{-x\varepsilon} = O(1)$. It is clear that as $\varepsilon \to 0$ the region in which the outer solution departs from the exact solution becomes arbitrarily close to $x = 0$ with a thickness $O(\varepsilon)$. This region is called the *boundary layer*.

The behavior of the exact solution and the zeroth-order term of the outer expansion are plotted in **Figure 1** for various values of ε.

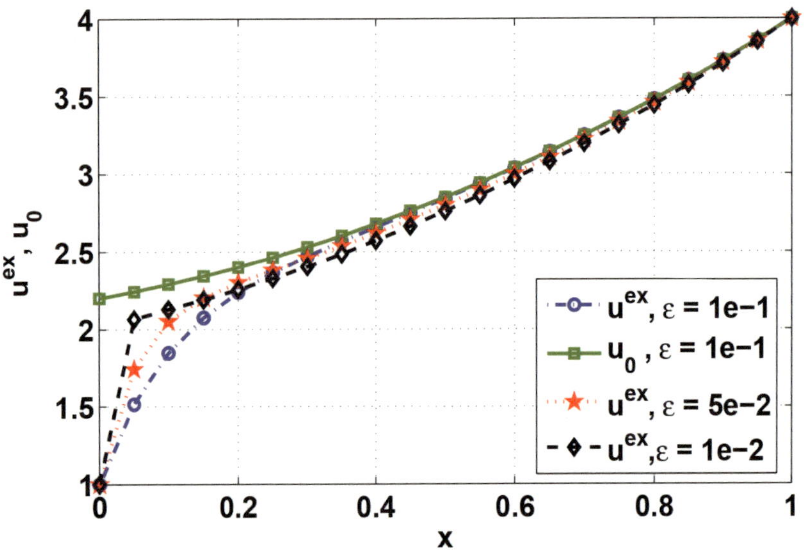

Figure 1.
Exact solution of (38) for various values of ε.

By differentiating the leading order term u_0^{ex}, of the exact solution, we have

$$u_0^{\text{ex}} = x^2 + x + 2 - e^{-x/\varepsilon}$$

$$\frac{du_0^{\text{ex}}}{dx} = 2x + 1 + \frac{1}{\varepsilon}e^{-x/\varepsilon}$$

$$\frac{d^2u_0^{\text{ex}}}{dx^2} = 2 - \frac{1}{\varepsilon^2}e^{-x/\varepsilon}$$

Outside the boundary layer, i.e., for $x = O(1)$, we have $e^{-x/\varepsilon} = o\left(\varepsilon^N\right), \forall N$, so $\varepsilon^{-1}e^{-x/\varepsilon}$ and $\varepsilon^{-2}e^{-x/\varepsilon}$ are also transcendentally small. Within the boundary layer when $x = O(\varepsilon)$, we have $e^{-x/\varepsilon} = O(1)$. The order of u_0^{ex} and its derivatives are given below:

	Outside BL	Inside BL
u_0^{ex}	$O(1)$	$O(1)$
$\frac{du_0^{\text{ex}}}{dx}$	$O(1)$	$O\left(\frac{1}{\varepsilon}\right)$
$\frac{d^2u_0^{\text{ex}}}{dx^2}$	$O(1)$	$O\left(\frac{1}{\varepsilon^2}\right)$

This indicates that x is the appropriate independent variable outside the boundary layer where u_0^{ex} and its derivatives are of $O(1)$ quantities. However, within the boundary layer the appropriately scaled independent variable is $s = x/\varepsilon$, then

$$\frac{du}{dx} = \varepsilon^{-1}\frac{dv}{ds}, \quad \frac{d^2u}{dx^2} = \varepsilon^{-2}\frac{d^2v}{ds^2},$$

so that within the boundary layer

$$\frac{du}{dx} = O(1), \quad \text{and} \quad \frac{d^2u}{dx^2} = O(1).$$

The variable $s = x/\varepsilon$ is called a *stretched variable*. The differential equations becomes

$$\frac{d^2v}{ds^2} + \frac{dv}{ds} = \varepsilon + 2\varepsilon^2 s. \tag{48}$$

We assume a boundary layer expansion, called the *inner expansion* of the form

$$v(s;\varepsilon) \sim v_0(s) + \varepsilon v_1(s) + \cdots. \tag{49}$$

The inner expansion will satisfy the boundary condition at $x = s = 0$ namely $v_0(s = 0) = 1$ giving $v_0(0) = 1$, and $v_n(0) = 0, n = 1, 2, \cdots$. Substituting (49) into the DE (48), we obtain the following set of equations:

$$\begin{cases} O(1): & \dfrac{d^2v_0}{ds^2}+\dfrac{dv_0}{ds}=0, \quad v_0(0)=1 \\ O(\varepsilon): & \dfrac{d^2v_1}{ds^2}+\dfrac{dv_1}{ds}=1, \quad v_1(0)=0 \\ O(\varepsilon^2): & \dfrac{d^2v_2}{ds^2}+\dfrac{dv_2}{ds}=2s, \quad v_1(0)=0 \\ O(\varepsilon^n): & \dfrac{d^2v_n}{ds^2}+\dfrac{dv_n}{ds}=0, \quad v_n(0)=0, \quad n=3,4,\cdots \end{cases} \tag{50}$$

with solutions

$$\begin{cases} v_0=A+(1-A)e^{-s} \\ v_1=B-Be^{-s}+s \\ v_2=C-Ce^{-s}+s^2-2s \\ v_n=D_n-D_ne^{-s}, \quad n=3,4,\cdots \end{cases} \tag{51}$$

The boundary condition at $x=1$ cannot be used to determine the constants appearing in these solutions because the DEs (50) are only valid in the boundary layer. The constants in (51) are determined by matching the inner and outer expansions. We shall first restrict our attention to matching the leading order expansions u_0 and v_0. The method which we shall apply is *Prandtl's matching condition.*

The leading order terms in the 'inner' and 'outer' expansions are to be matched at the 'edge of the boundary layer'. Of course there is no precise edge of the boundary layer, we simply know that it has thickness of order $O(\varepsilon)$. A plausible matching procedure would be to equate u_0 and v_0 at a value of x such that the region of rapid change has passed. We might choose to equate the terms at the point $x=5\varepsilon$. The leading order expansions are

$$u_0=x^2+x+2 \qquad v_0=A+(1-A)e^{-s}.$$

Equating at $x=5\varepsilon$ gives the following:

$$A=\frac{2+5\varepsilon+25\varepsilon^2-e^{-5}}{1-e^{-5}}.$$

If, instead we choose to match at $x=6\varepsilon$, then we obtain

$$A=\frac{2+6\varepsilon+36\varepsilon^2-e^{-6}}{1-e^{-6}}.$$

These two expressions differ in the argument of the exponential and differ algebraically with 5ε replaced by 6ε. The exponential functions are approaching transcendentally small values so that their contribution can be neglected. The algebraic difference is of $O(\varepsilon)$. Thus, the arbitrariness in the decision of the point at which we choose to equate the expansions leads to a difference of $O(\varepsilon)$. But we are only dealing with leading order expansions anyway. The difference between the exact solution and the leading order expansions will of $O(\varepsilon)$ so that an arbitrariness in v_0 and u_0 of $O(\varepsilon)$ is immaterial. Rather than choose between, for example, 5ε and 6ε as the value of x to evaluate u_0 we may take the value at $x=0$, since

$$u_0[x = O(\varepsilon)] = u_0(0) + O(\varepsilon),$$

where the remainder is uniformly $O(\varepsilon)$ since the gradient of u_0 is $O(1)$. For the inner expansion we are to ensure that the rapidly varying function has achieved its asymptotic value at the edge of the boundary layer. This means that the term $e^{-x/\varepsilon}$ should be replaced by zero. This can be achieved by taking the limit $s \to \infty$. Thus, rather than choosing a specific point to equate the inner and outer terms er are led to the following *Prandtl's matching condition*:

$$\lim_{x\to 0} u_0(x) = \lim_{s\to\infty} v_0(s). \tag{52}$$

The limit $s \to \infty$ may appear rather dangerous since although it certainly removes the exponential term it could lead to an algebraically unbounded term. For example, if $v_0 = As + (1 - A)e^{-s}$, then the first member would be unbounded as $s \to \infty$. This possibility can be eliminated since the inner expansion must be of a form which varies rapidly for $x = O(\varepsilon)$ but not for $x = O(1)$, i.e., not for $s \to \infty$. In practice, if the boundary layer has been properly located and the correct inner variable is used then Prandtl's matching condition is valid and elegantly avoids the need to choose an arbitrary 'edge' of the boundary layer.

Applying these conditions to the current example leads to

$$\lim_{x\to 0} \left(x^2 + x + 2\right) = \lim_{s\to\infty} \left[A + (1 - A)e^{-s}\right],$$

which yields $A = 2$. Thus the leading order terms in the expansion solutions are

$$\text{Outer region}: \quad u_0 = x^2 + x + 2, \quad \text{for } x = O(1)$$

$$\text{Inner region}: \quad v_0 = 2 - e^{-x/\varepsilon}, \quad \text{for } x = O(\varepsilon)$$

To prove that these are valid leading terms we consider u^{ex}:

$$\text{If } x = O(1), \quad \text{then } u_0^{\text{ex}} = x^2 + x + 2 + \text{T.S.T.}$$

$$\text{If } x = O(\varepsilon), \quad \text{then } u_0^{\text{ex}} = 2 - e^{-x/\varepsilon} + O(\varepsilon)$$

We conclude that the matching condition has correctly predicted the leading order terms.

4.1 Composite expansion

As single composite expression for these leading order terms can be constructed using the combination

$$u_0^{\text{comp}} = u_0 + v_0 - u_0^{\text{match}}, \tag{53}$$

where u_0^{match} is given by (52). Then,

$$\text{for } x = O(1), \quad v_0 = u_0^{\text{match}} + \text{T.S.T.}, \quad \text{so that } u_0^{\text{comp}} = u_0^{\text{match}} + \text{T.S.T.}$$

$$\text{for } x = O(\varepsilon), \quad u_0 = u_0^{\text{match}} + O(\varepsilon), \quad \text{so that } u_0^{\text{comp}} = v_0 + O(\varepsilon)$$

For the current example, $u_0^{\text{match}} = 2$, so the composite expansion is

$$u_0^{\text{match}} = x^2 + x + 2 - e^{-x/\varepsilon}. \tag{54}$$

Prandtl's matching condition can only be used for the leading order terms in the asymptotic expansions.

The outer, inner and composite expansions of the BVP (38) are presented in **Figures 2** and **3** for different values of ε. From these figures, one can easily identify the need and efficiency of the composite expansion.

4.2 Boundary layer location

Consider the following linear DE

$$\varepsilon \frac{d^2u}{dx^2} + a(x)\frac{du}{dx} + b(x)u = c(x), \quad x \in (x_1, x_2). \tag{55}$$

The following general statements can be made about the boundary layer location and the nature of the inner expansion.

Case I. If $a(x) > 0$ throughout (x_1, x_2), then the boundary layer will occur at $x = x_1$. The stretching transformation will be $s = (x - x_1)/\varepsilon$, and the one-term inner expansion will satisfy

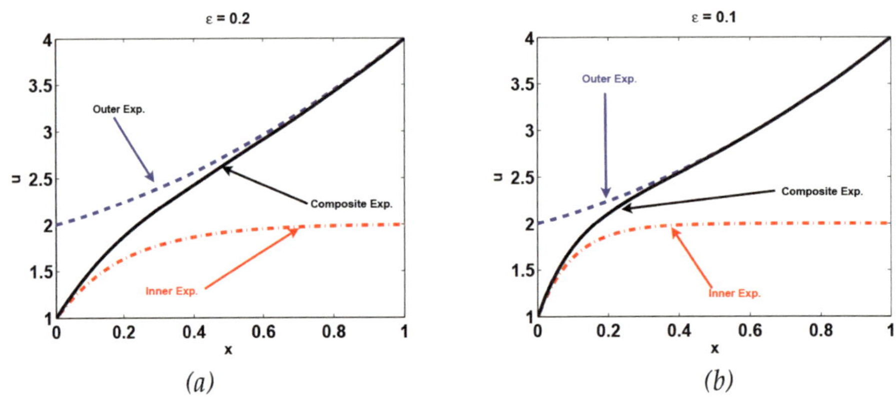

Figure 2.
Outer, inner and composite expansions. (a) For ε = 0.2; (b) For ε = 0.1.

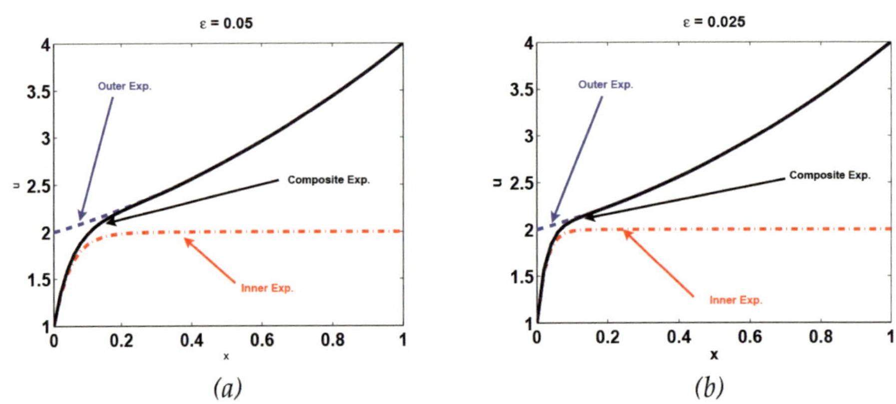

Figure 3.
Outer, inner and composite expansions. (a) For ε = 0.05; (b) For ε = 0.025.

$$\frac{d^2v_0}{ds^2} + a(x_1)\frac{dv_0}{ds} = 0.$$

The solution of this equation is

$$v_0 = A + Be^{a(x_1)(x-x_1)/\varepsilon},$$

where $A + B = u(x = x_1)$. The other condition to determine the constants A and B is obtained by matching with the value of the outer expansion at $x = x_1$.

Case II. If $a(x) < 0$ throughout (x_1, x_2), then the boundary layer will occur at $x = x_2$. The stretching transformation will be $s = (x_2 - x)/\varepsilon$, and the one-term inner expansion will involve the rapidly decaying function $e^{a(x_2)(x_2-x)/\varepsilon}$.

Case III. If $a(x)$ changes sign in the interval $x_1 < x < x_2$, then a boundary layer occurs at an interior point x_0, where $a(x_0) = 0$ and boundary layers may also occur at both ends x_1 and x_2.

4.3 Boundary layer thickness and the principle of least degeneracy

The boundary layers which we have met so far have all had thickness $O(\varepsilon)$. By this we mean that a variation of $O(\varepsilon)$ in the independent variable will encompass the region of rapid change in the dependent variable. The associated stretched independent variable s, appropriate for the boundary layer is related to x by a linear transformation involving division by ε.

There are practical situations where the boundary layer thickness will be of $O(\varepsilon^p)$. This means that if the boundary layer is located at $x = x_0$, then the appropriate stretching transformation is $s = (x - x_0)/\varepsilon^p$. More generally, the choice of the function $\delta(\varepsilon)$ to use in the stretching transformation $s = (x - x_0)/\delta(\varepsilon)$ is determined by the need to represent the region of rapid change correctly. We must ensure that the boundary layer solution contains rapidly varying functions. The form of the governing equation in the boundary layer region must have sufficient structure to allow such solutions.

Consider the example

$$\begin{cases} \varepsilon\dfrac{d^2u}{dx^2} + \dfrac{du}{dx} + u = x, \quad (0, 1) \\ u(0) = 1, \quad u(1) = 2. \end{cases} \tag{56}$$

Since the signs of the first and second derivatives are the same, and the boundary layer will occur at $x = 0$. We are not going to assume at the outset that the boundary layer thickness is $O(\varepsilon)$. Our intension is to deduce that the appropriate stretching variable is $s = x/\varepsilon$.

The one-term outer expansion u_0 satisfies $\frac{du_0}{dx} + u_0 = x,\ \ u_0(1) = 2$ The solution is

$$u_0(x) = 2e^{1-x} + x - 1. \tag{57}$$

To determine the inner expansion we first wrongly assume that the boundary layer thickness is $O(\varepsilon^{1/2})$. The stretching transformation $s = x/\varepsilon^{1/2}$ changes the original DE (56) into the following one:

$$\frac{d^2v}{ds^2} + \frac{1}{\varepsilon^{1/2}}\frac{dv}{ds} + v = \varepsilon^{1/2}s \tag{58}$$

If the appropriate stretching transformation has been used for the boundary layer then dv/ds and d^2v/ds^2 will be of $O(1)$ within it. The leading order expansion v_0 will satisfy the dominant part of (58), i.e., the component of $O(\varepsilon^{-1/2})$

$$\frac{dv_0}{ds} = 0, \quad v_0(0) = 1. \tag{59}$$

The solution is $v_0(s) = 1$. This of course does not have the rapidly varying behavior which we anticipate in the boundary layer. Prandtl's matching condition cannot be satisfied since

$$\lim_{x\to 0}\left(2e^{1-x} + x - 1\right) = 2e - 1 \neq \lim_{s\to\infty} v_0(s) = 1.$$

Thus, we reject the assumption of a boundary layer of thickness $O(\varepsilon^{1/2})$.

Next, suppose that the boundary layer thickness is $O(\varepsilon^2)$ and again we will discover that this is incorrect because the corresponding inner expansion cannot be matched to the outer expansion. Proceeding with the analysis we introduce the stretching transformation $s = x/\varepsilon^2$ which leads to the equation

$$\frac{1}{\varepsilon^3}\frac{d^2v}{ds^2} + \frac{1}{\varepsilon^2}\frac{dv}{ds} + v = \varepsilon^2 s.$$

Again we argue that if the appropriate stretching has been used then all derivatives are of $O(1)$ so that the governing equation for the leading term in $O(\varepsilon^{-3})$, namely

$$\frac{d^2v_0}{ds^2} = 0, \quad v_0(0) = 1. \tag{60}$$

The solution is $v_0(s) = 1 + As$, where the constant A is to be determined from matching. This solution is rapidly varying but the rapidity does not decay at the edge of the boundary layer (i.e., as $s \to \infty$). Indeed, we cannot match v_0 to the outer expansion because the term As becomes arbitrarily large as $s \to \infty$.

The correct choice of stretching transformation is $s = x/\varepsilon$ showing that the boundary layer thickness is $O(\varepsilon)$. The boundary layer equation becomes

$$\frac{1}{\varepsilon}\frac{d^2v}{ds^2} + \frac{1}{\varepsilon}\frac{dv}{ds} + v = \varepsilon s.$$

The dominant equation satisfied by v_0 is $O(1/\varepsilon)$, namely

$$\frac{d^2v_0}{ds^2} + \frac{dv_0}{ds} = 0, \quad v_0(0) = 1. \tag{61}$$

The solution is $v_0(s) = 1 - A + Ae^{-s}$. The last member provides the necessary rapid decay away from the point $x = s = 0$. Prandtl's matching condition requirest

$$\lim_{x\to 0}\left(2e^{1-x} + x - 1\right) = \lim_{s\to\infty}\left(1 - A + Ae^{-s}\right),$$

which leads to $A = 2 - 2e$, and

$$v_0(x) = 2e - 1 + 2(1 - e)e^{-x/\varepsilon}.$$

The one-term composite expansion is

$$u^{\text{comp}} = \left(2e^{1-x} + x - 1\right) + (2e - 1) + 2(1 - e)e^{-x/\varepsilon} - (2e - 1). \tag{62}$$

The leading order boundary layer equation associated with the stretching transformation $s = x/\varepsilon$, (61) involves more terms than (59), associated with $s = x/\varepsilon^{1/2}$, and (60) associated with $s = x/\varepsilon^2$. The extra term in (61) allows sufficient structure in the solution to produce the required boundary layer behavior. An aid for choosing the boundary layer thickness is to seek a stretching transformation which retains the largest number of terms in the dominant equation governing v_0. This referred to as the *principle of least degeneracy* by Van Dyke.

The composite expansion (62) can be verified by comparing with the exact solution of (56). The general solution of (56) is

$$u^{\text{ex}} = C_1 e^{m_1 x} + C_2 e^{m_2 x} + (x - 1),$$

where

$$m_1 = \frac{-1 + \sqrt{1 - 4\varepsilon}}{2\varepsilon}, \quad m_2 = \frac{-1 - \sqrt{1 - 4\varepsilon}}{2\varepsilon}.$$

We expand $\sqrt{1 - 4\varepsilon}$ using the binomial series, $\sqrt{1 - 4\varepsilon} = 1 - 2\varepsilon + O(\varepsilon^2)$, then

$$m_1 = -1 + O(\varepsilon), \quad \text{and} \quad m_2 = -\frac{1}{\varepsilon} + 1 + O(\varepsilon),$$

so that

$$u^{\text{ex}} = C_1 e^{-x} + C_2 e^{-x/\varepsilon} \cdot e^x + (x - 1) + O(\varepsilon). \tag{63}$$

Using the boundary conditions and by neglecting the transcendentally small term $e^{-1/\varepsilon}$, we have $C_1 = 2e, C_2 = 2(1 - e)$. Then, (63) becomes

$$u^{\text{ex}} = 2e^{1-x} + 2(1 - e)e^{-x/\varepsilon} \cdot e^x + (x - 1) + O(\varepsilon). \tag{64}$$

There is an apparent discrepancy between (64) and the composite expansion (62) in the coefficient of the $e^{-x/\varepsilon}$ term. There is an extra term only contributes in the boundary layer where $x = O(\varepsilon)$ so that the coefficient e^x may to leading order, be replaced by unity. Thus, the leading order composite expansion and the leading order term in the exact solution are in complete agreement.

4.4 Boundary layer of thickness of $O(\sqrt{\varepsilon})$

Consider the following two-point BVP:

$$\begin{cases} \varepsilon \dfrac{d^2 u}{dx^2} + x^2 \dfrac{du}{dx} - u = 0, & (0, 1) \\ u(0) = 1, \quad u(1) = 2. \end{cases} \tag{65}$$

We seek a one-term composite expansion for the above BVP. We will tentatively assume that a boundary layer occurs at $x = 0$ although the vanishing of the coefficient of the first derivative suggests the possibility of nonstandard behavior.

The one term outer expansion satisfies

$$x^2 \frac{du_0}{dx} - u_0 = 0, \quad u_0(1) = 2.$$

Its exact solution is $u_0(x) = 2e^{(1-1/x)}$.

Let us assume that the boundary layer thickness is of $O(\varepsilon^p)$, where p is to be determined from the principle of least degeneracy. The stretched variable is $s = x/\varepsilon^p$, and (65) becomes

$$\varepsilon^{1-2p} \frac{d^2 v}{ds^2} + \varepsilon^p s^2 \frac{dv}{ds} - v = 0.$$

The second-term is always dominated by the third, so the principle of degeneracy requires the first term to be of the same order as the third term (i.e., $O(1)$). Thus, $p = 1/2$, and the one-term inner expansion satisfies

$$\frac{d^2 v_0}{ds^2} + \frac{dv_0}{ds} = 0, \quad v_0(0) = 1.$$

The solution of the above problem is $v_0(s) = Ae^s + (1 - A)e^{-s}$. Prandtl's matching condition requires

$$\lim_{x \to 0} 2e^{(1-1/x)} = \lim_{s \to \infty} \left[Ae^s + (1 - A)e^{-s}\right]$$

which yields $A = 0$. This example is rather special in that A will be zero for all boundary conditions.

The on-term composite expansion is

$$u_0^{\text{comp}} = 2e^{(1-x)} + e^{-x/\sqrt{\varepsilon}}.$$

We conclude this example with the observation that a choice for the value of the index p other than $p = 1/2$ leads to boundary layer equations with insufficient structure to generate the required rapidly decaying behavior.

Thus, if $p > 1/2$, the dominant equation becomes

$$\frac{d^2 v_0}{ds^2} = 0, \quad v_0(0) = 1,$$

which gives $v_0(s) = 1 + As$. It is obvious that Prandtl's matching condition cannot be used to determine A. Whereas, if $p < 1/2$ the dominant equation degenerates to $v_0(s) = 0$ which does not satisfy the boundary condition at $s = 0$.

4.5 Interior layer

Consider the BVP:

$$\begin{cases} \varepsilon \dfrac{d^2 u}{dx^2} + x \dfrac{du}{dx} + xu = 0, \quad (-1, 1) \\ u(-1) = e, \quad u(1) = 2e^{-1}. \end{cases} \tag{66}$$

The coefficient of the first derivative (convective term) is positive in $(0, 1)$ which indicates the occurrence of a boundary layer at the left hand limit $x = 0$.

While the corresponding coefficient is negative in the range $-1 < x < 0$ indicates a boundary layer located at the right-hand limit which is again is $x = 0$. Thus, we are led to expr = ect two outer expansions for positive and negative x respectively and an inner expansion in the interior layer located at $x = 0$. We denote the leading term in the outer expansion for positive x by u_0^+, it satisfies

$$\frac{du_0^+}{dx} + u_0^+ = 0, \quad u_0^+(1) = 2e^{-1} \tag{67}$$

with the solution $u_0^+(x) = 2e^{-x}$.

The outer expansion for negative x, u_0^- satisfies

$$\frac{du_0^-}{dx} + u_0^- = 0, \quad u_0^+(-1) = e \tag{68}$$

with the solution $u_0^-(x) = e^{-x}$.

We suppose the boundary layer at $x = 0$ has thickness $O(\varepsilon^p)$ and determine the index p using the principle of least degeneracy. Let $s = x/\varepsilon^p$ so that the DE becomes

$$\varepsilon^{1-2p}\frac{d^2v}{ds^2} + s\frac{dv}{ds} + \varepsilon^p sv = 0.$$

The third term is dominated by the second term. The first term has the same order as the second term if $p = 1/2$. For this choice of p the leading term of the inner expansion v_0 satisfies

$$\frac{d^2v_0}{ds^2} + s\frac{dv_0}{ds} = 0.$$

Its solution can be given by

$$v_0(s) = B\,\mathrm{erf}\left(s/\sqrt{2}\right) + v_0(0),$$

Prandtl;s matching condition applied to the region $x > 0$ is

$$\lim_{s\to+\infty} v_0(s) = \lim_{x\to 0^+} u_0^+(x)$$

and corresponding for $x < 0$, we have

$$\lim_{s\to-\infty} v_0(s) = \lim_{x\to 0^-} u_0^-(x)$$

Using the limiting values $erf(\pm\infty) = \pm 1$ yields $v_0(0) = 1.5$ and $B = 0.5$. The leading order terms over the whole region are

$$\begin{aligned} &u_0^+(x) = 2e^{-x}, && x > O(\sqrt{\varepsilon}) \\ &v_0 = 0.5\,\mathrm{erf}\left(x/\sqrt{2\varepsilon}\right) + 1.5, && x = O(\sqrt{\varepsilon}) \\ &u_0^-(x) = e^{-x}, && x < -O(\sqrt{\varepsilon}) \end{aligned}$$

A composite expansion cannot be formed in the standard way when there is more than one outer solution. However, the behavior of v_0 for $|x| > O(\sqrt{\varepsilon})$ is as follows:

$$v_0\left[x > O(\sqrt{\varepsilon})\right] = 0.5 + 1.5 + \text{T.S.T}$$
$$v_0\left[x < -O(\sqrt{\varepsilon})\right] = -0.5 + 1.5 + \text{T.S.T}$$

Utilizing this enables a uniformly valid one-term composite expansion to be constructed which yields the correct coefficient of e^{-x} outside the boundary layer and the correct leading order behavior within the boundary layer. It is

$$u_0^{\text{comp}} = \left[0.5\,\text{erf}\left(x/\sqrt{2\varepsilon}\right) + 1.5\right]e^{-x}.$$

4.6 Nonlinear differential equation

Consider the following semilinear

$$\begin{cases} \varepsilon\dfrac{d^2u}{dx^2} + \dfrac{du}{dx} + u^2 = 0, \quad (0, 1) \\ u(0) = 2, \quad u(1) = 1/2. \end{cases} \tag{69}$$

The coefficient of the first and second order derivatives have the same sign, so the boundary layer will occur at the left boundary $x = 0$. The one-term outer expansion satisfies

$$\frac{du_0}{dx} + u_0^2 = 0, \quad u_0(1) = 1/2,$$

and the solution is $u_0(x) = 1/(1+x)$. The stretching transformation for the inner region will be $s = x/\varepsilon$ and therefore, the inner expansion satisfies

$$\frac{d^2v}{ds^2} + \frac{dv}{ds} + \varepsilon v^2 = 0, \quad v(0) = 2.$$

The one-term inner expansion v_0 satisfies the dominant part of this equation, i.e.,

$$\frac{d^2v_0}{ds^2} + \frac{dv_0}{ds} = 0, \quad v_0(0) = 2,$$

which gives $v_0(s) = A + (2 - A)e^{-s}$. Prandtl's matching condition yields $A = 1$, and the composite one-term uniformly valid expansion is

$$u_0^{\text{comp}} = \frac{1}{1+x} + e^{-x/\varepsilon}.$$

Next, consider the quasilinear problem

$$\begin{cases} \varepsilon\dfrac{d^2u}{dx^2} + 2u\dfrac{du}{dx} - 4u = 0, \quad (0, 1) \\ u(0) = 0, \quad u(1) = 4. \end{cases} \tag{70}$$

The nonlinearity is associated with the first derivative term. The location of the boundary layer depends on the relative sign of the first and second derivative coefficients. If we assume that the dependent variable is nonnegative throughout the interval $0 < x < 1$, then the boundary layer will occur at $x = 0$. The one-term outer expansion satisfies

$$2u_0\frac{du_0}{dx} - 4u_0 = 0, \quad u_0(1) = 4,$$

with the solution $u_0(x) = 2x + 2$.

Assuming that the boundary layer thickness is $O(\varepsilon)$, therefore, the dominant-order equation for the one-term inner expansion becomes

$$\frac{d^2v_0}{ds^2} + 2v_0\frac{dv_0}{ds} = 0, \quad v_0(0) = 0.$$

Its solution is $v_0(s) = a\tanh(as)$. Prandtl's matching condition yields $a = 2$. Thus, $v_0(s) = 2\tanh(2s)$, and the uniformly valid one-term composite expansion is

$$u_0^{\text{comp}} = 2x + 2 + 2\tanh(2s) - 2.$$

Application of perturbation techniques to partial differential equations, and other types of problems can be seen in the books [5, 6].

Conflict of interest

The authors declare no conflict of interest.

Notes/thanks/other declarations

I owe a great debt to my mentor Prof. S Natesan, Department of Mathematics, IIT Guwahati, who introduced me to this topic. The chapter was discussed during my stay at IIT Guwahati.

Author details

Jugal Mohapatra
Department of Mathematics, National Institute of Technology Rourkela, Odisha, India

*Address all correspondence to: jugal@nitrkl.ac.in

References

[1] A.B Bush. *Perturbation Methods for Engineers and Scientists*. CRC Press, Boca Raton, 1992.

[2] M.H. Holmes. *Introduction to Perturbation Methods*. Springer Verlag, Heidelberg, 1995.

[3] J. Kevorkian and J.D. Cole. *Perturbation Methods in Applied Mathematics*. Springer-Verlag, Heidelberg, 1981.

[4] R.E. O'Malley. *Singular Perturbation Methods for Ordinary Differential Equations*. Springer-Verlag, Heidelberg, 1991.

[5] J.A. Murdock. *Perturbations Theory and Methods*. John Wiley & Sons, New York, 1991.

[6] A.H. Nayfeh. *Introduction to Perturbation Techniques*. John Wiley & Sons, New York, 1993.

Chapter 9

Application of Perturbation Theory in Heat Flow Analysis

Neelam Gupta and Neel Kanth

Abstract

Many physical and engineering problems can be modeled using partial differential equations such as heat transfer through conduction process in steady and unsteady state. Perturbation methods are analytical approximation method to understand physical phenomena which depends on perturbation quantity. Homotopy perturbation method (HPM) was proposed by Ji Huan He. HPM is considered as effective method in solving partial differential equations. The solution obtained by HPM converges to exact solution, which are in the form of an infinite function series. Biazar and Eslami proposed new homotopy perturbation method (NHPM) in which construction of an appropriate homotopy equation and selection of appropriate initial approximation guess are two important steps. In present work, heat flow analysis has been done on a rod of length L and diffusivity α using HPM and NHPM. The solution obtained using different perturbation methods are compared with the solution obtained from most common analytical method separation of variables.

Keywords: heat conduction equation, homotopy perturbation method, new homotopy perturbation method, specific heat, diffusivity

1. Introduction

Partial differential equations play a dominant role in applied mathematics. The classical heat conduction equation is second order linear partial differential equation. The solutions of which are obtained by using various analytical and numerical methods [1–3]. This equation describes the heat distribution in each domain over some time. Jean-Joseph Fourier was the first to formulate and describe the heat conduction process [1, 4]. Perturbation methods depending upon small/large parameters have been encountered from past few years. Perturbation methods are analytical approximation method to understand physical phenomena which depends on perturbation quantity. But these methods do not provide an easy way to find out the rapid convergence of approximate series. Therefore, this method is simple, suitable and appropriate method to provide the rapid convergence of series [5–7]. The perturbation method along with the homotopy method has been employed to develop a hybrid method known as homotopy perturbation method (HPM) [1–4]. Ji-Huan was the first to introduce HPM. Homotopy perturbation method provides analytical approximation to linear/nonlinear problems without linearization or discretization. It helps in formulating simpler equations by breaking down the complex problems, which can be solved easily. Since HPM does not depend on small parameters, therefore drawbacks of the existing perturbation methods can be abolished [8–11]. The solution obtained by HPM converges to exact solution, which are in the form of an infinite function series. Various problems are modeled by linear and non-linear partial differential equations problems in the fields of physics, engineering etc. To solve such kind of partial differential equations (PDE), many methods are used to find the numerical or exact solutions. Homotopy perturbation method (HPM) is one of the methods used in recent years to solve various linear and non-linear PDE [12–15]. Initial and boundary value problems can be solved using HPM extensively. Many researchers and scientists show great interest in homotopy perturbation method. Huan was the first who described homotopy perturbation method. He showed that this method is a one of the powerful tools used to investigate various problems which are arising nowadays. HPM is used for solving linear and non-linear ordinary and partial differential equations [16].

In HPM, complex linear or non-linear problem can be continuously distorted into simpler ones. Perturbation theory and homotopy theory in topology is combined to develop homotopy perturbation method [1]. HPM is applicable to linear and non-linear boundary value problems. The solution obtained by HPM gives the solution approximately near to the universally accepted method of separation of variable [17–19].

Recently, Biazar and Eslami proposed the new homotopy perturbation method (NHPM). Construction of an appropriate homotopy equation and selection of appropriate initial approximation guess are two important steps of NHPM [19, 20]. The study reveals that with less computational work, we can construct proper homotopy by decomposition of source function in a correct way. New homotopy perturbation method is the most powerful tool which can be used to obtain analytical solution of various kinds of linear and nonlinear PDE's. This method is widely used by researchers to obtain solution of various functional Equations [20–22].

To develop this new technique, HPM is combined with the decomposition of source function. The decomposition of a source function is the basis of homotopy used in this method because convergence of a solution is affected by the decomposition of source functions [23]. Different kind of homotopy can be formed using various decomposition of a source functions. This study is aimed at constructing suitable homotopy by decomposition of a source function which requires less computational efforts and made calculations in simpler form unlike other perturbation methods. The obtained results directly imply the fact that NHPM is very influential as compared to HPM or any other perturbation technique. To establish exact solution of linear and non-linear problem with boundary and initial condition, new homotopy method is most appropriate method to apply [23].

The two most important steps in application of new homotopy perturbation method to construct a suitable homotopy equation and choose a suitable initial guess, we aim in this work to effectively employ the (NHPM) to establish exact solution for two-dimensional Laplace equation with Dirichlet and Neumann boundary condition, the difference between (NHPM) and standard (HPM) is starts from the form of initial approximation of the solution.

In this chapter, the semi analytic solution of one-dimensional heat conduction equation is obtained by means of homotopy perturbation method and new homotopy perturbation method. These methods are effectively applied to obtain the exact solution for the problem in hand which reveals the effectiveness and simplicity of the method. Numerical results have also been analyzed graphically to show the rapid convergence of infinite series expansion. The obtained analytic solution for one dimensional heat conduction equation with boundary and initial conditions using NHPM is same as the universally accepted exact solution. This tells us about the capability and reliability of this method. The solution obtained using NHPM is considered in the form of an infinite series. The convergence of solution to the exact solution is very rapid.

2. Heat conduction equation

The one-dimensional heat equation

$$\frac{\partial U}{\partial \theta} = \beta \frac{\partial^2 U}{\partial z^2} \tag{1}$$

with boundary conditions

$$U(0, \theta) = 0, U(1, \theta) = 0, \tag{2}$$

and initial condition

$$U(z, 0) = h(z), 0 \le z \le 1. \tag{3}$$

3. Basic idea of Homotopy perturbation method

First, we outline the general procedure of the homotopy perturbation method developed and advanced by He. We consider the differential Eq. [2]

$$A(u) - f\,(r) = 0, r \in \Omega \tag{4}$$

$$B\left(u, \frac{\partial u}{\partial x}\right) = 0, r \in \Gamma \tag{5}$$

where A is a general differential operator, linear or nonlinear, $f\,(r)$ is a known analytic function, B is a boundary operator and Γ is the boundary of the domain Ω. The operator A can be generally divided into two operators, L and N, where L is linear and N is a nonlinear operator. Eq. (4) can be written as

$$L(u) + N(u) - f\,(r) = 0 \tag{6}$$

Using the homotopy technique, we can construct a homotopy [1,2]

$$v(r,p) : \Omega \times [0,1] \to R \text{ which satisfies the relation}$$
$$H(v,p) = (1-p)[L(v) - L(u_0)] + p[A(v) - f\,(r)] = 0, r \in \Omega \tag{7}$$

Here $p \in [0, 1]$ is called the homotopy parameter and u_0 is an initial approximation for the solution of Eq. (4), which satisfies the boundary conditions. Clearly, from Eq. (7), we have

$$H(v, 0) = L(v) - L(u_0) \tag{8}$$

$$H(v, 1) = A(v) - f\,(r) \tag{9}$$

We assume that the solution of Eq. (7) can be expressed as a series in p as follows:

$$v = v_0 + pv_0 + p_2 v_2 + p_3 v_3 + \cdots \tag{10}$$

On setting $p = 1$, we obtain the approximate solution of Eq. (10) as

$$u = \lim_{p \to 1} v = v_0 + v_0 + v_2 + v_3 + \cdots \tag{11}$$

4. Basic idea of new Homotopy perturbation method

First, following homotopy is constructed for solving heat conduction equation using NHPM

$$(1-p)\left(\frac{\partial T}{\partial \theta} - U_0\right) + p\left(\frac{\partial T}{\partial \theta} - \beta \frac{\partial^2 T}{\partial z^2}\right) = 0 \tag{12}$$

Taking $L^{-1} = \int_{\theta_0}^{\theta} (.) d\theta$ i.e. inverse operator on Eq. (12), then

$$T(z,\theta) = \int_{\theta_0}^{\theta} U_0(z,\theta) d\theta - p \int_{\theta_0}^{\theta} \left(U_0 - \beta \frac{\partial^2 T}{\partial z^2}\right) d\theta + T(z,\theta_0) \tag{13}$$

Where $T(z,\theta_0) = U(z,\theta_0)$

Let the solution of Eq. (13) is given by

$$T = T_0 + pT_1 + p^2 T_2 + p^3 T_3 + \ldots \tag{14}$$

where $T_0, T_1, T_2, T_3, \ldots$ are to be determined.

Suppose solution given by Eq. (14) is the solution of Eq. (13). On comparing the coefficients of powers of p and equating to zero and using Eq. (14) in Eq. (13), following are obtained:

$$p^0 : T_0(z,\theta) = \int_{\theta_0}^{\theta} U_0(z,\theta) d\theta + T(z,\theta_0)$$

$$p^1 : T_1(z,\theta) = -\int_{\theta_0}^{\theta} \left(U_0(z,\theta) - \beta \frac{\partial^2 T_0}{\partial z^2}\right) d\theta$$

$$p^2 : T_2(z,\theta) = \int_{\theta_0}^{\theta} \left(\beta \frac{\partial^2 T_1}{\partial z^2}\right) d\theta$$

$$p^3 : T_3(z,\theta) = \int_{\theta_0}^{\theta} \left(\beta \frac{\partial^2 T_2}{\partial z^2}\right) d\theta$$

and so on ... (15)

Consider the initial approximation of Eq. (1) as

$$U_0(z,\theta) = \sum_{n=0}^{\infty} c_n(z) P_n(\theta), T(z,0) = U(z,0), P_k(\theta) = \theta^k, \tag{16}$$

where, $P_1(\theta), P_2(\theta), P_3(\theta), \ldots$ and $c_0(z), c_1(z), c_2(z), \ldots$ are specified functions and unknown coefficients respectively, depending on the problem.

Using Eq. (16) in (15), following are obtained:

$$T_0(z,\theta) = \left(c_0(z)\theta + c_1(z)\frac{\theta^2}{2} + c_2(z)\frac{\theta^3}{3} + c_3(z)\frac{\theta^4}{4} + \ldots\right) + U(z,0)$$

$$T_1(z,\theta) = \left(-c_0(z) - \beta\pi^2 \sin \pi z\right)\theta + \left(-\frac{1}{2}c_1(z) + \frac{1}{2}\beta c_0''(z)\right)\theta^2 + \left(-\frac{1}{3}c_2(z) + \frac{1}{3}c_1''(z)\right)\theta^3 + \dots$$

$$\text{and so on} \dots \tag{17}$$

Now solving the above equations in such a manner that, $T_1(z,\theta) = 0$. Therefore Eq. (17) reduces to

$$T_1(z,\theta) = T_2(z,\theta) = \dots = 0.$$

So $U(z,\theta) = T_0(z,\theta) = \sum_{n=0}^{\infty} c_n(z)P_n(\theta)$ is obtained solution which is found to be exactly same as the exact solution obtained through method of separation of variable.

If $U_0(z,\theta)$ is analytic at $\theta = \theta_0$,

$U_0(z,\theta) = \sum_{n=0}^{\infty} c_n(z)(\theta - \theta_0)^n$ is the taylor series expansion which can be used in Eq. (9).

5. Applications of Homotopy perturbation method and new Homotopy perturbation method

For understanding the application of HPM and NHPM, we will solve the one-dimensional heat equation given by

$$\frac{\partial U}{\partial \theta} = \beta \frac{\partial^2 U}{\partial z^2} \tag{18}$$

with boundary conditions

$$U(0,\theta) = 0, U(1,\theta) = 0, \tag{19}$$

and initial condition

$$U(z,0) = \sin \frac{2\pi z}{L}, 0 \le z \le L. \tag{20}$$

The homotopy for the diffusion equation given by (18) is obtained as follows [2].

$$\left(\frac{\partial v}{\partial \theta} - \frac{\partial u_0}{\partial \theta}\right) + \mathfrak{p}\left(\frac{\partial u_0}{\partial \theta} - \beta \frac{\partial^2 v}{\partial z^2}\right) = 0 \tag{21}$$

Let $u_0 = \sin \frac{2\pi z}{L} \cos \pi^2 \theta$ be the initial approximation, which satisfies boundary conditions given by (19).

Let solution of (18) has the following form

$$v = v_0 + \mathfrak{p}v_1 + \mathfrak{p}^2 v_2 + \mathfrak{p}^3 v_3 + \mathfrak{p}^4 v_4 + \dots \tag{22}$$

On substituting the value of v in Eq. (21) and comparing the coefficients of like powers of $\mathfrak{p}$ we obtain

$$\mathfrak{p}^0 : \frac{\partial v_0}{\partial \theta} = \frac{\partial u_0}{\partial \theta}$$

$$\mathfrak{p}^1 : \frac{\partial v_1}{\partial \theta} = \beta \frac{\partial^2 v_0}{\partial z^2}, v_1(0, \theta) = 0 = v_1(L, \theta)$$

$$\mathfrak{p}^2 : \frac{\partial v_2}{\partial \theta} = \beta \frac{\partial^2 v_1}{\partial z^2}, v_2(0, \theta) = 0 = v_2(L, \theta)$$

$$\mathfrak{p}^3 : \frac{\partial v_3}{\partial \theta} = \beta \frac{\partial^2 v_2}{\partial z^2}, v_3(0, \theta) = 0 = v_3(L, \theta)$$

$$\mathfrak{p}^n : \frac{\partial v_n}{\partial \theta} = \beta \frac{\partial^2 v_{n-1}}{\partial z^2}, v_n(0, \theta) = 0 = v_n(L, \theta) \tag{23}$$

On solving the system of Eq. (23) using Mathematica 5.2

$$v_0 = u_0 = sin\ \frac{2\pi z}{L}\ \ cos\ \pi^2\theta$$

$$\frac{\partial v_1}{\partial \theta} = -\frac{4\beta\pi^2}{L^2} v_0 \Rightarrow v_1 = -\frac{\beta sin\left[\frac{2\pi z}{L}\right] sin\left[\pi^2\theta\right]}{L^2} + sin\left[\frac{\pi z}{L}\right]$$

$$\frac{\partial v_2}{\partial \theta} = -\frac{4\beta\pi^2}{L^2} v_1 \Rightarrow v_2 = -\frac{\beta\left(L^2\pi^2\theta + \alpha cos\left[\pi^2\theta\right]\right) sin\left[\frac{2\pi z}{L}\right]}{L^4} + \frac{L^4\, sin\left[\frac{2\pi z}{L}\right] + \beta^2\, sin\left[\frac{2\pi z}{L}\right]}{L^4}$$

$$\frac{\partial v_3}{\partial \theta} = -\frac{4\beta\pi^2}{L^2} v_2 \Rightarrow v_3 = \frac{\beta\left(\pi^2\theta\left(-2L^4 + L^2\pi^2\beta\theta - 2\beta^2\right) + 2\beta^2\, sin\left[\pi^2\theta\right]\right) sin\left[\frac{2\pi z}{L}\right]}{2L^6} + sin\left[\frac{2\pi z}{L}\right]$$

$$\begin{aligned}\frac{\partial v_4}{\partial \theta} &= -\frac{4\beta\pi^2}{L^2} v_3 \Rightarrow v_4 \\ &= \frac{1}{6L^8}\left(\beta\left(\pi^2\theta\left(-6L^6 + 3L^4\pi^2\beta\theta - L^2\pi^4\theta^2\beta^2 + 3\beta^3\pi^2\theta\right)6\beta^3\cos\left[\pi^2\theta\right]\right) sin\left[\frac{2\pi z}{L}\right]\right) \\ &+ \frac{L^8\, sin\left[\frac{2\pi z}{L}\right] - \beta^4\, sin\left[\frac{2\pi z}{L}\right]}{L^8}\end{aligned}$$

$$\begin{aligned}\frac{\partial v_5}{\partial \theta} &= -\frac{4\beta\pi^2}{L^2} v_4 \Rightarrow v_5 \\ &= \frac{-1}{24L^{10}}\left(\beta\left(\pi^2\theta\left(24L^8 - 12L^6\pi^2\beta\theta + 4L^4\pi^4\theta^2\beta^2 - L^2\pi^6\theta^3\beta^3 + 4\left(-6 + \pi^4\theta^2\right)\beta^4\right)\right.\right. \\ &\left.\left. + 24\beta^4\, \sin\left[\pi^2\theta\right]\right) \sin\left[\frac{2\pi z}{L}\right]\right) + \sin\left[\frac{2\pi z}{L}\right]\end{aligned}$$

$$\begin{aligned}\frac{\partial v_6}{\partial \theta} = -\frac{\beta\pi^2}{L^2} v_5 \Rightarrow v_6 = \frac{-1}{120L^{12}}&\left(\beta\left(\pi^2\theta\left(120L^{10} - 60L^8\pi^2\beta\theta + 20L^6\pi^4\theta^2\beta^2\right.\right.\right. \\ &-5L^4\pi^6\theta^3\beta^3 + L^2\pi^8\theta^4\beta^4 - 5\pi^2\theta\left(-12 + \pi^4\theta^2\right)\beta^5) \\ &+120\beta^5\cos\left[\pi^2\theta\right]) \sin\left[\frac{2\pi z}{L}\right]) \\ &+ \frac{L^{12}\sin\left[\frac{2\pi z}{L}\right] + \beta^6\sin\left[\frac{2\pi z}{L}\right]}{L^{12}}\end{aligned}$$

$$\text{and so on} \ldots \tag{24}$$

The approximate solution of (1) by setting $\mathfrak{p} = 1$ in (23) is given by

$$u = \lim_{p \to 1} v = v_0 + v_1 + v_2 + v_3 + v_3 + \ldots \tag{25}$$

On substituting values of $v_i's$ in Eq. (25), solution is obtained in terms of a summation of infinite series which gives results near to the exact solution.

Now we will solve the Eq. (18) using NHPM. First of all, following homotopy is constructed for solving heat conduction equation using NHPM

$$(1-p)\left(\frac{\partial T}{\partial \theta}-U_0\right)+p\left(\frac{\partial T}{\partial \theta}-\beta\frac{\partial^2 T}{\partial z^2}\right)=0 \tag{26}$$

Taking $L^{-1}=\int_{\theta_0}^{\theta}(.)d\theta$ i.e. inverse operator on Eq. (26), then

$$T(z,\theta)=\int_0^\theta U_0(z,\theta)d\theta-p\int_0^\theta\left(U_0-\beta\frac{\partial^2 T}{\partial z^2}\right)d\theta+T(z,0). \tag{27}$$

Let the solution of the (27) is

$$T=T_0+pT_1+p^2T_2+p^3T_3+\dots, \tag{28}$$

where, $T_0, T_1, T_2, \dots$ are to be determined.

Suppose Eq. (25) is the solution of Eq. (24). Comparing the coefficients of powers of p and equating to zero and using Eq. (25) in Eq. (24), following are obtained:

$$\begin{aligned}
p^0 &: T_0(z,\theta)=\int_0^\theta U_0(z,\theta)d\theta+T(z,0)\\
p^1 &: T_1(z,\theta)=-\int_0^\theta\left(U_0(z,\theta)-\beta\frac{\partial^2 T_0}{\partial z^2}\right)d\theta\\
p^2 &: T_2(z,\theta)=\int_0^\theta\left(\beta\frac{\partial^2 T_1}{\partial z^2}\right)d\theta\\
p^3 &: T_3(z,\theta)=\int_0^\theta\left(\beta\frac{\partial^2 T_2}{\partial z^2}\right)d\theta\\
&\text{and so on.}
\end{aligned} \tag{29}$$

Consider initial approximation of Eq. (18) as

$$U_0(z,\theta)=\sum_{n=0}^{\infty}c_n(z)P_n(\theta), T(z,0)=U(z,0), P_k(\theta)=\theta^k, \tag{30}$$

where, $P_1(\theta), P_2(\theta), P_3(\theta), \dots$ and $c_0(z), c_1(z), c_2(z), \dots$ are specified functions and unknown coefficients respectively, depending on the problem.

Using Eq. (30) in (29), following are obtained:

$$\begin{aligned}
T_0(z,\theta)&=\left(c_0(z)\theta+c_1(z)\frac{\theta^2}{2}+c_2(z)\frac{\theta^3}{3}+c_3(z)\frac{\theta^4}{4}+\dots\right)+\sin\frac{2\pi z}{L}\\
T_1(z,\theta)&=\left(-c_0(z)-\frac{4\beta\pi^2}{L}\sin\frac{2\pi z}{L}\right)\theta+\left(-\frac{1}{2}c_1(z)+\frac{1}{2}\beta c_0''(z)\right)\theta^2\\
&\quad+\left(-\frac{1}{3}c_2(z)+\frac{1}{3}c_1''(z)\right)\theta^3+\dots\\
&\text{and so on}\dots
\end{aligned} \tag{31}$$

Now solving the above equations in such a manner that, $T_1(z,\theta) = 0$. Therefore Eq. (31) reduces to

$$c_0(z) = -\frac{2^2\beta\pi^2}{L}\sin\frac{2\pi z}{L}$$

$$c_1(z) = \frac{2^4\beta^2\pi^4}{L^2}\, sin\,\frac{2\pi z}{L}$$

$$c_2(z) = -\frac{2^6\beta^3\pi^6}{L^3}\, sin\,\frac{2\pi z}{L}$$

$$U(z,\theta) = T_0(z,\theta) = sin\,\frac{2\pi z}{L} + c_0(z)\theta + c_1(z)\frac{\theta^2}{2} + c_2(z)\frac{\theta^3}{3} + c_3(z)\frac{\theta^4}{4} + \ldots$$
$$= sin\,\frac{2\pi z}{L}\left[1 - \frac{2^2\beta\pi^2}{L}\theta + \frac{2^4\beta^2\pi^4}{L^2}\frac{\theta^2}{2} - \frac{2^6\beta^3\pi^6}{L^3}\frac{\theta^3}{3} + \ldots\right] = sin\,\frac{2\pi z}{L} e^{-\frac{4\beta\pi^2}{L}\theta} \tag{32}$$

which is same as the universally accepted exact solution for the problem which is shown in **Figure 1**.

The solution of one-dimensional heat conduction equation is solved using HPM and NHPM and then compared with the universally accepted exact solution obtained from method of separation of variable. **Figure 2** represents the comparison of solution of heat equation using HPM, NHPM and method of separation of variable. It is found that the solution obtained using HPM gives result near to the exact solution whereas solution using NHPM gives same results as the exact solution.

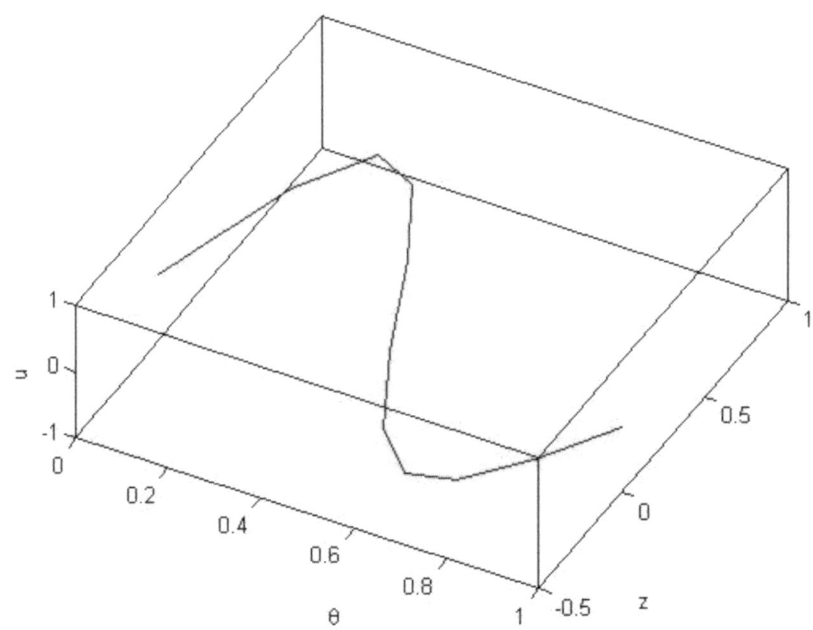

Figure 1.
Solution using NHPM.

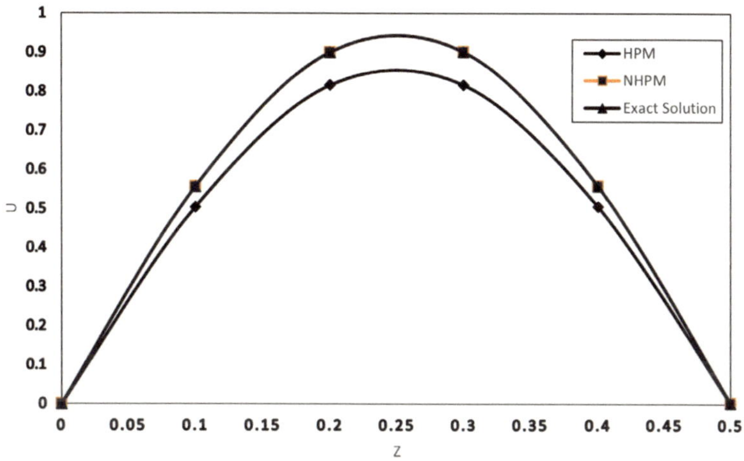

Figure 2.
Comparison of HPM, NHPM and the exact solution.

6. Conclusion

The analytical approximate solutions of one-dimensional heat conduction equation are obtained by applying new homotopy perturbation method and new homotopy perturbation method. It is found that new homotopy perturbation method (NHPM) converges very rapidly as compared to homotopy perturbation method (HPM) and other traditional methods. The exact solutions are obtained up to more accuracy using NHPM. An infinite convergent series solution for particular initial conditions are obtained using these methods which shows the effectiveness and efficiency of NHPM and HPM. The convergence rate of NHPM is much faster than traditional methods which directly indicates that this method is better than other methods. The solution of heat equation obtained by homotopy perturbation method and new homotopy perturbation method are exactly same and very close to the solution obtained by universally accepted and tested analytical method of separation of variables. If the initial guess in homotopy perturbation method is effective and properly chosen which satisfy boundary and initial condition, homotopy perturbation method provides solution with rapid convergence. It is illustrated that NHPM is very prominent, when accuracy has a vital role to play. The numerical results also reflect the remarkable applicability of NHPM to linear and non-linear initial and boundary value problems. NHPM provides the rapid convergence of the series solution for linear as well as non-linear problems with less computational work.

Author details

Neelam Gupta and Neel Kanth*
Jaypee University of Information Technology, Waknaghat, Solan, India

*Address all correspondence to: neelkanth28@gmail.com

References

[1] N. Gupta and N. Kanth (2019). Study of heat flow in a rod using homotopy analysis method and homotopy perturbation method. AIP Conference Proceedings, 2061(1) 020013 1–8

[2] D. Grover, V. Kumar, and D. Sharma, A Comparative Study of Numerical Techniques and Homotopy Perturbation Method for Solving Parabolic Equation and Nonlinear Equations, Int. J. Comput. Methods Eng. Sci. Mech., vol. 13, no. 6, pp. 403–407, 2012.

[3] J.-H. He, Homotopy perturbation technique, Comput. Methods Appl. Mech. Eng., vol. 178, no. 3, pp. 257–262, Aug. 1999.

[4] J.-H. He, Comparison of homotopy perturbation method and homotopy analysis method, Appl. Math. Comput., vol. 156, no. 2, pp. 527–539, Sep. 2004.

[5] J.-H. He, A coupling method of a homotopy technique and a perturbation technique for non-linear problems, Int. J. Non-Linear Mech., vol. 35, no. 1, pp. 37–43, Jan. 2000.

[6] J.-H. He, Some asymptotic methods for strongly nonlinear equations, Int. J. Mod. Phys. B, vol. 20, no. 10, pp. 1141–1199, Apr. 2006.

[7] S. Liang and D. Jeffrey, Comparison of Homotopy Analysis Method and Homotopy Perturbation Method through an Evolution Equation, Commun. Nonlinear Sci. Numer. Simul., vol. 14, pp. 4057–4064, Dec. 2009.

[8] A. Demir, S. Erman, B. Ozgur, E. Korkmaz, Analysis of the new homotopy perturbation method for linear and nonlinear problems. Boundary Value Problems, 2013(1), 1–11, 2013.

[9] J-H. He, Homotopy perturbation method: a new nonlinear analytical technique. Applied Mathematics and Computation, 135(1) 73–79, 2003.

[10] J. Biazar, M. Eslami, A new homotopy perturbation method for solving systems of partial differential equations. Computers and Mathematics with Applications, 62(1), 225–234, 2011.

[11] J-H. He, Homotopy perturbation method for solving boundary value problems. Physics Letters A. 350(1), 87–88, 2006.

[12] M. Mirzazadeh, Z. Ayati, New homotopy perturbation method for system of Burgers equations. Alexandria Engineering Journal. 55(3), 1619–1624, 2016.

[13] M. Elbadri, A New Homotopy Perturbation Method for Solving Laplace Equation. Advances in Theoratical and Applied Mathematics. 8 (3), 237–242, 2013.

[14] M. Elbadri, T.M. Elzaki, New Modification of Homotopy Perturbation Method and the Fourth - Order Parabolic Equations with Variable Coefficients. Pure Appl. Math. J. 4(6) 242–247, 2015.

[15] J.H. He, A coupling method of homotopy technique and a perturbation technique for non linear problems, Int. J. Non linear Mech., 35, pp. 37–43, 2000.

[16] J.H. He, Application of homotopy perturbation method to nonlinear wave equations, Chaos Solitons Fractals, 26, pp. 295–700,2005.

[17] A. Yıldırım, Solution of BVPs for fourth-order integro-differential equations by using homotopy perturbation method, Comput. Math. Appl., 56 (12), 3175–3180, 2008.

[18] S. Abbasbandy, Homotopy perturbation method for quadratic

Riccati differential equation and comparison with Adomian's decomposition method, Appl. Math. Comput., 172, pp. 485–490, 2006.

[19] S.J. Liao, An approximate solution technique not depending on small parameters: A special example, Int. J. Non linear Mech., 36, 371–180, 1995.

[20] N. Gupta and N. Kanth, Study of heat conduction inside rolling calender nip for different roll temperatures, Journal of physics: Conference series, 1276(1), 012044 1–9, 2019.

[21] N. Gupta and N. Kanth, Analytical Approximate Solution of Heat Conduction Equation Using New Homotopy Perturbation Method, Matrix Science Mathematic (MSMK), 3(2), 01–07, 2019.

[22] N. Gupta and N. Kanth, Numerical Solution of Diffusion Equation Using Method of Lines, Indian Journal of Industrial and Applied Mathematics, 10 (2), 194–203, 2019.

[23] N. Gupta and N. Kanth, Analysis of heat conduction inside the calender nip used in textile industry, AIP Conference Proceedings, 2214(1), 020008, 2020.

Chapter 10

Chaos and Complexity Dynamics of Evolutionary Systems

Lal Mohan Saha

Abstract

Chaotic phenomena and presence of complexity in various nonlinear dynamical systems extensively discussed in the context of recent researches. Discrete as well as continuous dynamical systems both considered here. Visualization of regularity and chaotic motion presented through bifurcation diagrams by varying a parameter of the system while keeping other parameters constant. In the processes, some perfect indicator of regularity and chaos discussed with appropriate examples. Measure of chaos in terms of Lyapunov exponents and that of complexity as increase in topological entropies discussed. The methodology to calculate these explained in details with exciting examples. Regular and chaotic attractors emerging during the study are drawn and analyzed. Correlation dimension, which provides the dimensionality of a chaotic attractor discussed in detail and calculated for different systems. Results obtained presented through graphics and in tabular form. Two techniques of chaos control, pulsive feedback control and asymptotic stability analysis, discussed and applied to control chaotic motion for certain cases. Finally, a brief discussion held for the concluded investigation.

Keywords: chaos, Lyapunov exponents, chaos indicator, bifurcation, topological entropy, correlation dimension

1. Introduction

Henri Poincaré, (1892–1908), [1], was first to acknowledge the possible existence of chaos in nonlinear systems while studying a 3-body problem comprising Sun, Moon and Earth. He noticed the dynamics of the system turned to be sensitive towards initial conditions, which was later termed as chaos. His results based on theoretical analysis and he could not demonstrate it because computers were not available at that time. Lorenz, a weather scientist, demonstrated existence of chaos by using a computer in 1963, [2], and in this way supported chaos theory of Poincaré. Thus, **Lorenz** provided the foundation of chaos theory and inspired a fundamental reappraisal of systems of nonlinearity in many disciplines of science, engineering, biological and medical sciences, atmospheric science, economics, social sciences and where not? In our everyday life, chaos happened frequently in various form like cyclone, tsunami, tornado, epidemics/pandemics etc. Spread of any uncontrollable form of disease in medical science is nothing but a chaotic and contagious nature of disease. Systematic studies in various areas resulting in numerous articles on chaos and nonlinear dynamics appeared in many well-reputed scientific journals, [3–19].

Most biological systems exhibit enormous diversity and structurally multicomponent resulting in ecological imbalance and disorder/disharmony in environment. Inspired by articles of Lotka, Volterra, and Allee, numerous articles appeared with diversity in assumptions depending of species and their living environmental conditions in predator-prey models, [20–44].

Real systems are mostly nonlinear and many of them are with multicomponent structure. Their individual elements possess individual properties. Such systems are termed as the complex system.

During evolution, a complex system exhibits chaos in some parameter space but also some other phenomena called complexity. This complexity is due to the interaction among multiple agents within the system displayed in the form of coexistence of multiple attractors, bistability, intermittency, cascading effects, exhibit of hysteresis properties etc. Thus, complexity can viewed as its systematic nonlinear properties and it is due to the interaction among multiple agents within the system. Foundation work and elaborate descriptions on complexity can viewed from some pioneer articles on complexity in nonlinear dynamics presented in [45–51]. Study of complexity means to know the results that emerging from a collection of interacting parts.

A dynamical system be chaotic then it must be (i) sensitive to initial conditions, (ii) topologically mixing and (iii) its periodic orbits must be dense. In chaotic systems, there exists a strange attractor, a chaotic set, which has fractal structure. Complex systems are also sensitive to their initial conditions and two complex systems that are initially very close together in terms of their various elements and dimensions can end up in distinctly different places. Wide discussions on complex system may found in some pioneer literatures, [14, 18, 45, 46, 48, 50, 51].

Chaos measured by Lyapunov exponents, (also called Lyapunov characteristic components or LCEs); LCE > 0 indicates existence of chaos and LCE < 0 indicates regularity, [52–62]. A complex system can better understood by measuring (i) chaos, (ii) Topological entropies and (iii) correlation dimension. Topological entropy, a non-negative number, provides a perfect way to measure complexity of a system. More topological entropy in any system signifies more complexity in it. Actually, it measures the evolution of distinguishable orbits over time, thereby providing an idea of how complex the orbit structure of a system is, [48–50, 61–69]. A system may be chaotic with zero topological entropy. In addition, a significant increase in topological entropy does not justify that it is chaotic. The book by Nagashima and Baba, [62], gives a very clear definition of topological entropy. The correlation dimension provides the dimensionality of the chaotic attractor. Correlation dimensions are non-integers and this is one reasons besides self-similarity that chaotic sets have fractal structure, [60, 68–73].

It emerges from a good number of recent researches that chaos appearing in dynamical system be controlled and suggested number techniques to control chaos, [74–88]. These techniques have some limitations depending on the models and nature of nonlinearity.

Objective of this article is to investigate the emergence of chaos and complexity in nonlinear dynamical systems through examples of nonlinear models. Numerical simulations carried out for bifurcation analysis, plotting of LCEs and topological entropies for different systems. Numerical calculations extended to obtain correlation dimensions for certain chaotic attractors emerging in different systems. The study further extended to explain different types of chaos controlling technique. Studies confined to one, two and three-dimensional systems only.

2. Dynamic models with chaos and complexity

2.1 One dimensional discrete models

2.1.1 Dynamics of laser map

A highly simplified type discrete nonlinear model for laser system, arising from Laser Physics, described in articles, [12, 50, 89–91]. The model describes evolution of certain Fabry-Perot cavity containing a saturable absorber and driven by an external laser represented by

$$x_{n+1} = Q - \frac{A\,x_n}{1 + x_n{}^2}, \forall \in \mathfrak{R}, n \in N \tag{1}$$

Here Q is the normalized input field and A is a parameter depends on the specifics of the parameters and $A > 0$. The fixed points of the map are the real root of equation

$$x^3 - Qx^2 + (1 + A)x - Q = 0 \tag{2}$$

This equation has either three real roots or one real and a pair of complex conjugate roots depending on parameter space (A, Q). Stability occur in the form of stability and bistability, [89].

Fixed Points and Bifurcations:

For Q fixed, $Q = 2.76$, and $A < 4.3793$, only one stable steady state solution exits and stable two cycle starts when A exceeds this value. Thus, approximately, $A = 4.3793$, is the bifurcation point. At value $A = 4.3$, the stable steady state solution is $x^* = 0.720533$.

Keeping $Q = 2.76$ and varying parameter A, bifurcation diagrams are drawn, **Figure 1**, for four different ranges of values of A. Similarly, keeping A fixed,

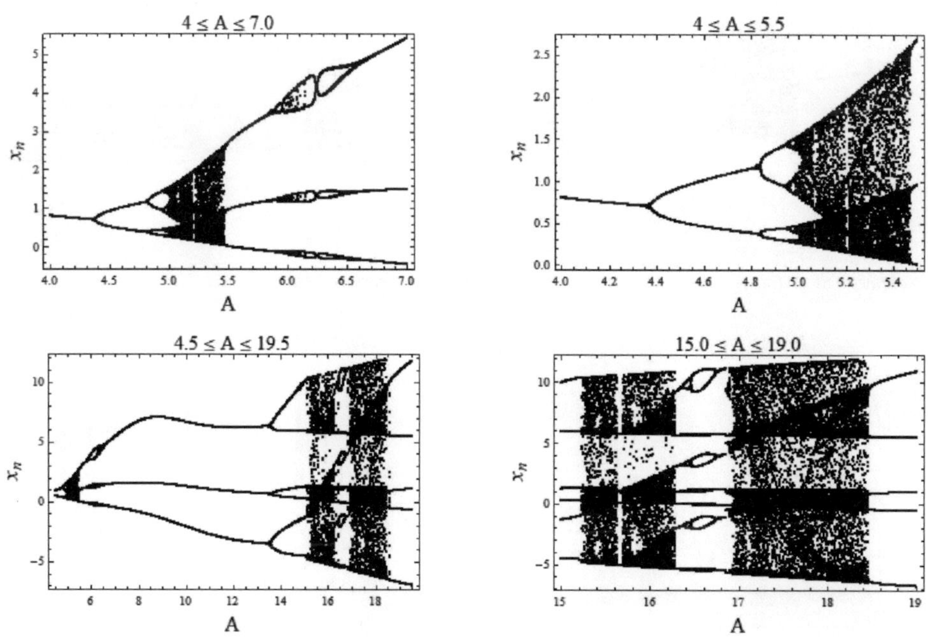

Figure 1.
Bifurcation diagrams of map (1) for four cases: when Q = 2.76 and parameter A varies.

$A = 5.4$ and varying Q in four different ranges, bifurcation diagrams are drawn, **Figure 2**. One observe clearly the appearance of periodic windows within chaotic region of bifurcations as an indication of intermittency and other complex phenomena. Periodic windows become gradually shorter and appearance become more frequent while moving forward in parameter space.

Both time series plots shown in **Figure 3** are for chaotic evolution of system (1) and correspond to parameters (a) (A, Q) = (5.3, 2.76), due to which an unstable fixed point obtained as $x^* = 0.58531$, and parameters (b)) (A, Q) = (5.4, 2.9), due to which an unstable fixed point obtained as $x^* = 0.572218$. For both cases, initial point taken is $x_0 = 0.5$ which lies nearby these points and so, also, unstable.

Calculations of Lyapunov Exponents, (LCEs):

Lyapunov exponents, LCEs, for map (1), calculated for four cases, **Figure 4**, positive LCEs appearing above zero line clearly indicate chaotic motion and those below this line indicate regular motion.

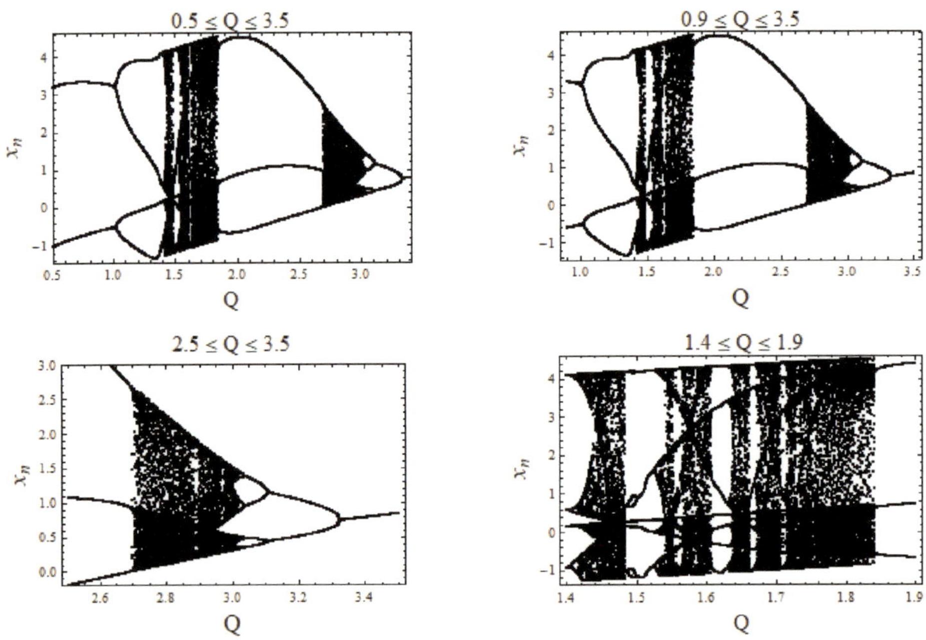

Figure 2.
Bifurcation diagrams of map (1) for four cases: when A = 5.4 and parameter Q varies.

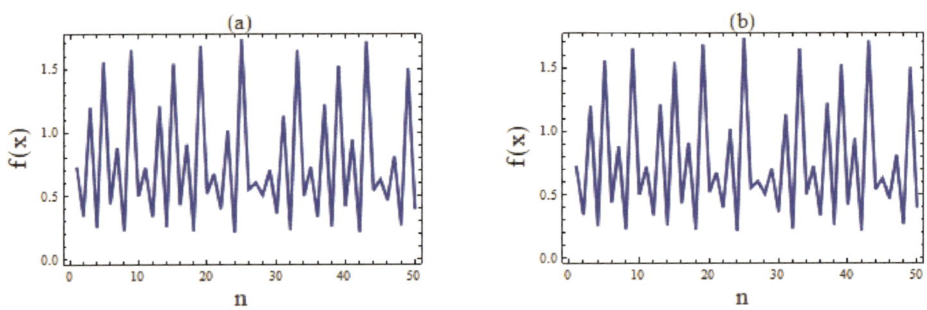

Figure 3.
Chaotic time series plots with initial value $x_0 = 0.5$: (a) A = 5.3, Q = 2.76 and (b) A = 5.4, Q = 2.9.

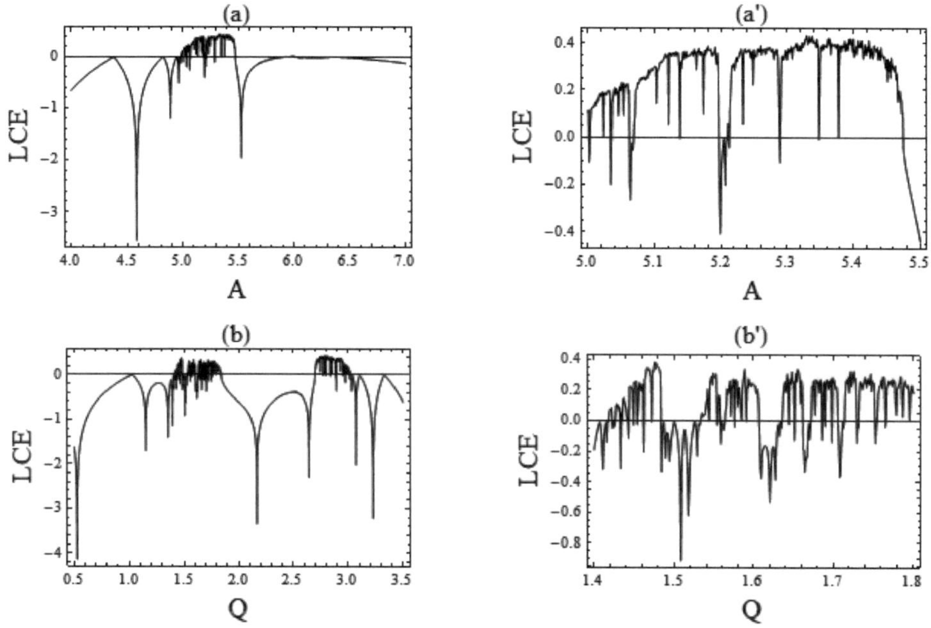

Figure 4.
Plots of LCEs: (a) for the upper row Q = 2.76, 4.0 ≤ A ≤ 5.5 and 5.0 ≤ A ≤ 7.0; (b) for the lower row A = 5.4, 0.5 ≤ Q ≤ 3.5 and 1.4 ≤ Q ≤ 1.8.

Topological Entropies:

Numerical calculations further proceeded to calculate topological entropies for system (1) and shown in **Figure 5**; where figures of upper row obtained by varying parameter A while keeping parameter Q = 2.76 and those of lower row obtained by varying parameter Q while keeping parameter A = 5.4.

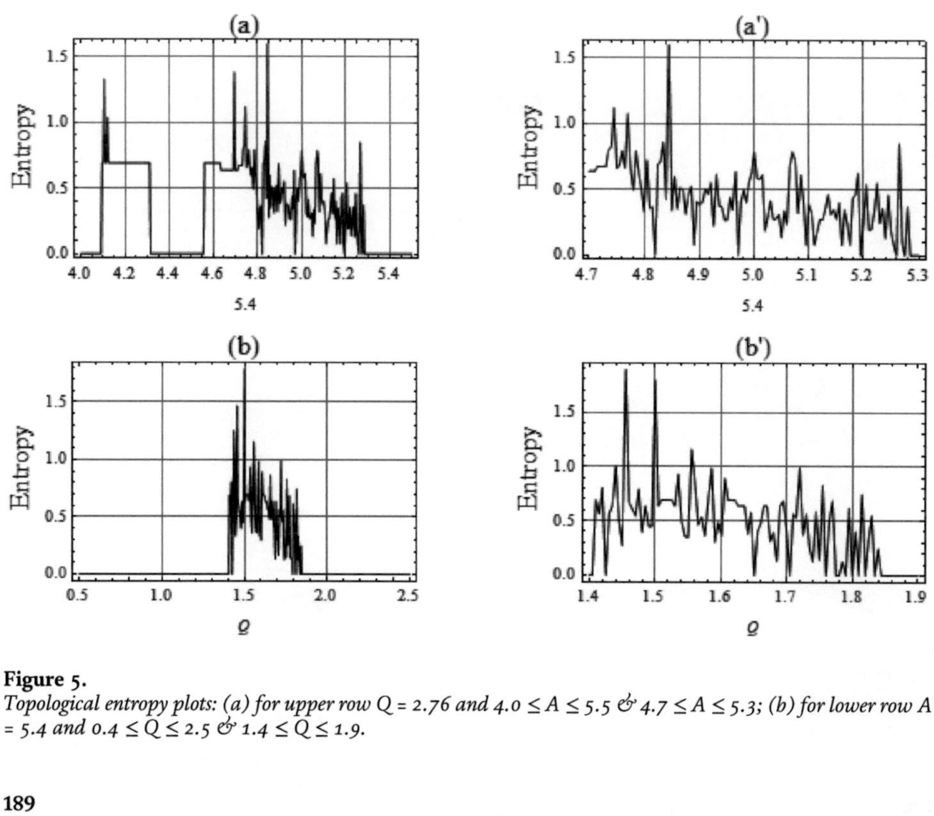

Figure 5.
Topological entropy plots: (a) for upper row Q = 2.76 and 4.0 ≤ A ≤ 5.5 & 4.7 ≤ A ≤ 5.3; (b) for lower row A = 5.4 and 0.4 ≤ Q ≤ 2.5 & 1.4 ≤ Q ≤ 1.9.

Correlation Dimension:

Extending further the numerical study, correlation dimensions of system (1) calculated for a chaotic attractor by using Mathematica codes, [73].

Consider an orbit $O(\mathbf{x}_1) = \{x_1, x_2, x_3, x_4 \dots \dots\}$, of a map $f : U \to U$, where Us an open bounded set in R^n. To compute correlation dimension of $O(x_1)$, for a given positive real number r, we form the correlation integral,

$$C(r) = \lim_{n\to\infty} \frac{1}{n\,(n+1)} \sum_{i\neq j}^{n} H\left(r - \|\mathbf{x_i} - \mathbf{x_j}\|\right) \tag{3}$$

Where,

$$H(x) = \begin{cases} 0, & x < 0 \\ 1, & x \geq 0 \end{cases}$$

is the unit-step function, (Heaviside function). The summation indicates number of pairs of vectors closer to r when$1 \leq i, j \leq n$ and $i \neq j$. $C(r)$ measures the density of pair of distinct vectors $\mathbf{x_i}$ and $\mathbf{x_j}$ that are closer to r.

The correlation dimension D_c of $O(\mathbf{x_1})$ is then defined as

$$D_c = \lim_{r\to 0} \frac{\log C(r)}{\log r} \tag{4}$$

To obtain D_c, $\log C(r)$ is plotted against $\log r$, **Figure 6**, and then we find a straight line fitted to this curve. The intercept of this straight line on y-axis provides the value of the correlation dimension D_C. Correlation dimensions of time series attractors, **Figure 3**, obtained as:

a. For first attractor, Q = 2.76, A = 5.3, a plot of the correlation integral curve is shown in **Figure 6**. Then, the linear fit of the correlation data used in this figure obtained as

$$y = 0.95661x + 0.687605$$

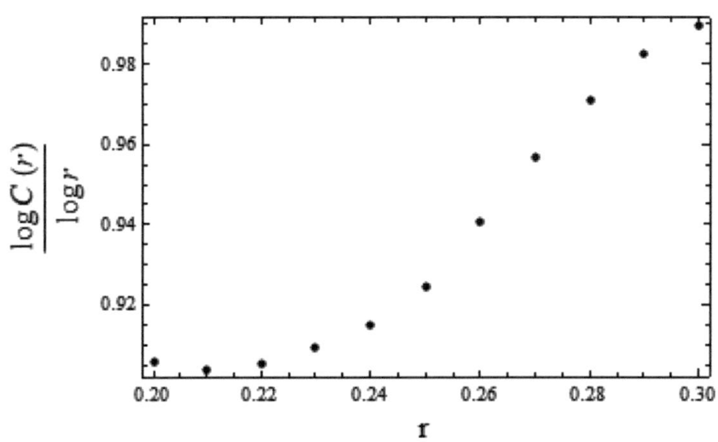

Figure 6.
Plot of correlation integral curve for A = 5.3, Q = 2.76 and x_o = 0.5.

The y-intercept of this straight line is 0.687605. Therefore the correlation dimension of the attractor in this case is $D_C = 0.69$.

b. In a similar way, correlation dimension for second attractor of **Figure 3**, A = 5.4 and Q = 2.9, as $D_c = 0.56$. Plots of correlation dimensions against parameters A, Q shown in **Figure 7**.

2.1.2 Dynamics of biological red cells model

The population of red blood cells in a healthy human being oscillates within a certain tolerance interval in normal circumstances. But, sometimes, in presence of a disease such as anemia, this behavior fluctuate dramatically. A discrete model of blood cell populations, Martelli, ([73], p: 35), presented here.

Let x_n, x_{n+1} representing quantities of cells per unit volume (in millions) at time n and $n + 1$, respectively and p_n, d_n are, respectively, the number of cells produced and destroyed during the nth generation then

$$x_{n+1} = x_n + p_n - d_n \tag{5}$$

Then, assuming that

$$d_n = a\, x_n, a \in [0, 1]$$
$$p_n = b(x_n)^r e^{-s x_n},$$

where b, r, s all positive parameters. With these our one-dimensional discrete model for blood cells populations comes as

$$x_{n+1} = (1 - a)\, x_n + b\, (x_n)^r e^{-s\, x_n} \tag{6}$$

The case $a = 1$, means that during the time interval under consideration all cells that were alive at time n are destroyed. In such a case, above models simply comes as

$$x_{n+1} = b(x_n)^r e^{-s\, x_n} \tag{7}$$

For $a = 0.8, b = 10$, $r = 6$ and $s = 2.5$, three fixed points $x^*{}_0 = 0, x^*{}_1 = 0.989813$, $x^*{}_2 = 3.53665$ obtained for system (6) of which only $x_0^* = 0$ is stable and other two are unstable. Chaotic motion observed for values of parameter $a = 0.8, b = 10, r = 6, s = 2.5$, as shown in the time series plot, **Figure 8**, with initial condition $x_0 = 1.5$.

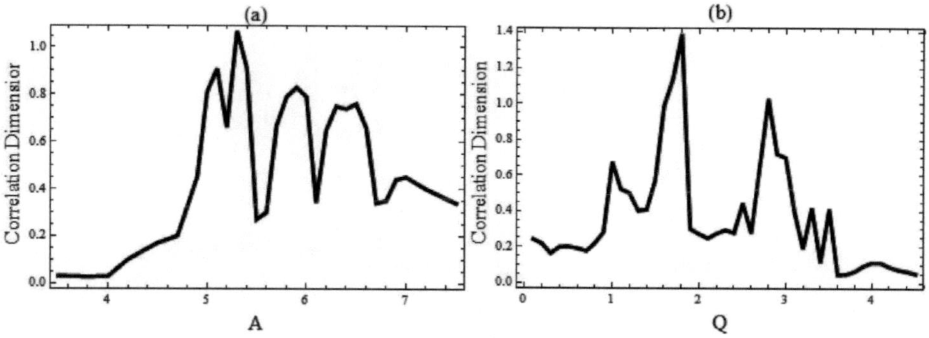

Figure 7.
Plots of correlation dimensions: (a) with Q = 2.76 and varying A, (b) with A = 5.4 and varying Q.

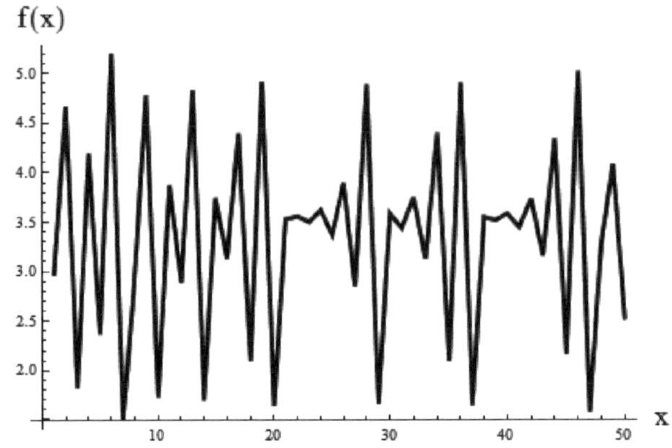

Figure 8.
Chaotic time series plot of map (6) for a = 0.8, b = 10, r = 6, s = 2.5 and $x_0 = 1.5$.

Interesting bifurcations observed for this map: For b = 1.1 × 10^6, r = 8, two bifurcation diagrams are drawn; (a) in one for s =16 and $0 \leq a \leq 1$, and (b) in another for $a = 0.8$ and $3.5 \leq s \leq 16.0$ and shown in **Figure 9**. In former case one finds initially period doubling bifurcation followed by loops before emergence of chaos. In later case, one finds some typical type of bifurcation showing chaos adding, folding and the bistability like phenomena. A magnification of right figure, **Figure 10**, for smaller range, $4.5 \leq s \leq 8.5$, justifying chaos adding behavior.

Regular and chaotic motion experienced through bifurcation diagrams, **Figures 9** and **10**, again confirmed by plots of Lyapunov exponents, **Figure 11**. This system, bears enough complexity and, as its measure, plot of topological entropies, **Figure 12**, obtained for values $r = 6$, $s = 16$ and b = 1.1 × 10^6 and $0 \leq a \leq 1$. Fluctuations in increase of topological entropies appear, approximately, in the region $0.25 \leq a \leq 0.95$ indicate existence of complexity.

The correlation dimension of its chaotic attractor for values $a = 0.78$, when $r = 6$, $s = 16$ and b = 1.1 × 10^6 is obtained as $D_c \cong 0.253$.

2.2 Two-dimensional models

2.2.1 Two-Gene Andrecut-Kauffman System

Chaos and complexity study of a discrete two-dimensional map for two-gene system, proposed by Andrecut and Kaufmann, investigated recently, [35, 71, 92]. The map used to investigate the dynamics of two-gene system for chemical

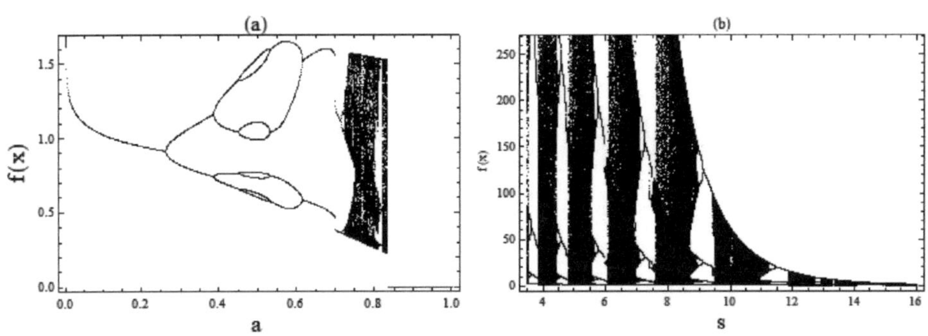

Figure 9.
Bifurcation plots of Blood Cell model for = 8 , b = 1.1 × 10^6 then for (a) s= 16 and $0 \leq a \leq 1$ and for (b) a = 0.1 and $3.5 \leq s \leq 16$.

DOI: http://dx.doi.org/10.5772/intechopen.94295

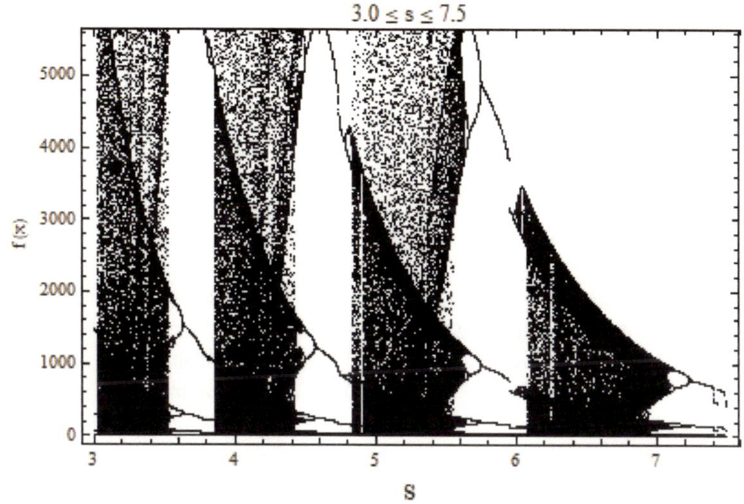

Figure 10.
Bifurcation of Blood Cell model when $3.0 \leq s \leq 7.5$ *and* $a = 0.8,\ r = 8,\ b = 1.1 \times 10^6$.

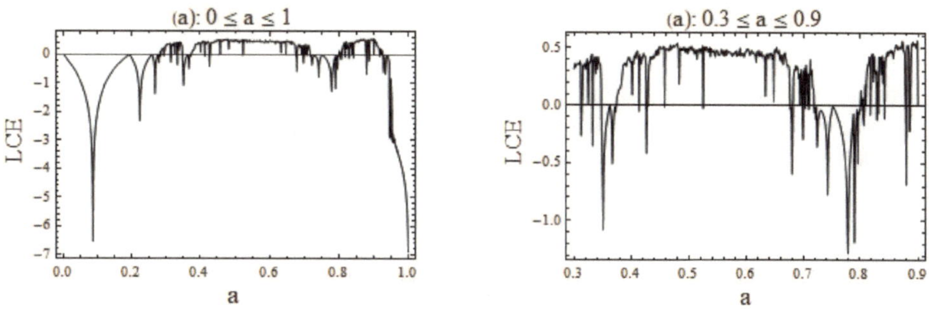

Figure 11.
LCE Plots for $= 6$, $s = 16$ *and* $b = 1.1 \times 10^6$, *negative and positive values of LCEs, respectively, below and above the zero line show the regular and chaotic zones of parameter space.*

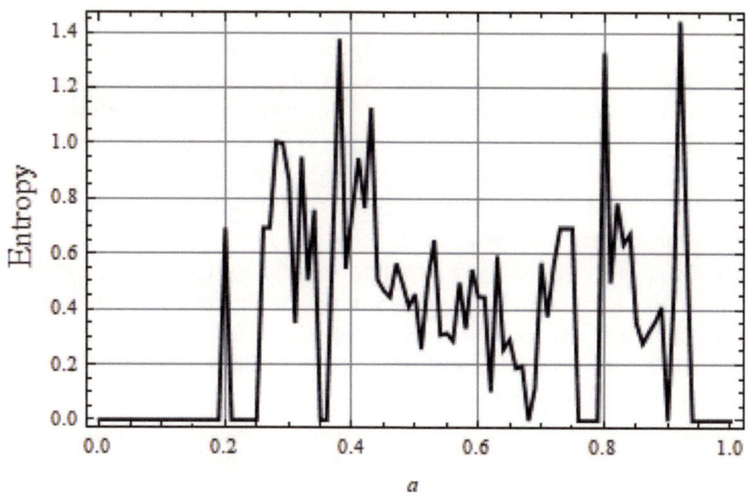

Figure 12.
Topological entropy plot for $r = 6, s = 16$ *and* $b = 1.1 \times 10^6$ *and* $0 \leq a \leq 1$.

reactions corresponding to gene expression and regulation. The discrete dynamic variables x_n and y_n describe the evolutions of the concentration levels of transcription factor proteins. The map represented by following pair of difference equations:

$$x_{n+1} = \frac{a}{1 + (1-b)\, {x^t}_n + b\, {y^t}_n} + c\, x_n$$

$$y_{n+1} = \frac{a}{1 + (1-b)\, {y^t}_n + b\, {x^t}_n} + d y_n \tag{8}$$

With parameter values $a = 25, b = 0.1, c = d = 0.18$ and $t = 3$, one obtains four different fixed points with coordinates (2.30409, 2.30409), (−2.52688, 2.44162), (2.44162, −2.52866), (−2.39464, −2.39464) and all are unstable.

For $c \neq d$ and when $a = 25, b = 0.1, c = 0.18, d = 0.42$, and $t = 3$, again, four unstable fixed points exists as (2.2832, 2.5413), (−2.5458, 2.6566), (2.4613, −2.7288), (−2.3744, −2.61705).Therefore, for all these the cases, orbit with initial point taken nearby any of the fixed points be unstable and may be chaotic also.

We intend to investigate certain dynamic behavior of system (8) for cases when $c = d$ and when $c \neq d$ of evolutions showing irregularities due to presence of chaos and complexity.

Numerical Simulations:

Drawing bifurcation diagrams and calculating Lyapunov exponents, topological entropy and correlation dimensions of the system for different cases have investigated performing numerical simulations. For values of the control parameters following ranges proposed: $a \in [0, 50]$, $c \in [-0.4, 0.4]$, $b = 0.1$, $d = 0.5, t = 3, 4, 5$.

Case 1: Taking $c = d$, bifurcation diagrams are drawn along the directions x and y, by varying c for cases t = 3, 4, 5 and certain fixed values of other parameters as shown in **Figure 13**. Then, plots of attractors have been obtained for parameters $a = 25, b = 0.1, t = 3$ and (i) for regular case $c = d = 0.32$ and (ii) for chaotic case $c = d = 0.18$ and shown in **Figure 14**. In each case when t = 3, 4, 5, bifurcations show period doubling leading to chaos and then to regularity. Also, bistability and folding nature of phenomena are appearing here.

Lyapunov Exponents & Topological Entropies:

For chaotic evolution, when $a = 25, b = 0.1, t = 3, c = d = 0.18$, Lyapunov exponents are obtained shown in **Figure 15**. Numerical investigations further proceeded for calculation of topological entropies. In **Figure 16**, plots of topological entropies are presented for t = 3, 4, 5 and for different ranges of parameter c. Analysis of these plots, gives an impression that for the case t = 3, system shows enough complexity in the range $0.05 \leq c \leq 0.23$. For the case t = 4, the system shows high complexity in the range $0 \leq c \leq 0.22$ and in case t = 5, high complexity appears in $0 \leq c \leq 0.44$.

Case II: When c and d are different, bifurcation diagrams, **Figure 17**, shows clear picture of complex nature of the system.

In **Figure 18**, plots of Lyapunov exponents, (LCE's), for chaotic evolution for different cases discussed above are shown in the upper row and plots of topological entropies are shown in the lower row for these cases. For all the plots, parameters a = 25 and b = 0.1 are common. Here, topological entropy plots are drawn for different ranges of parameter c.

When parameters c and d both were allowed to vary, one gets 3D plots for topological entropies as shown here in **Figure 19**.

Correlation dimensions:

Being one of the characteristic invariants of nonlinear system dynamics, the correlation dimension provides measure of dimensionality for the underlying attractor of the system. A statistical method used to determine correlation dimension. It is

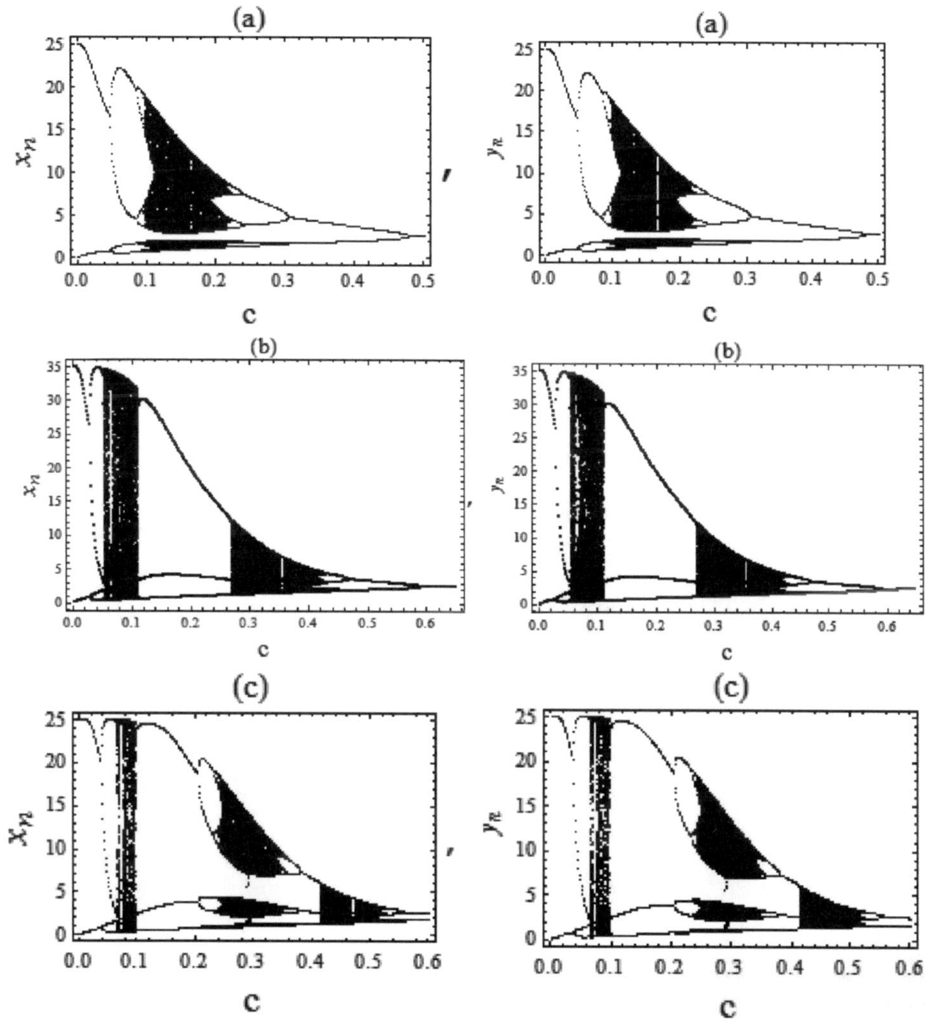

Figure 13.
Three cases of bifurcation scenarios of map (8) for parameters $c = d$: *(a)* $t = 3$, $a = 25, b = 0.1$ *and* $0 \le c \le 0.5$; *(b)* $t = 4$, $a = 35, b = 0.1$ *and* $0 \le c \le 0.65$; *(c)* $t = 5$, $a = 25, b = 0.1$ *and* $0 \le c \le 0.5$.

an efficient and practical method in comparison to others, like box counting etc. The procedure to obtain correlation dimension follows from steps of calculations in [73]:

For case $t = 3$ and $a = 25$, $b = 0.1$, $c = 0.28$, $d = 0.12$, correlation integral data calculated and its plot is obtained, **Figure 20**. The linear fit of correlation integral data obtained as

$$Y = 0.0581323x - 0.580866$$

The y-intercept of this straight line is 0.580866. Therefore the correlation dimension of the attractor in this case is, approximately, $D_c = 0.581$.

Computation of correlation dimension carried out for more cases for different set of values of parameters as shown in **Table 1**.

2.2.2 Complexities in micro-economic Behrens Feichtinger model

Investigation on microeconomic chaotic disturbances and certain measure to control chaos appeared in some recent articles, [72, 93–95], extended here for

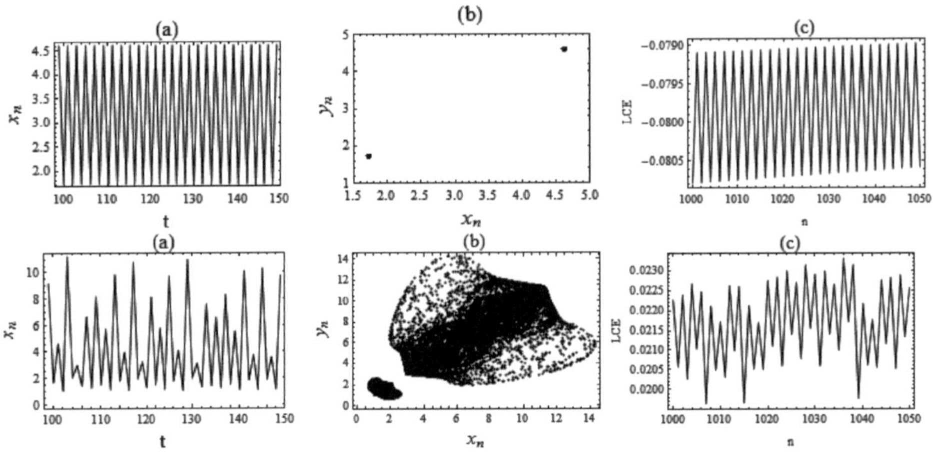

Figure 14.
Figures (a), (b), (c) correspond to time series, phase plane attractors and Lyapunov exponents; upper row is for regular case and the lower row is for chaotic case of map (8). Parameters values are taken as $a = 25, b = 0.1, t = 3$ *and (i) for regular case* $c = d = 0.32$ *and (ii) for chaotic case* $c = d = 0.18$.

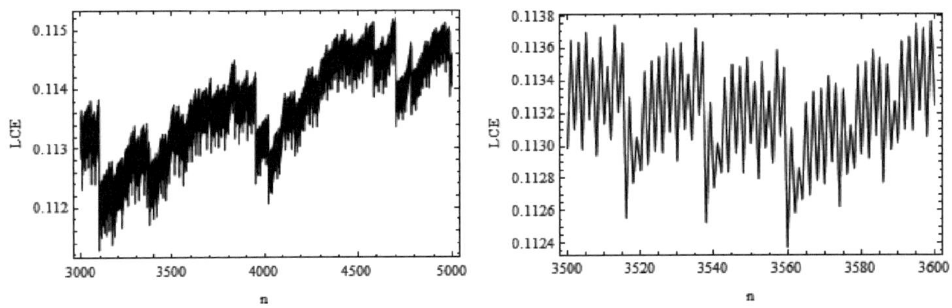

Figure 15.
Plots of Lyapunov exponents for chaotic evolution of map (8). Parameters are $a = 25, b = 0.1, t = 3, c = d = 0.18$ *and when evolving from initial point (2.1, 2.1).*

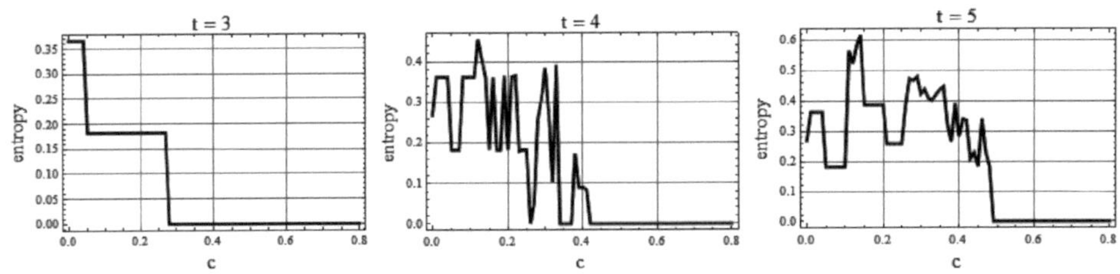

Figure 16.
Plots of topological entropy for map (8) when parameter $c = d$. *From left: (i)* $t = 3$, $a = 25, b = 0.1$ *and* $0 \leq c \leq 0.5$; *(ii)* $t = 4$, $a = 35, b = 0.1$ *and* $0 \leq c \leq 0.65$; *(iii)* $t = 5$, $a = 25, b = 0.1$ *and* $0 \leq c \leq 0.8$.

complexity analysis. The problem proposed as an micro economic model of two firms X and Y competing on the same market of goods having asymmetric strategies. The sales x_n and y_n of both firms are evolving in discrete time steps.

$$x_{n+1} = (1-\alpha)\, x_n + \frac{a}{1 + e^{[-c\,(x_n - y_n)]}}$$

$$y_{n+1} = (1-\beta)\, y_n + \frac{b}{1 + e^{[-c\,(X_n - Y_n)]}} \qquad (9)$$

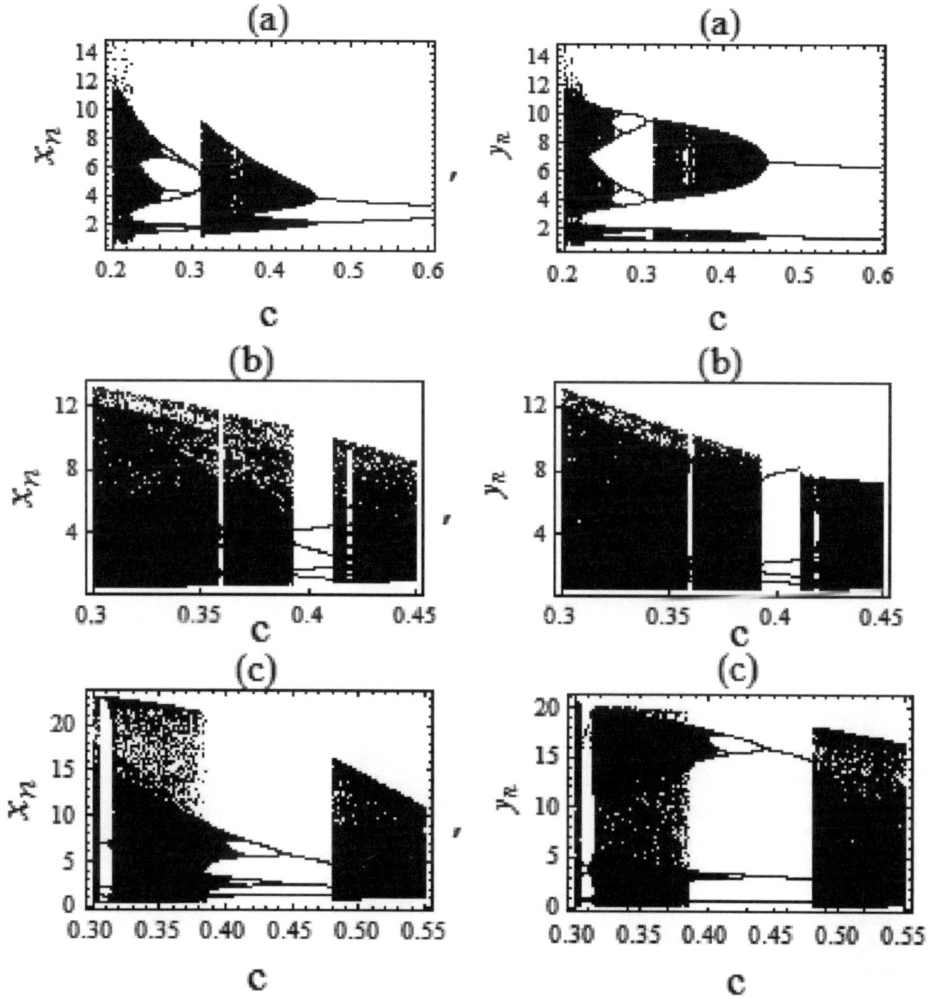

Figure 17.
Bifurcation plots when $c \neq d$ for different ranges of parameter c. Cases (a), (b), (c), corresponds to t = 3, t = 4, t = 5. Parameters are $a = 25, b = 0.1$ and $d = 0.20$ for plots (a) & (c) and d = 0.30 for plot (b).

where α, β $(0 < \alpha, \beta < 1)$ are the time rates at which the sales of both firm decays in the absence of investments. Parameters a, b describe the investment effectiveness of both the firms. Parameter c is an "elasticity" measure of the investment strategies. For parameter values α = 0.46, β = 0.7, a = 0.16, b = 0.9, c = 105, we have observed the chaotic attractor of this model.

Bifurcation Diagram:

Bifurcation diagrams for system (9) obtained for α = 0.46, β = 0.7, a = 0.16, b = 0.9 and by varying parameter c, $8 \leq c \leq 160$ and in close range, $6 \leq c \leq 8$, **Figure 21**. Then, again it obtained for values α = 0.46, β = 0.7, a = 0.16, b = 0.6, c = 110 and $0 \leq a \leq 0.4$, **Figure 22**. Appearance of period doubling followed by chaos visible from these figures.

Attractors:

Time series plots and a plot of chaotic attractor obtained for values a = 0.16, b = 0.9, c = 105, α = 0.46, β = 0.7 of system (9) shown in **Figure 23**. Plots shown in **Figure 24** are of LCEs for the chaotic motion.

Topological Entropies: Topological entropies calculated numerically and plotted. These are shown in **Figure 25**. One finds significant increase topological

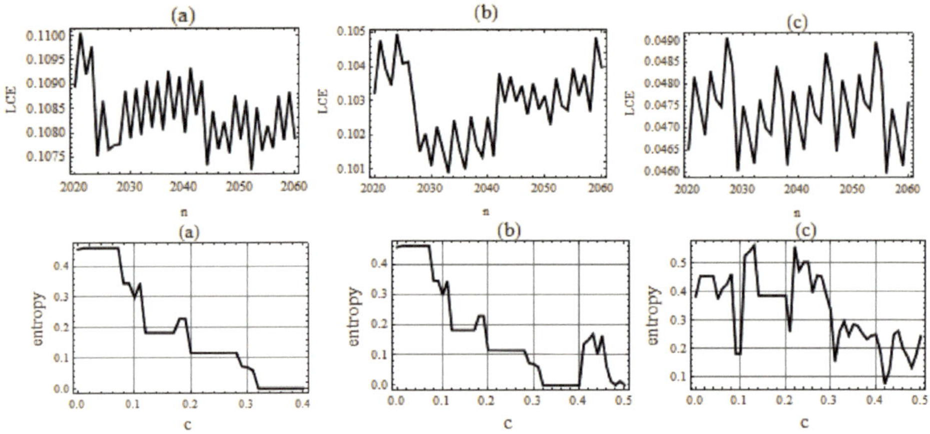

Figure 18.
Upper row plots are for LCE's and lower row plots are for topological entropies. Plots with (a), (b), (c) are respectively corresponds to the cases $t = 3, 4, 5$. *Parameters* $a = 25$, $b = 0.1$ *are common for all the plots. Then, for (b) & (c) LCE's plots,* $c = 0.2$, $d = 0.15$ *and that for plot (c) ,* $c = 0.28$, $d = 0.12$. *For lower row topological entropy plots, except parameter t, parameters* $a = 25$, $b = 0.1$, $d = 0.15$ *are common for all.*

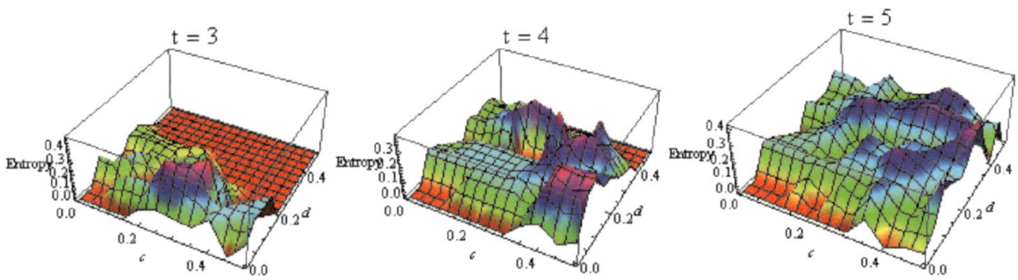

Figure 19.
3D plots for topological entropy variations. Parameters values are taken as $a = 25$, $b = 0.1$ *and then* $0 \le c \le 0.5$ *&* $0 \le d \le 0.5$.

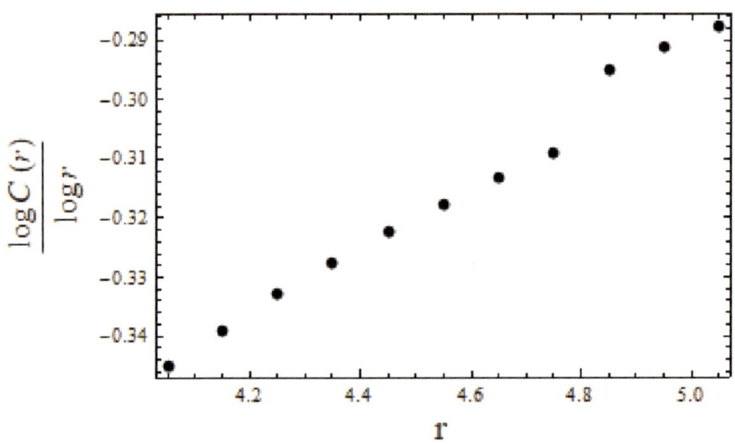

Figure 20.
Plot of correlation integral curve for $t = 3$ *and* $a = 25$, $b = 0.1$, $c = 0.28$, $d = 0.12$.

entropy where the system shows regularity, (e.g., $20 \le c \le 75$), and for values $\alpha = 0.46$, $\beta = 0.7$, $a = 0.16$ and $b = 0.9$. This shows presence of complexities though there is no chaos.

DOI: http://dx.doi.org/10.5772/intechopen.94295

Cases (t)/Parameters	a	b	c	d	Approximate D_c
t = 3	25	0.1	0.28	0.12	0.581
t = 4	25	0.1	0.18	0.18	0.645
t = 5	25	0.1	0.18	0.18	0.703
t = 4	25	0.1	0.28	0.12	0.676
t = 5	25	0.1	0.28	0.12	0.772
t = 3	35	0.1	0.2	0.2	0.877
t = 4	35	0.1	0.2	0.2	0.618
t = 5	35	0.1	0.2	0.2	1.264

Table 1.
Correlation Dimensions for different sets of parameters.

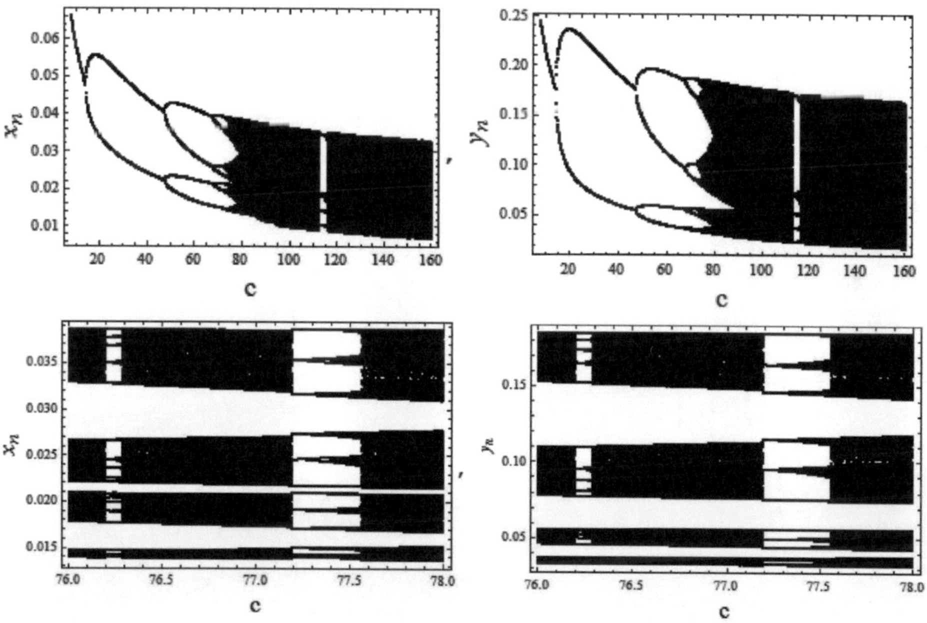

Figure 21.
Bifurcation diagrams of system (9) with respect to coordinates x and y. Lower plots are correspond to bifurcations in close range to indicate the appearance of periodic windows within bifurcation. $\alpha = 0.46$, $\beta = 0.7$, $a = 0.16$, $b = 0.9$, $8 \leq c \leq 160$ & $6 \leq c \leq 8$.

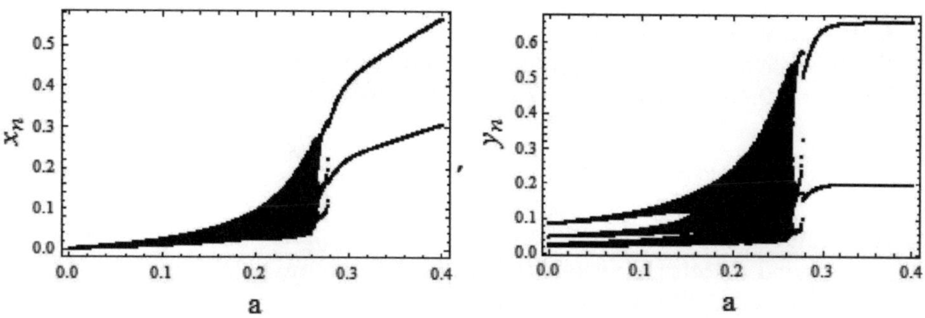

Figure 22.
Bifurcation of map (9) $\alpha = 0.46$, $\beta = 0.7$, $a = 0.16$, $b = 0.6$, $c = 110$ *and* $0 \leq a \leq 0.4$

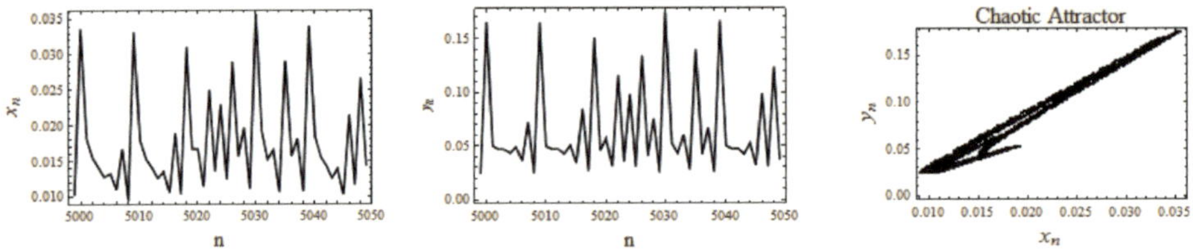

Figure 23.
Time series plots and chaotic attractor of the system (9) for a = 0.16, b = 0.9, c = 105, α = 0.46, β = 0.7 and initial condition (0.1, 0.1).

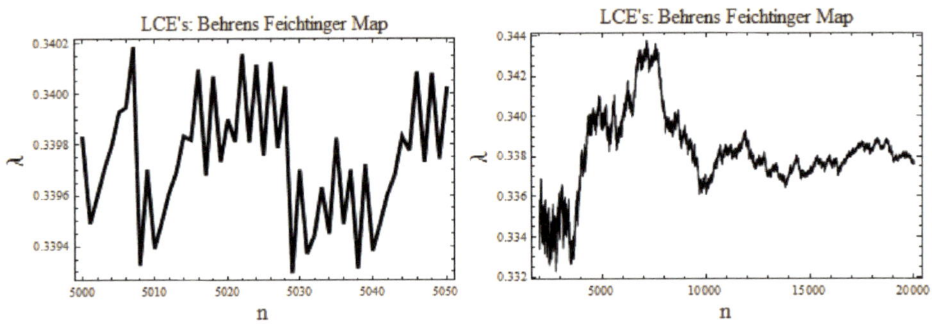

Figure 24.
Plots of Lyapunov exponents for chaotic evolution of the system (9) for a = 0.16, b = 0.9, c = 105, α = 0.46, β = 0.7.

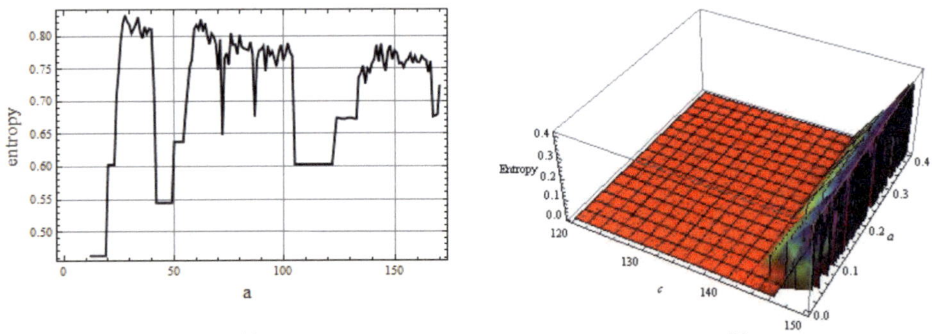

(a) (b)

Figure 25.
Plots of topological entropies: (a) left 2D plot is obtained for $12 \le c \le 170$ and values of a = 0.16, b = 0.9, α = 0.46 and β = 0.7 and (b) right 3D plot is for $120 \le c \le 150$ and $0 \le a \le 0.4$ keeping same values for α and β.

Correlation dimension:

Following steps used for map (8), correlation dimension of chaotic the attractor for values α = 0.46, β = 0.7, a = 0.16, b = 0.9, c = 105, obtained as D_c = 0.064

2.2.3 Continuous Volterra-Petzoldt Model

A continuous 2-dimensional Lotka – Volterra type predator– prey model of constant period chaotic amplitude, (UPCA model), proposed by Petzoldt, [96] based on works, [97, 98], written as

$$\frac{dx}{dt} = a\,x - \alpha_1 \frac{x\,y}{1 + k_1 x}$$

$$\frac{dy}{dt} = -b\,y + \alpha_1 \frac{x\,y}{1 + k_1 x} - \alpha_2 \frac{y\,z}{1 + k_2 y}$$

$$\frac{dz}{dt} = -c(z - w) + \alpha_2 \frac{y\,z}{1 + k_2 y} \qquad (10)$$

Bifurcation diagram for predator z while varying prey parameter b shown there, Petzoldt [86], is interesting. Periodic bifurcations and chaotic attractor of this model for different parameter space are presented in the figure, **Figure 26**.

Plots of time series for x(t), for cases of chaos, are given in **Figure 27** and that of Lyapunov exponents, (LCEs), of chaotic attractors shown in last two plots in **Figure 28**.

In conclusion, one observes that the system (10) evolve into chaos after period doubling phenomena.

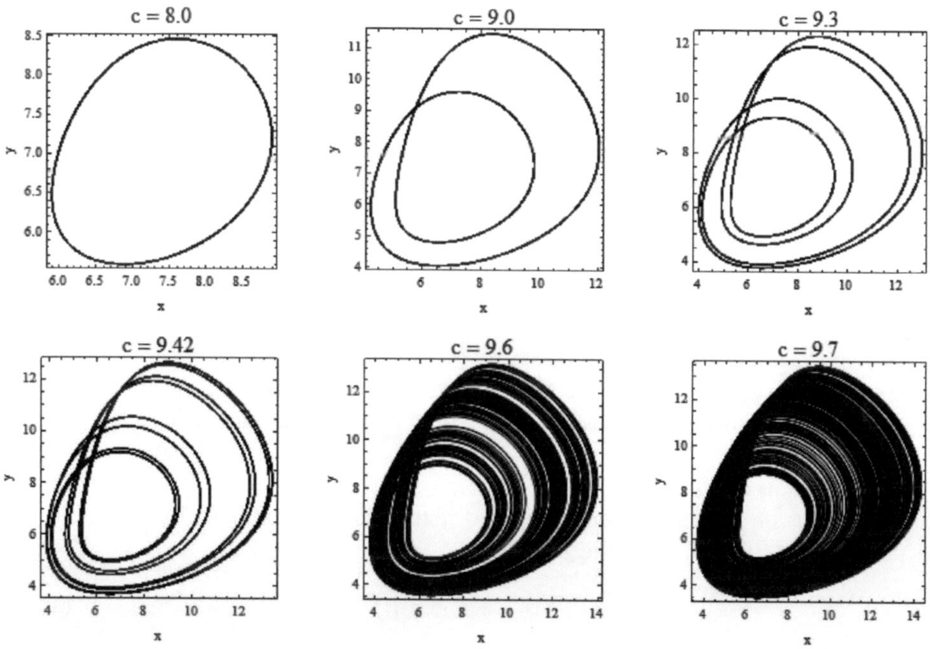

Figure 26.
Periodic bifurcations and chaotic attractor formations of Volterra – Petzoldt model for different values of c fixed parameters $a = 1$, $b = 1$, $\alpha_1 = 0.205$, $\alpha_2 = 1$, $k_1 = 0.05$, $k_2 = 0$, $w = 0.006$.

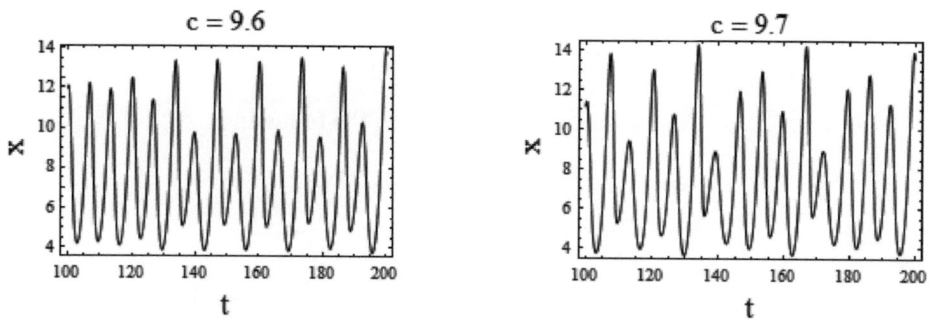

Figure 27.
Plots of time series curves for x(t) for chaotic evolutions for values of c. Other parameters are same as in ***Figure 26****.*

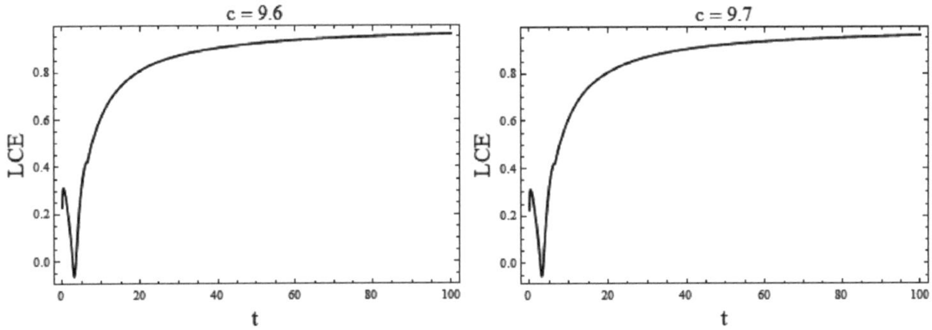

Figure 28.
*Plots of LCEs of chaotic attractors of model (1) for values of c. Other parameters are same as in **Figure 26**.*

3. Chaos control technique

As nonlinear systems are hardly comparable in the sense that behavior of one nonlinear system hardly match with another nonlinear system so the chaotic evolutions. So controlling chaos to bring any chaotic system to regularity may differ from one nonlinear system to another nonlinear system. Different types of controlling chaos technique discussed in recent literatures, [75–88].

Following two chaos controlling technique discussed here:

3.1 Asymptotic Stability Method

Asymptotic stability analysis to stabilize unstable fixed point and to control chaotic motion appeared in some recent researches, [83–85]. Though this method has some limitations, it is perfect way to control chaos in models where it can be applicable.

Description of the Method:

Dynamics of the actual map $\mathbf{X}_{n+1}$ and that of the desired map $\mathbf{Y}_{n+1}$ can be explained by following mapping:

$$\mathbf{X}_{n+1} = \mathbf{F}(\mathbf{x}_n, \mathbf{p}) \tag{11}$$

$$\mathbf{Y}_{n+1} = \mathbf{F}(\mathbf{y}_n, \mathbf{p}*) \tag{12}$$

Also, the neighborhood dynamics of $\mathbf{X}_{n+1}$ and $\mathbf{Y}_{n+1}$ can be represented by the relation:

$$\mathbf{X}_{n+1} = \mathbf{A}_R\mathbf{X}_n + \mathbf{B}_R\,\mathbf{p}$$
$$\mathbf{Y}_{n+1} = \mathbf{A}_D\mathbf{Y}_n + \mathbf{B}_D\,\mathbf{p}*$$

Matrices $\mathbf{A}_R$, $\mathbf{A}_D$, $\mathbf{B}_R$, $\mathbf{B}_D$ can be obtained from the following:

$$\mathbf{A}_R = \mathbf{D}_{Xn}\,\mathbf{F}(\mathbf{X}_n, \mathbf{p}), \mathbf{A}_D = \mathbf{D}_{Yn}\,\mathbf{F}(\mathbf{Y}_n, \mathbf{p}^*)$$
$$\mathbf{B}_R = \mathbf{Dp}\,\mathbf{F}(\mathbf{X}_n, \mathbf{p}), \mathbf{B}_D = \mathbf{Dp}*\mathbf{F}(\mathbf{Y}_n, \mathbf{p}^*)$$

Here,

$$\mathbf{X}_{n+1} = \begin{pmatrix} x_{n+1} \\ y_{n+1} \end{pmatrix} \mathbf{Y}_{n+1} = \begin{pmatrix} x^*_{n+1} \\ y^*_{n+1} \end{pmatrix}$$

Let a, b are two parameters of the system and (x^n, y^n) be any unstable fixed point of above system for given values of a and b. Then, our objective is to obtain two new values for a and b so that this unstable point becomes stable. For this, we need the Jacobian matrices defined by

$$\mathbf{J} = \begin{pmatrix} \frac{\partial \mathbf{f}}{\partial \mathbf{x}} & \frac{\partial \mathbf{f}}{\partial \mathbf{y}} \\ \frac{\partial \mathbf{g}}{\partial \mathbf{x}} & \frac{\partial \mathbf{g}}{\partial \mathbf{y}} \end{pmatrix}, \quad \mathbf{J}^* = \begin{pmatrix} \frac{\partial \mathbf{f}}{\partial \mathbf{a}} & \frac{\partial \mathbf{f}}{\partial \mathbf{b}} \\ \frac{\partial \mathbf{g}}{\partial \mathbf{a}} & \frac{\partial \mathbf{g}}{\partial \mathbf{b}} \end{pmatrix}$$

The control input parameter matrix p*can be given by

$$\mathbf{P}^* = \mathbf{C_R}\,\mathbf{X_n} + \mathbf{C_M}\,\mathbf{p} - \mathbf{C_D}\mathbf{Y_n} \tag{13}$$

Then, using (11)-(13), one obtains the following error equation:

$$\mathbf{e_{n+1}} = (\mathbf{A_R} - \mathbf{B_D}\,\mathbf{C_R})\,\mathbf{e_n} + \{\mathbf{A_R} - \mathbf{A_D} + \mathbf{B_D}(\mathbf{C_D} - \mathbf{C_R})\}\mathbf{Y_n} + (\mathbf{B_R} - \mathbf{B_D}\mathbf{C_M})\mathbf{p} \tag{14}$$

And $\mathbf{e_n} = \mathbf{X_n}\text{-}\mathbf{Y_n}$.

Note that in equation (13) and (14) the coefficient matrices $\mathbf{C_R}$, $\mathbf{C_D}$ and $\mathbf{C_M}$ are to be determined so that if the error vector $\mathbf{e_n} = \mathbf{X_n}\text{-}\mathbf{Y_n}$ is initialized as $\mathbf{e_0} = \mathbf{0}$, then it will be zero for all n future times. For asymptotic stability, we must have $\mathbf{e_n} \to \mathbf{0}$ as $n \to \infty$, then equation (14) implies

$$\mathbf{A_R} - \mathbf{A_D} + \mathbf{B_D}\,(\mathbf{C_D} - \mathbf{C_R}) = \mathbf{0} => \mathbf{B_D}(\mathbf{C_D} - \mathbf{C_R}) = \mathbf{A_D} - \mathbf{A_R} \tag{15}$$

$$\text{And } \mathbf{B_R} - \mathbf{B_D}\,\mathbf{C_M} = \mathbf{0} => \mathbf{B_D}\mathbf{C_M} = \mathbf{B_R} \tag{16}$$

The necessary and sufficient condition for $\mathbf{e_n} \to \mathbf{0}$ as $\mathbf{n} \to \infty$ is

$$\mathbf{A_R} - \mathbf{B_D}\,\mathbf{C_R} = -\mathbf{I} \tag{17}$$

From these, one can obtain matrices $\mathbf{C_M}$, $\mathbf{C_D}$, $\mathbf{C_R}$ and then control parameter matrix $\mathbf{P}^*$ from (13).

A necessary and sufficient condition for the existence of matrices $\mathbf{C_M}$, $\mathbf{C_D}$, $\mathbf{C_R}$, given by:

$$\mathbf{Rank}\,(\mathbf{B_D}) = \mathbf{Rank}\,(\mathbf{B_D}, \mathbf{A_D} - \mathbf{A_R}) = \mathbf{Rank}\,(\mathbf{B_D}, \mathbf{B_R})$$

3.2 Applications

3.2.1 Chaos Control in a 2–Dimensional Prey-Predator map

Considered a prey-predator model where both species evolve with logistic rule and also influencing each other, [30], written as

$$\begin{aligned} x_{n+1} &= a\,x_n(1 - x_n) - b\,x_n\,y_n \\ y_{n+1} &= c\,y_n(1 - y_n) + b\,x_n\,y_n \end{aligned} \tag{18}$$

For $a = 3.7$, $b = 3.5$, $c = 0.2$, one obtains four fixed points obtained as: (0, 0), (0, −4.0), (0.72973, 0) & (0.25712, 0.49961) of which (0.25712, 0.49961) is unstable. So, the orbits originating nearby it would also be unstable and unpredictable & may be chaotic. Nearby this unstable fixed point, we assume a desired initial point as (0.3, 0.5). With this as initial point together with parameters $a = 3.7$, $b = 3.5$,

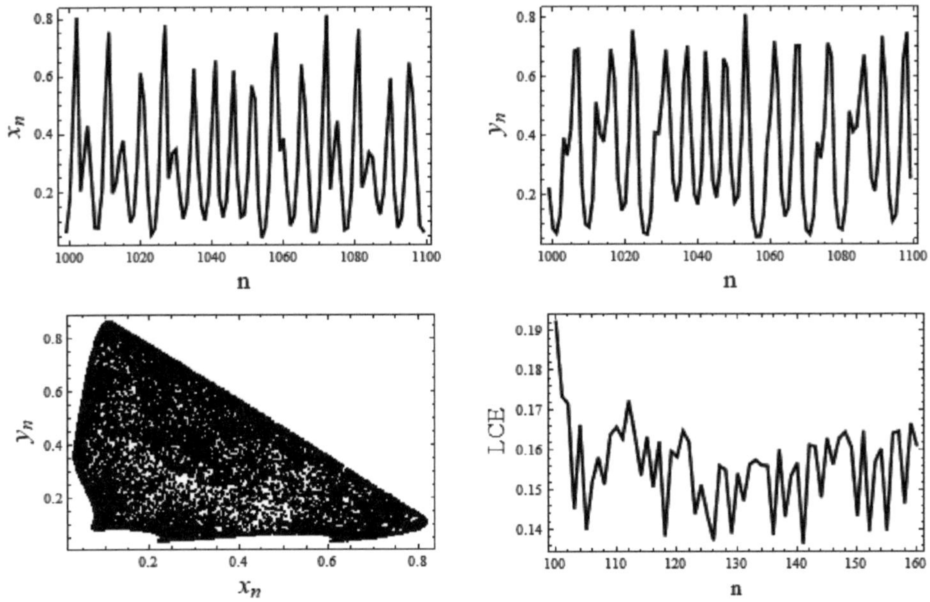

Figure 29.
Time series graphs, attractor and LCE plots of the unstable system.

$c = 0.2$, time series, attractor and LCE plots are obtained and shown by **Figure 29**. Clearly the system (18) is showing chaos at (0.3, 0.5) with $a = 3.7$, $b = 3.5$, $c = 0.2$.

Then, applying asymptotic stability discussed above for the map (18). For fixed value c = 0.2, unstable fixed point obtained as (0.25712, 0.49961). Nearby this point take initial point (0.3, 0.5) and $\boldsymbol{p}^* = \begin{pmatrix} a \\ b \end{pmatrix} = \begin{pmatrix} 3.7 \\ 3.5 \end{pmatrix}$. When above-mentioned method applied, one obtains matrices:

$$A_R = \begin{bmatrix} 0.048652 & -0.899924 \\ 1.74865 & 0.900078 \end{bmatrix} \qquad A_D = \begin{bmatrix} -0.27 & -1.05 \\ 1.75 & 1.05 \end{bmatrix}$$

$$B_R = \begin{bmatrix} 0.19101 & -0.128462 \\ 0 & 0.128462 \end{bmatrix} \qquad B_D = \begin{bmatrix} 0.21 & -0.15 \\ 0 & 0.15 \end{bmatrix}$$

$$C_M = \begin{bmatrix} 0.90957 & 0 \\ 0 & 0.85641 \end{bmatrix} \qquad C_R = \begin{bmatrix} 3.79669 & -4.76117 \\ 11.6577 & -0.66615 \end{bmatrix}$$

$$C_D = \begin{bmatrix} 2.28571 & -4.7619 \\ 11.6667 & 0.333333 \end{bmatrix} \qquad p* = \begin{pmatrix} 3.91525 \\ 2.99538 \end{pmatrix}$$

For the case when $c = 0.2$; new values of a and b; $a = 3.91525$, $b = 2.99538$ along with initial point (0.3, 0.5)a phase plot and a plot of Lyapunov exponents (LEC), are given in **Figure 30**.

3.2.2 Food chain model

Next, we have considered three dimensional food chain model, [23], written as

$$x_{n+1} = a\, x_n(1 - x_n) - b\, x_n y_n$$
$$y_{n+1} = c\, x_n\, y_n - d\, y_n\, z_n$$

$$z_{n+1} = r\, y_n z_n \tag{19}$$

For values $a = 4.1$, $b = 3.7$, $c = 3$, $d = 3.5$, $r = 3.8$ five fixed points exist for system (19) given by: $P_0(0, 0, 0)$, $P_1(0, 0.2632, 0.2857)$, $P_2(0.518614, 0.263158, 0.158812)$, $P_3(0.7561, 0, 0)$ and $P_4(0.3333, 0.4685, 0)$. Then, by stability analysis it has obtained that the fixed points $P_2(0.518614, 0.263158, 0.158812)$ and $P_4(0.3333, 0.4685, 0)$ are unstable. Then, taking nearby P_2, a desired initial point P^* (0.5, 0.3, 0.2), chaotic attractors drawn, **Figure 31**.

In the process of stabilizing the desired point (0.5, 0.3, 0.2), calculations performed to replace parameters $a = 4.1$, $d = 3.5$ and $r = 3.8$ to earlier case of map (18). After obtaining all concerned matrices, replacement matrix obtained as

$$p^* = \begin{pmatrix} a \\ d \\ r \end{pmatrix} = \begin{pmatrix} 4.1035 \\ 1.05194 \\ 1.02707 \end{pmatrix}$$

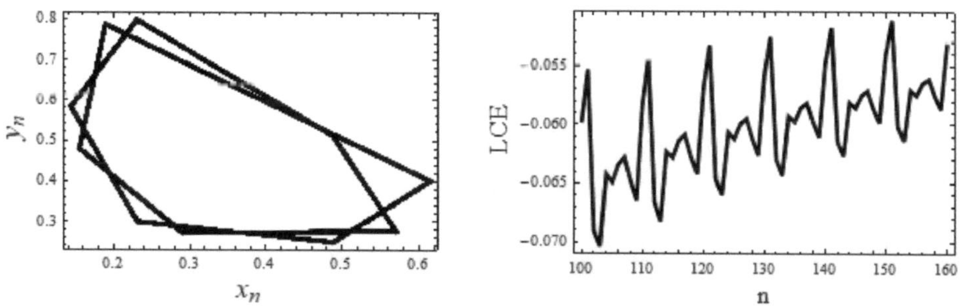

Figure 30.
Phase plot and LCE plot of controlled system when c = 0.2, a = 3.91525, b = 2.99538.

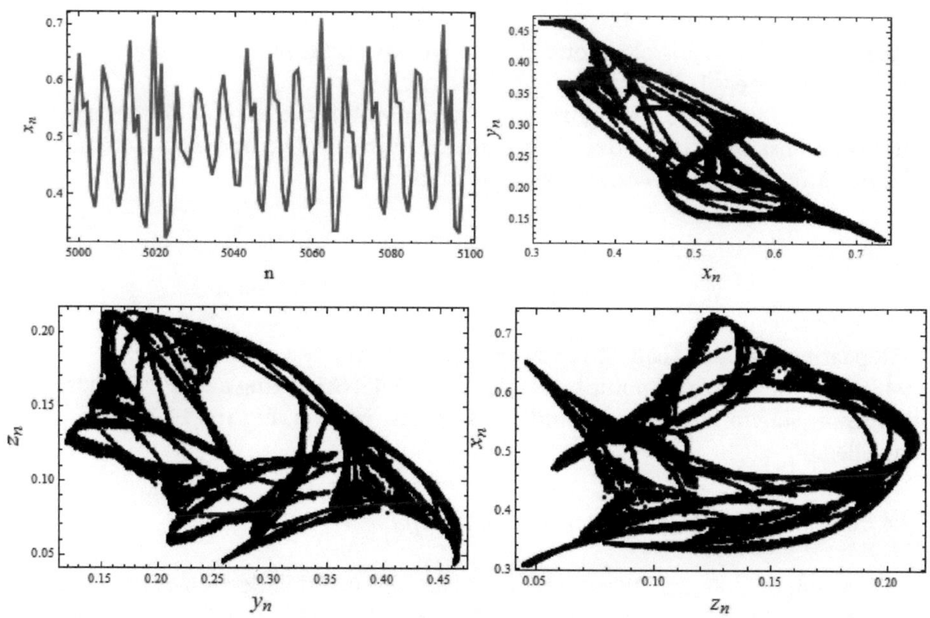

Figure 31.
Time series and attractors of unstable system.

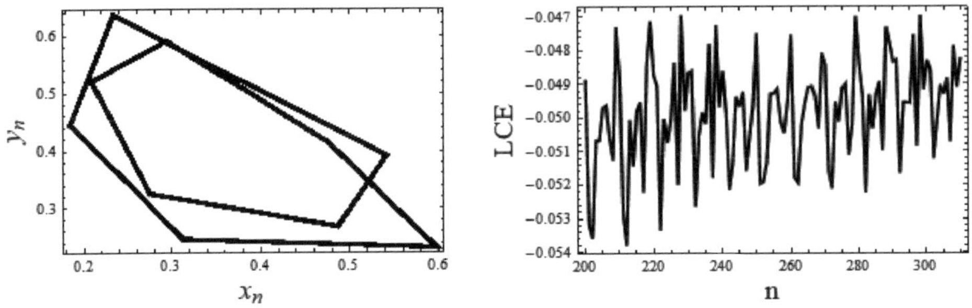

Figure 32.
Phase plot and LCE plot of map (19) showing regular motion and chaos is controlled.

At these new parameter values of a, d and r, the phase plot and the plot of Lyapunov exponents of map (19) obtained, **Figure 32**. These show chaotic motion controlled and the system returns to regularity.

3.2.2.1 Pulsive Feedback Technique to Chaos Control

Pulsive chaos control technique is discussed in detail in recent articles, [86–88]. As an application of this technique let us consider a simple 2 – dimension discrete time Burger's map

3.2.3 Controlling Chaos in 2-D Burger's Map

$$\begin{aligned} x_{n+1} &= (1-a)\,x_n - y_n^2 \\ y_{n+1} &= (1+b)\,y_n + x_n\,y_n \end{aligned} \tag{20}$$

where a and b are non-zero parameters . This map evolve chaotically when a= 0.9, b=0.856. To control chaotic motion we have used pulsive feedback control technique, Litak et al. [86] by

Here (−0.9, 0.948683) is an unstable fixed point of the original Burger's map. It has been observed that above chaotic motion is controlled and display regular behavior after re-writing equations (1) as follows:

$$\begin{aligned} x_{n+1} &= (1-a)\,x_n - y_n^2 + \in (x+0.9) \\ y_{n+1} &= (1+b)\,y_n + x_n\,y_n + \in (y-0.948683) \end{aligned} \tag{21}$$

Repeating stability analysis for system (2) with the fixed point (−0.9, 0.948683), one finds this point be stable if $\varepsilon < 0.45$. So, taking $\varepsilon = 0.435$, phase plot obtained as shown in **Figure 34**, indicates chaotic motion, **Figure 33**, is now controlled.

3.2.4 Controlling Chaos in Volterra-Petzoldt Map

Evolution of Volterra-Petzoldt map already discussed in Section 2, Eq. (10). For parameters $a = 1$, $b = 1$, $c = 9.7$, $\alpha_1 = 0.205$, $\alpha_2 = 1$, $k_1 = 0.05$, $k_2 = 0$, $w = 0.006$, this map shows chaotic motion. An unstable equilibrium solution P* (19.5374, 9.64328, 1.02602) exists in this case.

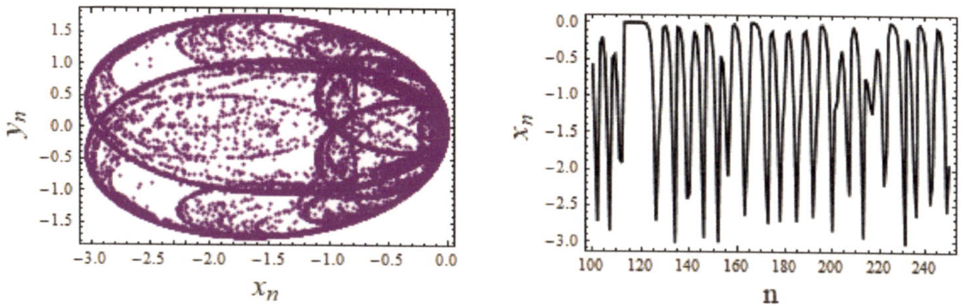

Figure 33.
Chaos in Burgerger's map for a = 1, b = 0.9.

Applying the method of pulsive feedback, and re-writing eq. (10) as

$$\frac{dx}{dt} = a\,x - \alpha_1 \frac{x\,y}{1 + k_1 x} + \in (x - 19.5374)$$
$$\frac{dy}{dt} = -b\,y + \alpha_1 \frac{x\,y}{1 + k_1 x} - \alpha_2 \frac{y\,z}{1 + k_2 y} + \in (y - 9.64328)$$
$$\frac{dz}{dt} = -c(z - w) + \alpha_2 \frac{y\,z}{1 + k_2 y} + \in (z - 1.02602) \tag{22}$$

Then, using stability analysis, for stabilize the above unstable point P*, one obtains the parameter $\varepsilon = -0.45$.

4. Discussions

Regular and chaotic evolutions observed in some 1-3 dimensional discrete and continuous nonlinear models, which have applications in different areas of science. Presence of complexity in these systems viewed by indications of significant increase in topological entropies in certain parameter spaces. More increase in topological entropy in a system signified the system is more complex. Bifurcation phenomena for different systems show interesting properties like bistability, folding, intermittency, chaos adding etc. which are not common to all nonlinear systems. Proper numerical simulations performed for each system to obtain regular and chaotic attractors, Lyapunov exponents (LCEs) as a measure of chaos, (evolution is regular if LCE < 0 and chaotic if LCE > 0), topological entropies and correlation dimensions for chaotic attractors. It appears from the plots of topological entropies that obtained for discrete models that complexity exists even in absence of chaos. Correlation dimensions obtained for chaotic attractors are non-integers because these attractors bear fractal properties. A chaotic attractor is composed of complex pattern and so, in a variety of nonlinear evolving systems measurement of topological entropy is equally important, [63–67].

To control chaotic motion, techniques of asymptotic stability analysis and that of pulsive feedback control applied here. Pulsive control technique applied to Volterra-Petzoldt map (10) and to Burger's map (20), show chaos successfully controlled and systems returned to regularity, **Figures 34** and **35**. Application of Pulsive control method perfectly controlled chaotic motions in systems (10), (20) shown here. Chaos is also controlled by this method for system (10), [72]. Asymptotic stability analysis method applied to a prey-predator system and to a food chain model, respectively, to maps (18) and (19), and chaos effectively controlled shown,

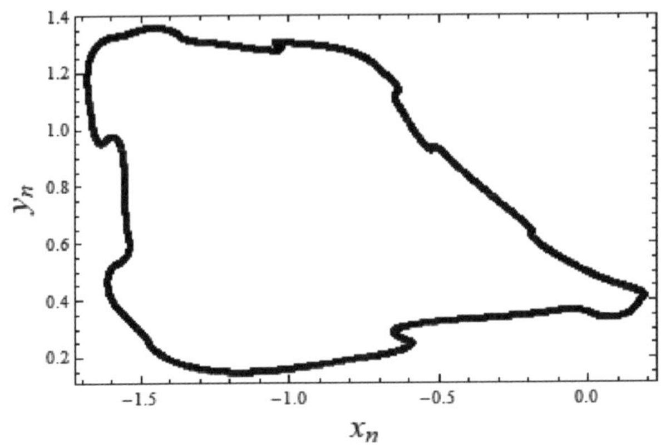

Figure 34.
Plot of regular attractor for a = 1, b = 0.9 and ε = 0.435.

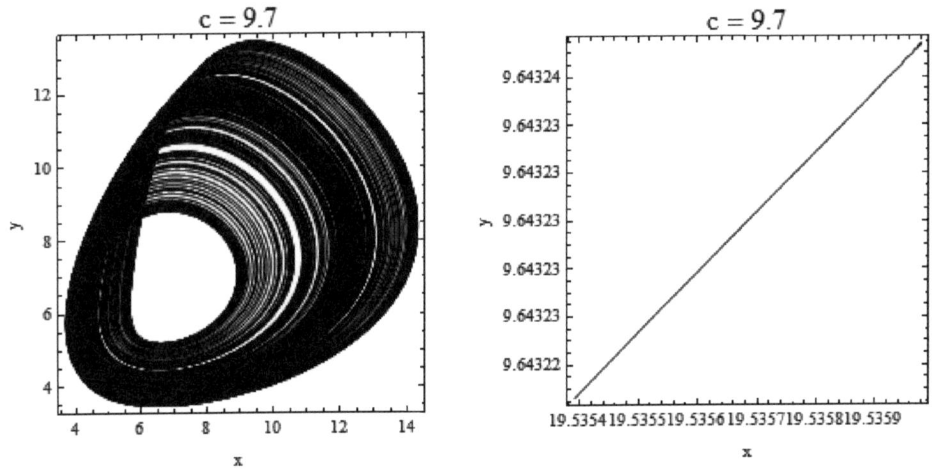

Figure 35.
Plots of chaotic attractor changing into regular attractor by application of pulsive feedback technique.

respectively, through figures, **Figures 30** and **32**. Asymptotic stability analysis technique has some limitations explained in the articles where this method proposed, [83, 84]. Though there are many ways to control chaos in dynamical systems, [74], both the techniques applied here are perfect and very effective in controlling chaos, especially in real systems.

Acknowledgements

The author wishes to present his sincere gratitude to Professor M.K. Das of Institute of Informatics & Communication, University of Delhi South Campus, for his all support and help in preparation of this article.

DOI: http://dx.doi.org/10.5772/intechopen.94295

Author details

Lal Mohan Saha
Department of Mathematics, Shiv Nadar University, Gautam Budha Nagar, India

*Address all correspondence to: lmsaha.msf@gmail.com

References

[1] Poincaré, H. (1957): Les Methodes Nouvelles de la Mecanique Celeste. Dover, New York (1957) Paris, 1899; reprint.

[2] Lorenz EN. Deterministic non-periodic flow. Journal of the Atmospheric Sciences. **1963**;**20**(2): 130-141

[3] Sharkovskii, A. N. (1964).: Co-existence of cycles of a continuous mapping of the line into itself. *Ukrainian Math. J.* **16**: 61–71.

[4] Smale, S.: Differentiable dynamical systems. Bull. Amer. Math. with Soc., 1967; 73 (6), 747–817.

[5] Hènon M. Numerical study of quadratic area-preserving maoings. Quart. J. Math. 1969;27:291-311

[6] Ruelle D, Takens F. On the nature of turbulence. Commun. Math. Phys. 1971; **20**:167-192

[7] Guckenheimer, J., Holmes, P. (1971): Nonlinear Oscillations Dynamical Systems, and Bifurcations of Vector Field. Applied Mathematical Sciences, Book Series, Springer.

[8] May RM. Stability and Complexity in Model Ecosystem. Princeton N. J: Princeton University Press; 1974

[9] Li T-Y, Yorke JA. Period Three Implies Chaos. *The American Mathematical Monthly*. 1975. DOI: https://doi.org/10.2307/2318254

[10] May RM. Simple mathematical models with very complicated dynamics. Nature. 1976;**261**:459-467

[11] Rössler OE. An equation for hyperchaos. Physics Letters A. 1979;**71**: 155-157

[12] K. Ikeda, H. Daido and O. Akimoto: Chaotic behavior of transmitted light from a ring cavity. Phys. Rev. Lett., 1980; 45 (9), 709 – 712.

[13] Feigenbaum MJ. Universal behavior in nonlinear systems. Physica. 1983;**7D**: 16-39

[14] Gleick, J. (1987). "Chaos: Making a New Science".

[15] F. C. Moon (1987): Chaotic Vibrations., John Wiley & Sons New York 1987

[16] Devaney RL. An Introduction to Chaotic Dynamical System. Reading: Addison-Wesley; 1989

[17] Ueda Y. Randomly transitional phenomena in the system governed by Duffing's equation. *J. Stat. Phys.* 1979;**20**: 181-196

[18] Stewart I. Does God Play Dice? Penguin Books; 1989

[19] Tanaka Y and Saha LM. (2012): Nonlinear Behaviors of Pulsating Stars with Convective Zones. PASJ**:** Publ. Astron. Soc. Japan 2012;**64**, L8-1- 4.

[20] Lotka AJ. *Elements of Physical Biology*. Baltimore MD: Williams and Wilkins; 1925

[21] Volterra, V. (1931): *.Lecons sur la Thorie Mathmatique de la Lutte pour la Vie*, Gauthiers-Viallars, Paris.

[22] Allee WC. Animal aggregations. *The Quarterly Review of Biology*. 1927;**2**: 367-398

[23] Allee WC, Bowen E. Studies in animal aggregations: mass protection against colloidal silver among goldfishes. Journal of Experimental Zoology. 1932;**61**(2):185-207

[24] Elsadany AA. Dynamical complexities in a discrete-time food chain. *Computational Ecology and Software*. 2012;**2**(2):124-139

[25] Smith M. Mathematical Ideas in Biology. Cambridge: Cambridge University Press; 1968

[26] Freedman HI. Deterministic Mathematical Models in Population Ecology. Marcel Dekker; 1980

[27] J.R. Beddington, C.A. Free and J.H. Lawton. Dynamic complexity in predator–prey models framed in difference equations, Nature. 1975;**255**: 58-60.

[28] Abrams PA, Ginzburg LR. The nature of predation: prey dependent, ratio dependent or neither? Trends in Ecology & Evolution. 2000;**15**(8): 337-341

[29] Grafton, R.Q., Silva-Echenique J. (1994). Predator–Prey Models: Implications for Management. *Atlantic Canada Economics Association Papers 23*, pp. 61–71.

[30] Kaitala V, Heino M. Complex non-unique dynamics in ecological interactions. Proc R Soc London B. 1996; **263**:1011-1015

[31] Quentin Grafton R, Silva-Echenique J. How to manage nature? Strategies, predator-prey models, and chaos. Marine Resource Economics. 1997; **12**(2):127-143

[32] Yakubu A-A. Prey dominance in discrete predator-prey systems with a prey refuge. Mathematical Biosciences. 1997;**144**:155-178

[33] Xiao Y, Cheng D, Tang S. Dynamic complexities in predator–prey ecosystem models with age-structure for predator. *Chaos, Solitons and Fractals*. 2002;**14**:1403-1411

[34] Liu X, Xiao D. Complex dynamic behaviors of a discrete-time predator-prey system. Chaos, Solitons & Fractals. 2007;**32**:80-94

[35] Andrecut, M. and Kauffman, S. A. Chaos in a Discrete Model of a Two-Gene System. Physics Letters A. 2007; 367, 281-287.

[36] Canan C¸ elik, Oktay Duman. Allee effect in a discrete-time predator–prey system. Chaos, Solitons & Fractals, 2009; 40, 1956–1962.

[37] Hadeler KP, Freedman HI. Predator-prey populations with parasitic infection. Journal of Mathematical Biology. 1989;**27**(6):609-631

[38] Danca M, Codreanu S, Bako B. Detailed analysis of a nonlinear predator-prey model. Journal of Biological Physics. 1997;**23**(1):11-20

[39] Wan-Xiong, Yan-Bo-Zhang and Chang-zhong Liu. Analysis of a discrete-time predator-prey system with Allee effect. Ecological Complexity, 2011, 8: 81 – 85.

[40] Zhao M, Yunfei D. Stability of a discrete-time predator-prey system with Allee effect. Nonlinear Analysis and Differential Equations. 2016;**4**(5):225-233

[41] Xian-wei X-l F, Jing Z-j. Dynamics in a discrete-time predator-prey system with Allee effect. Acta Mathematicae Applicatae Sinica. 2013;**20**:143-164

[42] Stephens PA, Sutherland WJ, Freckleton RP. What is the Allee effect? Oikos. 1999;**87**:185-190

[43] Tang S, Chen LA discrete predator–prey system with age-structure for predator and natural barriers for prey. *Mat. Model. Numer. Anal.* 2001;**35**: 675-690

[44] Tang S, Chen L. Density-dependent birth rate, birth pulses and their population dynamic consequences. J. Math. Biol. 2001;**44**(2):185-199

[45] W. Weaver,(1948): "Science and complexity," American Scientist, vol. 36, no. 4, p. 536.

[46] Gribbin, J (2004) Deep Simplicity: Chaos, Complexity and the Emergence of Life. Penguin Press Science.

[47] Simon HA. The architecture of complexity. Proceedings of the American Philosophical Society. 1962; **106**(6):467-482

[48] Gribble, S. 1995, Topological Entropy as a Practical Tool for Identification and Characterization of Chaotic System. Physics 449 Thesis.

[49] Iwai K. Continuity of topological entropy of one dimensional map with degenerate critical points. *J. Math. Sci. Univ. Tokyo*. 1998;**5**:19-40

[50] Hefferman DM. Multistability, intermittency and remerging Feigenbaum trees in an externally pumped ring cavity laser system. *Phys. Lett.* **A**. 1985;**108**:413-422

[51] Walby S. Complexity theory, systems theory, and multiple intersecting social inequalities. Philosophy of the Social Sciences. 2007; **37**(4):449-470

[52] Benettin G, Galgani L, Giorgilli A, Strelcyn JM. Lyapunov Characteristic Exponents for smooth dynamical systems and for Hamiltonian systems; a method for computing all of them. Part 1 & 2: Theory. *Meccanica*. 1980;**15**:9-30

[53] Katok A. Lyapunov exponents, entropy and periodic orbits for diffeomorphisms. *Publ. Math. IHES*. 1980;**51**:137-174

[54] P. Grassberger and Itamar Procaccia. "Measuring the Strangeness of Strange Attractors". Physica D: Nonlinear Phenomena, 1983;9 (1–2): 189–208.

[55] Grassberger P, Procaccia I. Characterization of Strange Attractors. *Physical Review Letters*. 1983;**50**(5): 346-349

[56] Bryant P, Brown R, Abarbanel H. Lyapunov exponents from observed time series. Physical Review Letters. 1990;**65**(13):1523-1526

[57] Brown R, Bryant P, Abarbanel H. Computing the Lyapunov spectrum of a dynamical system from an observed time series. *Phys. Rev. A*. 1991;**43**: 2787-2806

[58] Abarbanel HDI, Brown R, Kennel MB. Local Lyapunov exponents computed from observed data. *Journal of Nonlinear Science*. 1992;**2**(3):343-365

[59] Skokos C. The Lyapunov characteristic exponents and their computation. *Lect. Notes. Phys.* 2009; **790**:63-135

[60] Syta A, Litak G, Budhraja M, Saha LM. Detection of the chaotic behavior of a bouncing ball by 0 – 1 test. **Chaos, Solitons & Fractals**. 2009;**42**: 1511-1517

[61] Adler RL, Konheim AG, McAndrew MH. Topological entropy. Trans. Amer. Math. Soc. 1965;**114**:309-319

[62] Nagashima H, Baba Y. Introduction to Chaos: Physics and Mathematics of Chaotic Phenomena. Overseas Press India Private Limited; 2005

[63] Bowen R. Topological entropy for noncompact sets. Trans. Amer. Math. Soc.1973;184:125-136

[64] Holmes P. 'Strange' phenomena in dynamical systems and their physical

implications. *App. Math. Modelling*. 1977;**7**(1):362-366

[65] P. Holmes (1979) A nonlinear oscillator with a strange attractor, *Phil. Trans. Roy. Soc.Lond.***A 292**(1394): 419 – 448.

[66] Balmforth NJ, Spiegel EA, Tresser C. Topological entropy of one dimenonal maps: approximations and bounds. *Phys. Rev. Lett.* 1994;**72**:80-83

[67] Stewart L, Edward ES. Calculating topological entropy. J. Stat. Phys. 1997; **89**:1017-1033

[68] Yuasa M, Saha LM. Indicators of chaos. Science and Technology, Kinki University. Japan. No. 2008;**20**:1-12

[69] Saha LM, Prasad S and Yuasa M Measuring Chaos: Topological Entropy and Correlation Dimension in Discrete Maps, Science and Technology, Vol. 24, 2012, pg. 10 – 23.

[70] DeCoster GP, Mitchell DW. The efficacy of the correlation dimension technique in detecting determinism in small samples. *Journal of Statistical Computation and Simulation*. 1991;**39**: 221-229

[71] Saha LM, Sharma R. Dynamics of Two-Gene Andrecut-Kauffman System: Chaos and Complexity. Accepted. **Italian Journal of Pure and Applied Mathematica (IJPAM)**. 2018;**41**:405-413

[72] Saha LM, Das MK. Complexities in Micro-Economic Behrens Feichtinger Model. **Indian Journal of Industrial and Applied Mathematics**. 2016;**7**(2): 127-135

[73] M. Martelli (1999) *Introduction to Discrete Dynamical Systems and Chaos*, Wiley- Interscience

[74] Guanrong Chen and Xiaoning Dong (1998): From Chaos to Order: Methodologies, Perspectives and Applications. World Scientific, Singapore, New Jersey, London, Hong Kong.

[75] Auerbach D, Grebogi C, Ott E, Yorke JA. Controlling chaos in high dimensional systems. Phys. Rev. Lett. 1992;**69**:3479-3482

[76] Erjaee GH, Atabakzade MH, Saha LM. Interesting synchronization-like behavior. **Int. Jour. Bifur.. Chaos**. 2004;**14**(**4**):1447-1453

[77] Carroll TL, Pecora LM. Cascading synchronized chaotic systems. Physica D. 1993;**67**:126-140

[78] Chen G. Optimal control of chaotic systems. Int'l J. of Bifur. Chaos. 1994;**4**: 461-463

[79] Ott E, Grebogi C, Yorke JA. Controlling chaos. Phys. Rev. Lett. 1990; **64**:1196-1199

[80] Pan S, Yin F. Using chaos to control chaotic systems. Phys. Lett. A. 1997;**231**: 173

[81] Pyragas K. Continuous control of chaos by self-controlling feedback. Phys. Lett. A. 1992;**170**:421-428

[82] Shinbrot T, Grebogi C, Ott E, Yorke JA. Using small perturbations to control chaos. Nature. 1993;**363**:411-417

[83] Erjaee GH. On the asymptotic stability of a dynamical system. *IJST, Transaction A*. 2002;**26**(A1):131-135

[84] Saha LM, Erjaee GH and Budhraja M. Controlling chaos in 2-dimensional systems, **Iranian Jour. Sci. Tech.**, Trans. A, 2004;**28**, No.A2, 219 – 226.

[85] Saha LM, Das MK, Bhardwaj R. Asymptotic stability analysis applied to price dynamics. **Ind. J, Industrial and Appl, Math. (IJIAM)**. 2018;**9**(2): 186-195

[86] Litak G, Ali M, Saha LM. Pulsating feedback control for stabilizing unstable periodic orbits in a nonlinear oscillator with a non-symmetric potential. **Int. J. Bifur. Chaos**. 2007;**17**:2797-2803

[87] Litak G, Borowiec M, Ali M, Saha LM, Friswell MI. Pulsive feedback control of a quarter car forced by a road profile. **Chaos Soliton and Fractals**. 2007;**33**:1672-1676

[88] G. Litak, L. M. Saha and M. Ali (2010): Continuous and Pulsive Feedback Control of Chaos, **Recent Progress in Controlling Chaos**, by Miguel A. F. Sanjuan and Celso Grebogi, World Scientific (eBooks), p. 337 – 369.

[89] O'Cairbre F, O'Farrell AG, O'Reilly A. Bistability, bifurcation and chaos in a laser system. *Int. Jour. Bifurcation and Chaos*. 1995;**5**(4):1021-1031

[90] Bonifacio R, Lugiato LA. Bistable absorption in a ring cavity, Lett.l. Nuovo Cimento. 1978;**21**(15):505-510

[91] Benefacio R, Lugato LA. Theory of optical bistability. In: *Dissipative Systems in Quantum Optics*. Ed. R. Benefacio: Springer-Verlag; 1982. pp. 61-92

[92] de Souza SLT, Lima AA, Caldas IL, Medrano-T RO, Guimarães-Filho ZO. Self-similarities of periodic structures for a discrete model of a two-gene system. Physics Letter A. 2012;**376**:1290-1294

[93] Holyst JA, Hagel T, Haag G, Weidlich W. How to control chaotic economy? J. Evol. Econ. 1996;**6**:31-42

[94] Behrens DA. Feichtinger G., Prskawetz A. Complexity dynamics and control of arms race. European Journal of Operational Research, 1997;100: 192-215

[95] Perc M. Microeconomic uncertainties cooperative allince and social welfare. Economics Letters. 2007;**95**:104-109

[96] Thomas Petzoldt (2003): R as a simulation platform in ecological modelling. R. News, Vol. 3/3, 8 – 16.

[97] B. Blasius, A. Huppert and L. Stone. Complex dynamics and phase synchronization in spatially extended ecological systems. Nature, 1999;399: 354 – 359.

[98] Blasius B, Stone L. Chaos and phase synchronization in ecological systems. Int. J. Bifur. And chaos. 2000;**10**: 2361-2380

Chapter 11

Green's Function Method for Electromagnetic and Acoustic Fields in Arbitrarily Inhomogeneous Media

Vladimir P. Dzyuba and Roman Romashko

Abstract

An analytical method based on the Green's function for describing the electromagnetic field, scalar-vector and phase characteristics of the acoustic field in a stationary isotropic and arbitrarily inhomogeneous medium is proposed. The method uses, in the case of an electromagnetic field, the wave equation proposed by the author for the electric vector of the electromagnetic field, which is valid for dielectric and magnetic inhomogeneous media with conductivity. In the case of an acoustic field, the author uses the wave equation proposed by the author for the particle velocity vector and the well-known equation for acoustic pressure in an inhomogeneous stationary medium. The approach used allows one to reduce the problem of solving differential wave equations in an arbitrarily inhomogeneous medium to the problem of taking an integral.

Keywords: inhomogeneous media, Green's function, electromagnetic field, acoustic field, analytical method

1. Introduction

The chapter discusses the procedure for using the Green's function for the analytical description of electromagnetic and acoustic fields in a stationary isotropic and arbitrarily inhomogeneous medium. In the case of the electromagnetic field, the wave equation for the electric vector of the electromagnetic field in the inhomogeneous medium with conductivity, dielectric and magnetic permeability is used. In the case of the acoustic field, the wave equation proposed by the author for the vector of particle velocity and the well-known equation for acoustic pressure in an inhomogeneous stationary medium are used. Using the Green's function and the method of successive approximations makes it possible to achieve the required accuracy of calculating the electric and magnetic vectors of the electromagnetic field, as well as to calculate the vectors of complex intensity and intensity, density of energy, acoustic pressure and the particle velocity vector of the acoustic field in media with arbitrary spatial variability of the parameters. The approach used allows one to reduce the problem of solving differential wave equations in an arbitrarily inhomogeneous medium to integration. The chapter is divided into two parts. At the beginning of each part, the corresponding wave equations are derived and next, a method of using the Green's function is described and analytical expressions describing the fields are formulated. At the beginning we will describe the method as applied to the electromagnetic field, and then as applied to the acoustic field.

Research and modeling of the electromagnetic field in spatially inhomogeneous natural and composite media are actively developing in various fields of science and technology, ranging from systems of underground and underwater electromagnetic communication to photonics, metamaterials and metasurfaces [1]. Such a wide field of scientific research requires methods of mathematical modeling of the properties of the electromagnetic field in media with different spatial scales of conductivity, magnetic and dielectric permittivity. At present, analytical methods are applicable in a very limited range of environments. Among the methods of mathematical modeling of the electromagnetic field in the frequency range from fractions of the hertz to optical, various numerical methods and technologies are used [2, 3]. Numerical modeling uses a variety of methods and technologies, for example, parallel computing which are used in electrodynamic modeling programs. Among them, there are also direct and universal methods for solving boundary problems. The drawback of these methods is a large expenditure of computer resources, which leads to a significant simplification of physical models of the environment and mathematical approximations. There is a third class of methods, in which, at the initial stage, analytical methods are used, for example, the Green's function method, which brings the problem to a form that can be solved by fairly simple numerical methods. Below we will use exactly this approach using the Green's function. Green's function is actively used in a wide range of problems [4–6] of describing electromagnetic and other physical fields in various multilayer, chiral and anisotropic media, including inhomogeneous ones. The proposed procedure is also applicable to media with boundaries and arbitrary dependence on the coordinates of conductivity, magnetic and dielectric permittivity. The source of the field in the environment can be the electric current or an external field. The electric current can be located also inside the medium and outside it. The problem of descriptions the electric vector in an inhomogeneous medium by using the Green's function is formulated as the integral equation with its subsequent solution by the method of successive approximations. This procedure uses the equation for the vector of electric field strength in an inhomogeneous medium, with a certain conductivity, magnetic permeability, and dielectric constant.

The acoustic energy flux density vector (intensity vector), basically, until the beginning of the second half of the 20th century was only of theoretical interest. The second half of the 20th century brought about reliable means of synchronous measurement, practically at a single point, of the acoustic pressure and the components of particle velocity vector necessary to determine the intensity vector of acoustic field [7–14]. However, this did not lead to a significant increase in the number and quality of theoretical research methods and modeling of the intensity vector in an inhomogeneous medium. For the complete theoretical description of the acoustic field, knowledge of its acoustic pressure and the particle velocity vector is required. These two quantities make it possible to find the field of the acoustic intensity vector, to describe the energy and phase structure of the acoustic field. Knowledge of these quantities is useful for solving fundamental and applied problems of acoustic tomography and sounding of the geosphere, applied and fundamental hydroacoustics, creation of acoustic metamaterials, technical and architectural acoustics, noise control, etc. [15–17]. The acoustic pressure $\vec{P}_a\left(\vec{\mathbf{r}},t\right)$ and the particle velocity vector $\vec{\mathrm{V}}\left(\vec{\mathrm{r}},\mathrm{t}\right)$ are interrelated. This connection is obvious for a plane wave and, in the approximation of a continuous medium, has the following form: $\vec{V}\left(\vec{r},t\right) = -\frac{1}{\rho_0\left(\vec{r}\right)}\int \vec{P}_a\left(\vec{\mathbf{r}},t\right)dt$, where $\rho_0\left(\vec{r}\right)$ is the density of the medium unperturbed by the acoustic field at the point $\vec{\mathbf{r}}$ and t is time, and ∇ is the Nabla operator. This relationship largely determined the development of the theory of sound as a scalar field of acoustic pressure. Currently, there are several directions for the development of methods of calculation and theoretical analysis of the characteristics of the intensity vector. In the first direction, the relationship between the acoustic pressure and the particle velocity vector is used. This approach is applicable when there are mathematical expressions for the acoustic pressure field. As a rule, this is only possible in a homogeneous medium or for simple waveguides [18]. The second direction requires the use of the continuity equation and the equation of state of the inhomogeneous medium, as well as dynamic equations of motion of elementary volumes or particles of the inhomogeneous medium, for example: the Euler or Navier-Stokes equations. These equations are viewed as a system of equations for determining the pressure and the particle velocity vector. This approach is used to model the propagation of waves in various environments, including plasma and stellar atmospheres [19–22]. These equations are widely known, but to find analytical wave solutions of such systems given an arbitrary dependence of the density and speed of sound on the coordinates is a very difficult task. The use of the acoustic energy transfer equation is the third approach [9]. This approach allows one to describe the energy structure of the acoustic field which makes it possible to study the statistical characteristics of the complex intensity vector in a Gaussian delta-correlated inhomogeneous medium with refraction [9]. It is a very difficult task to find solutions to the transport equation in an inhomogeneous media. In turn, numerical methods for modeling metamaterials and propagation of acoustic waves in a medium are usually limited to specific problems [23–25]. None of the listed approaches, including numerical ones, provides the possibility of a complete theoretical description of the characteristics of the acoustic field and their evolution during field propagation in an arbitrary inhomogeneous medium. One of the promising directions is to use two wave equations in an inhomogeneous medium: equations for the acoustic pressure and equations for the particle velocity vector. We use this very approach. It is based on the proposed by authors wave equation for the particle velocity vector and the well-known equation for acoustic pressure in an inhomogeneous stationary medium. The proposed wave equation for the vector of the particle velocity of the acoustic field in a stationary inhomogeneous and isotropic medium is much more complicated than for the acoustic pressure. This makes it difficult to find the analytical solution for inhomogeneous media with an arbitrary spatial dependence of the density of the medium and the speed of sound in it. However, in an inhomogeneous medium, in which the field of the acoustic intensity vector is weakly vortex, the use of the Green's tensor together with the method of successive approximations makes it possible to find analytical solutions for an arbitrary spatial dependence of the speed of sound and density of the inhomogeneous medium.

2. Electromagnetic field

By an inhomogeneous medium, we mean a medium in which the conductivity $\boldsymbol{\sigma}\left(\vec{\mathbf{r}}\right)$, dielectric $\boldsymbol{\varepsilon}\left(\vec{\mathbf{r}}\right)$ and magnetic $\boldsymbol{\mu}\left(\vec{\mathbf{r}}\right)$ permittivity, and the current density $\vec{\mathbf{J}}\left(\vec{\mathbf{r}}\right)$ have an arbitrary, but differentiable, in the ordinary and in the generalized sense, dependence on coordinates points of the medium. Below we will not point out the explicit dependence of these and other quantities on time and coordinates, where this will not lead to misunderstanding. In an isotropic inhomogeneous medium, ε, μ, σ are scalar functions of coordinates. To derive the wave equation of the electromagnetic field in such a medium, we use the following well known fundamental and material Maxwell equations in a continuous isotropic and stationary medium:

$$\begin{array}{l} 1.\ \nabla\times\vec{H}=\vec{J}+\frac{\partial\vec{D}}{\partial t};\quad 2.\ \nabla\times\vec{E}=-+\frac{\partial\vec{B}}{\partial t};\quad 3.\ \vec{J}=\sigma\vec{E};\\ 4.\ \nabla\vec{J}=-\frac{\partial}{\partial t}(\rho_f+\rho_{ext});\quad 5.\ \vec{J}=\vec{J}_f+\vec{J}_{ext};\quad 6.\ \vec{D}=\varepsilon\vec{E};\\ 7.\ \vec{B}=u\vec{H};\quad 8.\ \varepsilon=\varepsilon_0\varepsilon_r;\quad 9.\ \mu=\mu_0\mu_r, \end{array} \tag{1}$$

where ρ_f is the density of free charges of the medium, and ρ_{ext} is the density of external charges introduced into the medium, $\boldsymbol{\varepsilon_r}$ and $\boldsymbol{\mu_r}$ are the relative dielectric and magnetic permeability of the medium, and $\vec{\mathbf{J}}_{ext}$ is the current density created by free and external charges. The wave equation for the electric vector in an inhomogeneous medium can be obtained, as for a homogeneous medium, excluding the vector of magnetic field strength from the system of Maxwell's equations. For this, we use the well-known vector analysis formulas [26] and take the rotor from the 2nd equation in system (Eq. (1)):

$$\nabla\times\nabla\times\vec{\mathbf{E}}=\nabla\cdot\left(\nabla\cdot\vec{\mathbf{E}}\right)-\Delta\cdot\vec{\mathbf{E}}=-\frac{\partial}{\partial t}\nabla\times\vec{\mathbf{B}} \tag{2}$$

In this case, the source of electromagnetic field is the electric current density $\vec{\mathbf{J}}$, so the divergence of the vector $\nabla\cdot\vec{\mathbf{E}}$, we need to associate with a current density in the medium. For this, we use (Eq. (5)) of the Maxwell system of equations (Eq. (1)) and obtain $\nabla\cdot\vec{E}=\frac{1}{\sigma}\left[\left(\nabla\cdot\vec{J}-\vec{E}(\nabla\cdot\sigma)\right)\right]$. Using the vector analysis formulas, we find the following expression:

$$\begin{aligned}\nabla\left(\nabla\cdot\vec{\mathbf{E}}\right)=\ &\vec{\mathbf{E}}\frac{(\nabla\cdot\sigma)^2}{\sigma^2}-[(\nabla\cdot\ln\sigma)\nabla]\vec{\mathbf{E}}-\left(\vec{\mathbf{E}}\cdot\nabla\right)(\nabla\cdot\ln\sigma)-(\nabla\cdot\ln\sigma)\times\nabla\times\vec{\mathbf{E}}+\\ &+\left(\nabla\cdot\vec{\mathbf{J}}\right)\left(\nabla\cdot\frac{1}{\sigma}\right)+\frac{1}{\sigma}\nabla\left(\nabla\cdot\vec{\mathbf{J}}\right)]\end{aligned} \tag{3}$$

The expression for the rotor of the magnetic field induction vector has the form

$$\nabla\times\vec{\mathbf{B}}=\nabla\times\left(\mu\vec{\mathbf{H}}\right)=\mu\left(\nabla\times\vec{\mathbf{H}}\right)+(\nabla\cdot\mu)\times\vec{\mathbf{H}} \tag{4}$$

Using equations 1 and 2 of the system of Maxwell equations and expressions (Eq. (2)) and (Eq. (3)), we find that $-\frac{\partial}{\partial t}\left(\nabla\times\vec{\mathbf{B}}\right)=-\mu\sigma\frac{\partial}{\partial t}\vec{\mathbf{E}}-\mu\frac{\partial}{\partial t}\vec{\mathbf{J}}-\varepsilon\mu\frac{\partial^2}{\partial t^2}\vec{\mathbf{E}}+(\nabla\cdot\ln\mu)\times\left(\nabla\times\vec{\mathbf{E}}\right)$, and the desired equation for an electric vector with a field source, in which the charge flux from the volume occupied by the current is not equal to zero, has the following form:

$$\varepsilon\mu\frac{\partial^2}{\partial t^2}\vec{\mathbf{E}}+\mu\sigma\frac{\partial}{\partial t}\vec{\mathbf{E}}-\Delta\vec{\mathbf{E}}-(\nabla\cdot\ln\mu\sigma)\times\left(\nabla\times\vec{\mathbf{E}}\right)+\vec{\mathbf{E}}\left(\frac{\nabla\cdot\sigma}{\sigma}\right)^2-[(\nabla\cdot\ln\sigma)\nabla]\vec{\mathbf{E}}-\left(\vec{\mathbf{E}}\cdot\nabla\right)(\nabla\cdot\ln\sigma)=$$
$$=-\mu\frac{\partial}{\partial t}\vec{\mathbf{J}}-\left(\nabla\cdot\vec{\mathbf{J}}\right)\left(\nabla\cdot\frac{1}{\sigma}\right)-\frac{1}{\sigma}\nabla\left(\nabla\cdot\vec{\mathbf{J}}\right) \tag{5}$$

If there is no injection of external charges into the medium, the value $\nabla\cdot\vec{\mathbf{J}}$ is equal to zero and $\nabla\vec{\mathbf{J}}=\nabla\vec{\mathbf{J}}_{ext}=-\frac{\partial}{\partial t}\rho_{ext}$ otherwise. When deriving (Eq. (5)), no conditions on the field frequency were used. Therefore, the equation is valid up to frequencies that correspond to wavelengths λ larger than the sizes of atoms or molecules. The smallness of the ratio of the first and second terms of equation (Eq. (5)) corresponds to the condition of quasi-stationarity of the electromagnetic field. For a monochromatic field with the angular frequency ω, the modulus of their ratio is equal $\frac{\varepsilon}{\sigma}\omega$ and small under conditions of high conductivity, low dielectric constant, or low the angular frequency. In this case, the propagation of the field in the medium will have a predominantly diffusion character and will be described by the following equation:

$$\mu\sigma\frac{\partial}{\partial t}\vec{\mathbf{E}}-\Delta\vec{\mathbf{E}}-(\nabla\cdot\ln\mu\sigma)\times\left(\nabla\times\vec{\mathbf{E}}\right)+\vec{\mathbf{E}}\left(\frac{\nabla\cdot\sigma}{\sigma}\right)^2-[(\nabla\cdot\ln\sigma)\nabla]\vec{\mathbf{E}}-\left(\vec{\mathbf{E}}\cdot\nabla\right)(\nabla\cdot\ln\sigma)=$$
$$=-\mu\frac{\partial}{\partial t}\vec{\mathbf{J}}-\left(\nabla\cdot\vec{\mathbf{J}}\right)\left(\nabla\cdot\frac{1}{\sigma}\right)-\frac{1}{\sigma}\nabla\left(\nabla\cdot\vec{\mathbf{J}}\right) \tag{6}$$

When $\frac{\varepsilon}{\sigma}\omega>>1$ the field propagation in the media is of the wave-type mainly. At the present time, there are no methods for finding exact solutions of equations of the type (Eq. (5)) and (Eq. (6)). Solutions satisfying a given accuracy can be obtained in two ways. The first is to use numerical methods. The second, which we will follow, consists in passing from the differential equation (5) to the integral equation using the tensor Green's function of the Helmholtz equation for the Fourier - the spectrum of the vector of the electric field strength. The solution to an integral equation can be written in the form of a sequence of approximate solutions, in which each subsequent term is more accurate. It is important that such a procedure for finding a solution is applicable for arbitrary differentiable, both in the usual and in the generalized sense, dependences of σ, ε, and μ on coordinates. For this, we express the vector of the electric field $\vec{\mathbf{E}}\left(\vec{\mathbf{r}},t\right)$ and the current density $\vec{\mathbf{J}}\left(\vec{\mathbf{r}},t\right)$ through their Fourier spectra $\vec{\mathbf{E}}\left(\vec{\mathbf{r}},\omega\right)$ and $\vec{\mathbf{J}}\left(\vec{\mathbf{r}},\omega\right)$

$$\vec{\mathbf{E}}\left(\vec{\mathbf{r}},t\right)=\frac{1}{2\pi}\int_{-\infty}^{+\infty}\vec{\mathbf{E}}\left(\vec{\mathbf{r}},\omega\right)e^{i\omega t}d\omega,\ \vec{\mathbf{J}}(\mathbf{r},t)=\frac{1}{2\pi}\int_{-\infty}^{+\infty}\vec{\mathbf{J}}\left(\vec{\mathbf{r}},\omega\right)e^{i\omega t}d\omega. \tag{7}$$

For high frequencies, when the dependence of ε and μ from the field frequency cannot be neglected, but spatial dispersion and nonlinear effects can be neglected, $\vec{\mathbf{J}}\left(\vec{\mathbf{r}},\omega\right)=\sigma\left(\omega,\vec{r}\right)\vec{\mathbf{E}}\left(\omega,\vec{r}\right),\ \vec{D}\left(\omega,\vec{r}\right)=\varepsilon\left(\omega,\vec{r}\right)\vec{\mathbf{E}}\left(\omega,\vec{r}\right)$ и $\vec{B}\left(\omega,\vec{r}\right)=\mu\left(\omega,\vec{r}\right)\vec{H}\left(\omega,\vec{r}\right)$. Spatial dispersion plays a minor role in comparison with temporal dispersion and is significant in media with the mean free path of the charge or its diffusion much longer than the field wavelength. Below we will not indicate the dependence of the conductivity and permeability on frequency. Let's consider conductivity and permittivity as a sum of a constant and a space-dependent variable:

$$\sigma\left(\vec{\mathbf{r}}\right) = \sigma_c + \sigma_1\left(\vec{\mathbf{r}}\right), \mu\left(\vec{\mathbf{r}}\right) = \mu_c + \mu_1\left(\vec{\mathbf{r}}\right), \varepsilon\left(\vec{\mathbf{r}}\right) = \varepsilon_c + \varepsilon_1\left(\vec{\mathbf{r}}\right) \quad (8)$$

Let's introduce the following notations:

$$\left(\omega^2\varepsilon\left(\vec{r}\right)\mu\left(\vec{r}\right) - i\omega\mu\left(\vec{r}\right)\sigma\left(\vec{r}\right) - \left(\frac{\nabla \cdot \sigma\left(\vec{r}\right)}{\sigma\left(\vec{r}\right)}\right)^2\right) = k^2\left(\vec{r}, \omega\right),$$

$$\vec{\mathbf{f}}_{ext}\left(\omega, \vec{\mathbf{r}}\right) = i\omega\mu\left(\vec{\mathbf{r}}\right)\vec{\mathbf{J}}\left(\vec{\mathbf{r}}, \omega\right) + \left(\nabla\vec{\mathbf{J}}\left(\vec{\mathbf{r}}, \omega\right)\right)\left(\frac{\nabla\sigma_1\left(\vec{\mathbf{r}}\right)}{\sigma^2\left(\vec{\mathbf{r}}\right)}\right) + \frac{1}{\sigma\left(\vec{\mathbf{r}}\right)}\nabla\left(\nabla \cdot \vec{\mathbf{J}}\left(\vec{\mathbf{r}}, \omega\right)\right), \quad (9)$$

$\vec{f}\left(\omega, \vec{r}\right) = -\left(\frac{\nabla\mu_1\left(\vec{r}\right)}{\mu\left(\vec{r}\right)} + \frac{\nabla\sigma_1\left(\vec{r}\right)}{\sigma\left(\vec{r}\right)}\right) \times \left(\nabla \times \vec{E}\left(\vec{r}, \omega\right)\right) - \left(\frac{\nabla\sigma_1\left(\vec{r}\right)}{\sigma\left(\vec{r}\right)}\nabla\right)\vec{E}\left(\vec{r}, \omega\right) - \left(\vec{E}\left(\vec{r}, \omega\right) \cdot \nabla\right)\left(\frac{\nabla\sigma_1\left(\vec{r}\right)}{\sigma\left(\vec{r}\right)}\right)$. Substituting expressions (Eq. (8)) and (Eq. (9)) into equation (Eq. (5)), we arrive at the following equation:

$$\Delta\vec{\mathbf{E}}\left(\vec{\mathbf{r}}, \omega\right) + k^2\left(\vec{\mathbf{r}}, \omega\right)\vec{\mathbf{E}}\left(\vec{\mathbf{r}}, \omega\right) = \vec{\mathbf{f}}\left(\vec{\mathbf{r}}, \omega\right) + \vec{\mathbf{f}}_{ext}\left(\vec{\mathbf{r}}, \omega\right) \quad (10)$$

Eq. (10) must be supplemented with boundary conditions. In an inhomogeneous medium, the interface between the media can be considered as an inhomogeneity with its dependence on coordinates, described by the corresponding functions, for example: Heaviside step function, etc. Therefore, the boundary conditions will be the conditions at infinity, where the field and its divergence must be equal to zero. In Eq. (10) the field source is not only the external currents (term $\vec{\mathbf{f}}_{ext}\left(\omega, \vec{\mathbf{r}}\right)$ but also the heterogeneity of the environment). These sources are described by $\vec{\mathbf{f}}\left(\omega, \vec{\mathbf{r}}\right)$. At present, there are no methods for the analytical solution of equations similar to (10) with an arbitrary dependence of the term $\vec{\mathbf{f}}\left(\omega, \vec{\mathbf{r}}\right)$on coordinates. Nevertheless, using the Green's functions of the vector Helmholtz equation in a homogeneous isotropic medium we can reformulate (10) into the integral with respect to the vector $\vec{\mathbf{E}}\left(\vec{\mathbf{r}}, \omega\right)$, the solution of which can be found in an iterative way, for example, by the method of successive approximations.

Using (Eq. (8)) one can formulate $k^2\left(\vec{\mathbf{r}}, \omega\right)$as the sum of independent on coordinates $\mathbf{k}_c^2$ function on coordinates and $\mathbf{k}_1^2\left(\vec{\mathbf{r}}, \boldsymbol{\omega}\right)$

$$k^2\left(\vec{r}\right) = \left(\omega^2\varepsilon_c\mu_c - i\omega\mu_c\sigma_c\right) + \omega^2\left(\varepsilon_c\mu_1\left(\vec{r}\right) + \varepsilon_1\left(\vec{r}\right)\mu_c + \varepsilon_1\left(\vec{r}\right)\mu\left(\vec{r}\right)\right)$$
$$-i\omega\left(\sigma_c\mu_1\left(\vec{r}\right) + \sigma_1\left(\vec{r}\right)\mu_c + \sigma_1\left(\vec{r}\right)\mu\left(\vec{r}\right)\right) - \left(\frac{\nabla \cdot \sigma_1\left(\vec{r}\right)}{\sigma\left(\vec{r}\right)}\right)^2 = k_c^2 + k_1^2(r) \quad (11)$$

Thus equation (Eq. (10)) can be written as:

$$\Delta\vec{\mathbf{E}}\left(\vec{\mathbf{r}},\omega\right)+k_c^2\vec{\mathbf{E}}\left(\vec{\mathbf{r}},\omega\right)=\vec{\mathbf{f}}_l\left(\omega,\vec{\mathbf{r}}\right)+\vec{\mathbf{f}}_{ext}\left(\omega,\vec{\mathbf{r}}\right),\tag{12}$$

where $\vec{f}_1\left(\omega,\vec{r}\right)=\vec{f}\left(\omega,\vec{r}\right)-k_1^2\left(\vec{r},\omega\right)\vec{E}\left(\vec{r},\omega\right)$.

Formally, we can consider Eq. (12) as an inhomogeneous Helmholtz equation and using the Green's function for it, we can rewrite Eq. (12) in the form of an integral equation. For vector fields, the Green's function [7–10] is a tensor of the second rank. In an orthogonal coordinate system, Eq. (5) decomposes into a system of three scalar equations for the projections of the field E (r, ω) on the coordinate axis. This simplifies the form of the Green's tensor and it has only diagonal elements that are not equal to zero. It can be represented as the vector

$\vec{G}\left(\vec{\mathbf{r}}-\vec{\mathbf{r}}_1\right)=\sum_{i=1}^{3}\vec{n}_iG_i\left(\vec{\mathbf{r}}-\vec{\mathbf{r}}_1\right)$, where $G_i\left(\vec{\mathbf{r}}-\vec{\mathbf{r}}_1\right)$ is the components of which are the Green's functions of the one-dimensional Helmholtz equation and $\mathbf{n}_i$ are the unit vectors of the coordinate axes. Let the area Ω in which we describe the field be large enough so that on its borders the field and its derivatives can be equated to zero. Using the Green tensor, we can rewrite Eq. (5) for the electric vector at the point $\vec{r}\in\Omega$ of the in the form of the following integral equation

$$\vec{\mathbf{E}}\left(\vec{\mathbf{r}},\omega\right)=\sum_{i=1}^{3}\mathbf{n}_i\int_\Omega G_i\left(\vec{\mathbf{r}}-\vec{\mathbf{r}}_1\right)\left[\mathbf{f}^i{}_{ext}\left(\omega,\vec{\mathbf{r}}_1\right)+\mathbf{f}^i{}_1\left(\omega,\vec{\mathbf{r}}_1\right)\right]d\vec{\mathbf{r}}_1\tag{13}$$

where $\left(\vec{\mathbf{r}},\ \vec{\mathbf{r}}_1\right)\in\Omega$, $\mathbf{f}^i{}_{ext}(\omega,\mathbf{r}_1)$ and $\mathbf{f}^i{}_1(\omega,\mathbf{r}_1)$ are the projections of vectors $\vec{\mathbf{f}}_{ext}\left(\omega,\vec{\mathbf{r}}\right)$ and $\vec{\mathbf{f}}_1\left(\omega,\vec{\mathbf{r}}\right)$ on to the coordinate axes. Integration is performed over the volume occupied by inhomogeneities, which are secondary sources of the field. In practice, the volume should be chosen such that secondary and higher order sources make a noticeable contribution to the field. Due to the rapid decrease in the amplitude of the Green's function and, especially with a strong absorption of the electromagnetic field by the medium, the region of integration can be about 1/α where α is the absorption coefficient of the field.

The steps for finding $\vec{\mathbf{E}}\left(\vec{\mathbf{r}},\omega\right)$ by the method of successive approximations can be as follows. We find the zeroth approximation $\vec{\mathbf{E}}_0\left(\vec{\mathbf{r}},\omega\right)$ for the field, which is valid in a homogeneous medium with parameters σ_c, μ_c, ε_c

$$\vec{\mathbf{E}}_0\left(\vec{\mathbf{r}},\omega\right)=\sum_{i-1}^{3}\vec{\mathbf{n}}_i\int_{\Omega_1}G\left(\vec{\mathbf{r}}-\vec{\mathbf{r}}_1\right)\left[\left(\vec{\mathbf{n}}_i\cdot\vec{\mathbf{f}}_{ext}\left(\omega,\vec{\mathbf{r}}_1\right)\right)\right]d\vec{\mathbf{r}}_1.\tag{14}$$

Integration is performed over the volume Ω_1 occupied by the external current (primary source of the field). This solution describes the primary field created by an external current. Using obtained by Eq. (14) expression $\vec{\mathbf{E}}_0\left(\vec{\mathbf{r}},\omega\right)$ and expression (9), we find $\mathbf{f}^i{}_1\left(\omega,\vec{\mathbf{r}}_1\right)$. Using (Eq. (13)) and integrating, we obtain a more accurate first $\vec{\mathbf{E}}_1\left(\vec{\mathbf{r}},\omega\right)$ approximation for $\vec{\mathbf{E}}\left(\vec{\mathbf{r}},\omega\right)$, which takes into account the influence of medium inhomogeneities on the field. To find the second approximation, it is necessary to substitute $\vec{\mathbf{E}}_1\left(\vec{\mathbf{r}},\omega\right)$ into $\mathbf{f}^i{}_1\left(\omega,\vec{\mathbf{r}}_1\right)$ and using (Eq. (13)) to obtain the second more accurate approximation. Similarly, more accurate solutions are obtained that take into account multiple field scattering by medium inhomogeneities. At these stages, the integration is performed over the volume occupied by inhomogeneities, which are secondary sources of the field. If the source of the field in an inhomogeneous medium is an external field with an electric vector $\vec{\mathbf{E}}_{ext}\left(\vec{\mathbf{r}},\omega\right)$ it should be used as the vector $\vec{\mathbf{E}}_0\left(\vec{\mathbf{r}},\omega\right)$.

For determining the magnetic field component one uses Maxwell's equations and writes them in terms of magnetic and electric fields Fourier spectrums: $\nabla\times\vec{\mathbf{E}}\left(\vec{\mathbf{r}},\omega\right)=-\omega\vec{B}\left(\vec{r},\omega\right)$. Substituting (Eq. (13)) in this equation one obtains

$$\vec{\mathbf{H}}\left(\vec{\mathbf{r}},\omega\right)=i\frac{1}{\omega\mu\left(\vec{\mathbf{r}}\right)}\int_{\Omega}\sum_{i=1}^{3}\vec{\mathbf{n}}_i\times\nabla G\left(\vec{\mathbf{r}}-\vec{\mathbf{r}}_1\right)\left[\mathbf{f}^i{}_{ext}\left(\omega,\vec{\mathbf{r}}_1\right)+\mathbf{f}^i{}_1\left(\omega,\vec{\mathbf{r}}_1\right)\right]d\vec{\mathbf{r}}_1. \quad (14a)$$

Using the Green's function, as the experience of its use shows (e.g. [7]) in such tasks, significantly reduces the requirements for computing resources and reduces the computation time. Note that the proposed procedure can be effective in simulating the optical properties of metamaterials, nanocomposites, and nanostructures.

3. Acoustic field

The wave equation for acoustic pressure $\mathrm{P}_a\left(\vec{r},\mathrm{t}\right)$ in a continuous inhomogeneous motionless and stationary medium is well known [16, 17]

$$\frac{1}{\mathrm{c}^2\left(\vec{r}\right)}\frac{\partial^2}{\partial \mathrm{t}^2}\mathrm{P}_a\left(\vec{r},\mathrm{t}\right)+\Delta\mathrm{P}_a\left(\vec{r},\mathrm{t}\right)+\left[\nabla\mathrm{P}_a\left(\vec{r},\mathrm{t}\right)+\vec{\mathbf{f}}\left(\vec{r},\mathrm{t}\right)\right]\nabla\ln\rho_0\left(\vec{r}\right)=\nabla\vec{f}\left(\vec{r},\mathrm{t}\right). \quad (15)$$

where $\vec{f}\left(\vec{r},\mathrm{t}\right)$ is the density of volumetric external forces that are the source of the acoustic field.

Eq. (1) is obtained by excluding the particle velocity vector from the linearized Euler equations, continuity and state of the medium. If we exclude acoustic pressure from these equations, then we get the equation for the vector of the particle velocity of the acoustic field. For this, we differentiate the equation of state in taking into account the smallness of the acoustic pressure, perturbation of the density of the medium by the $\rho_a\left(\vec{r},\mathrm{t}\right)$ in comparison with the background values $\rho_0\left(\vec{r}\right)$ and $\mathrm{P}_0\left(\vec{r}\right)$ the medium. In the inhomogeneous medium, the equation of state $\mathrm{P}_c\left[\rho\left(\vec{r},\mathrm{t}\right)\right]$ describes the relationship of the instantaneous local value of pressure and density of the medium. Therefore, it is necessary to use the total time derivative when differentiating the equation of state. Using it, we find in the linear approximation

$$\frac{\mathrm{d}}{\mathrm{dt}}\mathrm{P}_c\left[\rho\left(\vec{r},\mathrm{t}\right)\right]=\mathrm{c}^2\left(\vec{r},\right)\left[\frac{\partial}{\partial\mathrm{t}}\rho_a\left(\vec{r},\mathrm{t}\right)+\nabla\rho_0\left(\vec{r},\right)\vec{V}\left(\vec{r},\mathrm{t}\right)\right], \quad (16)$$

where $C(\vec{r},) = \sqrt{\frac{\partial P_c[\rho(\vec{r},t)]}{\partial\rho(\vec{r},t)}}$ is the local phase speed of sound for the acoustic pressure wave. We used the expansion $\rho(\vec{r},t) = \rho_0(\vec{r},t) + \rho_a(\vec{r},t)$ and the condition $\nabla\rho_a(\vec{r},t)V(\vec{r},t) << \nabla\rho_0(\vec{r})V(\vec{r},t)$. With the help of expression (2) and representation, $P_c[\rho(\vec{r},t)] = P_0(\vec{r}) + P_a(\vec{r},t)$ the equation of continuity$\frac{\partial}{\partial t}\rho(\vec{r},t) + \nabla[\rho(\vec{r},t)\vec{V}(\vec{r},t)] = 0$ is reduced to a linearized form

$$\frac{1}{\rho_0(\vec{r})\, c^2(\vec{r})}\frac{\partial}{\partial t}P(\vec{r},t) + \nabla V(\vec{r},t) = 0 \tag{17}$$

In the inhomogeneity of the medium $\nabla\vec{V}(\vec{r},t) \neq 0$, even in the approximation of an incompressible medium. Let us Take the time derivative on the linearized Euler equation $\rho_0(\vec{r})\frac{\partial}{\partial t}\vec{V}(\vec{r},t) + \nabla P(\vec{r},t) + \vec{f}(\vec{r},t) = 0$ and take the gradient of the equation of continuity (Eq. (17)). We exclude the acoustic pressure from the obtained expressions and find the equation for the particle velocity vector

$$\frac{1}{c^2(\vec{r})}\frac{\partial^2}{\partial t^2}\vec{V}(\vec{r},t) - \Delta\vec{V}(\vec{r},t) - \nabla\ln[\rho_0(\vec{r})c^2(\vec{r})]\nabla\vec{V}(\vec{r},t) - \nabla\times\nabla\times\vec{V}(\vec{r},t) = \\ = -\frac{1}{\rho_0(\vec{r})c^2(\vec{r})}\frac{\partial}{\partial t}\vec{f}(\vec{r},t) \tag{18}$$

When $\vec{f}(\vec{r},t) = 0$ then $\nabla\times\nabla\times\vec{V}(\vec{r},t) = -\nabla\times\left(\left(\nabla\ln\rho_0(\vec{r})\right)\times\vec{V}(\vec{r},t)\right)$ and can transformed equation (Eq. (18)) to the following form, which is valid in the absence of external forces

$$\frac{1}{c^2(\vec{r})}\frac{\partial^2}{\partial t^2}\vec{V}(\vec{r},t) - \Delta\vec{V}(\vec{r},t) - \nabla\ln\left[\rho_0(\vec{r})c^2(\vec{r})\right]\nabla\vec{V}(\vec{r},t) + \\ + \nabla\times\left(\left(\nabla\ln\rho_0(\vec{r})\right)\times\vec{V}(\vec{r},t)\right) = 0 \tag{19}$$

Eqs. (18) and (19) are much more complicated than equation (Eq. (15)) due to the third and fourth vortex term. The estimate of their ratio is

$$\left|\frac{\nabla\ln\left(\rho_0(\vec{r})c^2(\vec{r})\right)\nabla\vec{\mathbf{V}}(\vec{r},t)}{\nabla\times\left(\left(\nabla\ln\rho_0(\vec{r})\right)\times\vec{V}(\vec{r},t)\right)}\right| \sim \left|\frac{\nabla\ln\left(\rho_0(\vec{r})c^2(\vec{r})\right)}{\nabla\ln\rho_0(\vec{r})}\right| \tag{20}$$

In areas of the medium where this ratio is greater than unity, the fourth term in equations (Eqs. (18) and (19)) can be neglected. As a rule, these are media with a large relative gradient of the speed of sound. One of the examples of such media can be the marine environment, in which the local gradient of the speed of sound is determined less by the change in water density than salinity and temperature [12, 24]. Directly near the surface and the ocean floor or the interface between the media, the relative density gradient of the medium can be large. In these regions, the field of the particle velocity vector and acoustic intensity can have a significant rotational (vortex) component.

At present, the solution of these equations is possible only by numerical methods. If the fourth term in the equations is small, the equations for the vector of particle velocity and acoustic pressure allows one to find analytical expressions connecting the phases and moduli of vector of complex intensity and particle velocity vector, pressure, density of acoustic energy with the density of the medium and the speed of sound in it. Let us do it for equation 19. Using both scalar function $\Psi\left(\vec{r},t\right)$ and vector $\vec{U}\left(\vec{r},t\right)$

$$P\left(\vec{r},t\right)=\sqrt{\frac{Z_p\left(\vec{r}\right)}{Z_p^0}}\Psi\left(\vec{r},t\right)\ \vec{V}\left(\vec{r},t\right)=\sqrt{\frac{Z_v\left(\vec{r}\right)}{Z_v^0}}\vec{U}\left(\vec{r}',t\right), \tag{21}$$

we rewrite equations (Eqs. (15) and (19)) in the following form:

$$\frac{1}{c^2(r)}\frac{\partial^2}{\partial t^2}\Psi\left(\vec{r},t\right)-\Delta\Psi\left(\vec{r},t\right)+\left[\frac{3\nabla Z_p\left(\vec{r}\right)\nabla Z_p\left(\vec{r}\right)}{4Z_p^2\left(\vec{r}\right)}-\frac{\Delta Z_p\left(\vec{r}\right)}{2Z_p\left(\vec{r}\right)}\right]\Psi\left(\vec{r},t\right)=0 \tag{22}$$

$$\frac{1}{c^2(r)}\frac{\partial^2}{\partial t^2}\vec{U}\left(\vec{r},t\right)-\Delta\vec{U}\left(\vec{r},t\right)+\left[\frac{3\nabla Z_v\left(\vec{r}\right)\nabla Z_v\left(\vec{r}\right)}{4Z_v^2\left(\vec{r}\right)}-\frac{\Delta Z_v\left(\vec{r}\right)}{2Z_v\left(\vec{r}\right)}\right]\vec{U}\left(\vec{r},t\right)=0, \tag{23}$$

where $Z_p\left(\vec{r}\right)=\rho_0\left(\vec{r}\right)$ and $Z_v\left(\vec{r}\right)=\frac{1}{\rho_0\left(\vec{r}\right)c^2\left(\vec{r}\right)}$, and $Z_p^0=\rho_0\left(\vec{r}_0\right)$, $Z_v^0=\frac{1}{\rho_0\left(\vec{r}_0\right)c^2\left(\vec{r}_0\right)}$ are values $Z_p\left(\vec{r}\right)$ and $Z_v\left(\vec{r}\right)$ at some point in space $\vec{r}_0$. For the spectral components $\Psi\left(\vec{r},\omega\right)$ and $\vec{U}\left(\vec{r},\omega\right)$ using the Fourier transform of equations (Eqs. (22) and (23)) with respect to the time variable, we obtain the following equations

$$\Delta\Psi\left(\vec{r},\omega\right)+k_\psi^2\left(\vec{r}\right)\Psi\left(\vec{r},\omega\right)=0, \tag{24}$$

$$\Delta\vec{U}\left(\vec{r},\omega\right)+k_U^2\left(\vec{r}\right)\vec{U}\left(\vec{r},\omega\right)=0, \tag{25}$$

where $k_\psi^2\left(\vec{r}\right)=\frac{\omega^2}{c^2\left(\vec{r}\right)}-\frac{3}{4}\left[\frac{\nabla\rho_0\left(\vec{r}\right)}{\rho_0\left(\vec{r}\right)}\right]^2+\frac{\Delta\rho_0\left(\vec{r}\right)}{2\rho_0\left(\vec{r}\right)}$

$$\begin{aligned}k_U^2\left(\vec{r}\right)&=\frac{\omega^2}{c^2\left(\vec{r}\right)}+\frac{5}{4}\left[\frac{\nabla\rho_0\left(\vec{r}\right)}{\rho_0\left(\vec{r}\right)}\right]^2+\frac{\nabla\rho_0\left(\vec{r}\right)\nabla c\left(\vec{r}\right)}{\rho_0\left(\vec{r}\right)c\left(\vec{r}\right)}\\&+3\left[\frac{\nabla c\left(\vec{r}\right)}{c\left(\vec{r}\right)}\right]^2-\frac{\Delta\rho_0\left(\vec{r}\right)}{\rho_0\left(\vec{r}\right)}-2\frac{\Delta c\left(\vec{r}\right)}{c\left(\vec{r}\right)}\end{aligned} \tag{26}$$

From expressions (Eq. (26)) it follows that the gradient of the speed of sound affects only the vibrational speed in a medium with a small swirl of the particle velocity field. The acoustic pressure depends only on the gradient of the density of the medium and does not depend on the gradient of the speed of sound. This situation is valid for media in which the density gradient of the medium is less than the gradient of the speed of sound. These differences form the phase difference between the acoustic pressure and the vibrational velocity vector during the propagation of an acoustic wave in an inhomogeneous medium. Different field reactions $\vec{V}\left(\vec{r},\mathrm{t}\right)$ and $P\left(\vec{r},\mathrm{t}\right)$ to the density gradient of the medium and the gradient of the sound speed form the phase difference between the acoustic pressure $\Phi_p\left(\vec{\mathrm{r}},\mathrm{t}\right)$ and the particle velocity $\Phi_v\left(\vec{\mathrm{r}},\mathrm{t}\right)$ vector during the propagation of an acoustic wave in an inhomogeneous medium. Solutions to equations (Eqs. (24) and (25)) can be found using the method of successive approximations. For this, we represent these equations in the following form:

$$\Delta\Psi\left(\vec{r},\omega\right)+\mathrm{k}_0^2\Psi\left(\vec{r},\omega\right)=\mathrm{k}_{1\Psi}^2\left(\vec{r}\right)\Psi\left(\vec{r},\omega\right) \tag{27}$$

$$\Delta\vec{U}\left(\vec{r},\omega\right)+\mathrm{k}_0^2\vec{U}\left(\vec{r},\omega\right)=\mathrm{k}_{1_v}^2\left(\vec{r}\right)\overleftarrow{U}\left(\vec{r},\omega\right) \tag{28}$$

where $\mathrm{k}_0^2=\frac{\omega^2}{\mathrm{c}^2\left(\vec{r}_0\right)}$, $\mathrm{k}_{1\Psi}^2\left(\vec{r}\right)=\mathrm{k}_0^2-\mathrm{k}_\psi^2\left(\vec{r}\right)$ and $\mathrm{k}_{1U}^2\left(\vec{r}\right)=\mathrm{k}_0^2-\mathrm{k}_U^2\left(\vec{r}\right)$

Similarly to the case of an electromagnetic field in the inhomogeneous medium, using the scalar $\mathrm{G}\left(\vec{r}-\vec{r}_1\right)$ and vector $\vec{G}\left(\vec{r}-\vec{r}_1\right)$ Green's functions of the Helmholtz equation for a homogeneous unbounded medium. Eqs. (2) and (13) and Eq. (2.14) can be rewritten in the form of the following integral equations:

$$\Psi\left(\vec{r},\omega\right)=\Psi_0\left(\vec{r},\omega\right)+\int_\Omega \mathrm{G}\left(\vec{r}-\vec{r}_1\right)\mathrm{k}_{1\Psi}^2\left(\vec{r}_1\right)\Psi\left(\vec{r}_1,\omega\right)d\vec{r}_1 \tag{29}$$

$$U_i\left(\vec{r},\omega\right)=U_{0i}\left(\vec{r},\omega\right)+\int_\Omega G_i\left(\vec{r}-\vec{r}_1\right)\mathrm{k}_{1U}^2\left(\vec{r}_1\right)U_i\left(\vec{r}_1,\omega\right)d\vec{r}_1 \tag{30}$$

Here $U_i\left(\vec{r}_1,\omega\right)$ is the projections of the vector onto the coordinate axes and $\Psi_0\left(\vec{r},\omega\right)$ and $U_{0i}\left(\vec{r},\omega\right)$ are the solutions of equations (Eq. (27)) and (Eq. (28)) with the right-hand side equal to zero, and $G_i\left(\vec{r}-\vec{r}_1\right)$ are the components of the vector Green's function. The steps for finding a solution to equations (2.15) and (2.16) by the method of successive refinements can be as follows:

1. We find $\Psi_0\left(\vec{r},\omega\right)$ and $U_{0i}\left(\vec{r},\omega\right)$ which are the zeroth approximation for the field, valid in a homogeneous medium

2. We find an explicit form of dependence $\mathrm{k}_{1\Psi}^2\left(\vec{r}\right)=\mathrm{k}_0^2-\mathrm{k}_\psi^2\left(\vec{r}\right)$ and $\mathrm{k}_{1U}^2\left(\vec{r}\right)=\mathrm{k}_0^2-\mathrm{k}_U^2\left(\vec{r}\right)$ on coordinates

3. Using $\Psi_0\left(\vec{r},\omega\right)$ and $U_{0i}\left(\vec{r},\omega\right)$ and expressions for $\mathrm{k}_{1\Psi}^2\left(\vec{r}\right)$ and $\mathrm{k}_{1U}^2\left(\vec{r}\right)$with the help of (Eqs. (29) and (30)), we obtain more accurate first approximations that take into account single scattering of the primary field.

4. To obtain the second more accurate approximation, it is necessary to substitute the first approximations in expressions (Eqs. (29) and (30)) and obtain the second more accurate approximation of the solution.

Similarly, you can get solutions that are more accurate. Integration is performed over the volume of the inhomogeneous medium, the inhomogeneities of which will be secondary, etc. field sources. In a real situation, the volume should be chosen such that its secondary sources make a noticeable contribution to the field.

Let us consider an example that shows how the parameters of an inhomogeneous medium affect the characteristics of the acoustic field. We represent the acoustic pressure, and the vector of the particle velocity of the monochromatic acoustic field by the frequency ω in the following form

$$P\left(\vec{r},t\right) = P_0\left(\vec{r}\right)\exp i\left[\omega t - \Phi_p\left(\vec{r}\right)\right] \text{ and } \vec{\mathbf{V}}\left(\vec{r},t\right) = \vec{\mathbf{V}}_0\left(\vec{r}\right)\exp i\left[\omega t - \Phi_v\left(\vec{r}\right)\right].$$

Complex intensity vector will be written as $\vec{\mathbf{I}}\left(\vec{r}\right) = P\left(\vec{r},t\right)\vec{\mathbf{V}}^*\left(\vec{r},t\right) = \vec{\mathbf{I}}_0\left(\vec{r}\right)\exp i\left[\Phi_v\left(\vec{r}\right) - \Phi_p\left(\vec{r}\right)\right]$. In a medium without absorption of the acoustic energy, the phases $\Phi_p\left(\vec{r}\right)$ and $\Phi_v\left(\vec{r}\right)$, respectively are equal to the phases $\Psi\left(\vec{r},\mathrm{t}\right)$ and $\vec{U}\left(\vec{r},\mathrm{t}\right)$. The wave vector of a wave is normal to its phase surface and is determined by the wave phase gradient and the wave number by the modulus of this gradient. For the wavenumbers of the pressure $\mathrm{k}_p\left(\vec{r}\right)$ and the particle velocity vector $\mathrm{k}_v\left(\vec{r}\right)$, we can take, respectively, the quantities $k_\psi\left(\vec{r}\right)$ and $k_U\left(\vec{r}\right)$ if the inequalities $\left|\frac{\Delta\psi_0\left(\vec{r}\right)}{\psi_0\left(\vec{r}\right)}\right| << \left|k_\psi^2\left(\vec{r}\right) - \left(\nabla\Phi_p\left(\vec{r}\right)\right)^2\right|$ and $\left|\frac{\Delta U_0\left(\vec{r}\right)}{U_0\left(\vec{r}\right)}\right| << \left|\mathrm{k}_U^2\left(\vec{r}\right) - \left(\nabla\Phi_v\left(\vec{r}\right)\right)^2\right|$. The refractive indices of the medium for the acoustic pressure and the particle velocity relative to the point $\vec{r}_0$ are different and accordingly, equal $n_p\left(\vec{r}\right) = \frac{k_p\left(\vec{r}\right)}{k_0}$ and $n_v\left(\vec{r}\right) = \frac{k_v\left(\vec{r}\right)}{k_0}$, where $\mathrm{k}_0 = \frac{\omega}{\mathrm{c}\left(\vec{r}_0\right)}$ and $\mathrm{C}\left(\vec{r}_0\right)$ is the phase velocity of sound for the acoustic pressure wave at the point $\vec{r}_0$. The phase velocities of the acoustic pressure wave and the particle velocity vector become different, which leads to the inequality of the phases of the acoustic pressure and the particle velocity vector when the acoustic wave propagates in an inhomogeneous medium. The absorption of acoustic energy by the medium is taken into account by assuming the speed of sound and the density of the medium to be complex quantities, in which the imaginary part is responsible for the absorption of the energy of the acoustic field. In this case, the phases and acquire an additive equal to the phases of the values $\sqrt{\frac{Z_p\left(\vec{r}\right)}{Z_p^0}}$ and $\sqrt{\frac{Z_v\left(\vec{r}\right)}{Z_v^0}}$. To avoid cumbersome expressions, we restrict ourselves to the first approximation. The proposed example is often implemented in real measurements of the characteristics of the acoustic field. In practice, as a rule, the projections of the vibrational velocity vector and, accordingly, the intensity vector are measured in an orthogonal coordinate system, for example, in a Cartesian one. Consider the projection of a vector $\vec{U}\left(\vec{r},\omega\right)$ on the OX axis. In this case, the projection will be a function of only the x coordinate, and the Y and Z coordinates will act as parameters and determine the straight line parallel to the OX axis, along which the observation point x changes. The wave numbers $\mathrm{k}_{1\Psi}^2\left(\vec{r}\right)$ and $\mathrm{k}_{1U}^2\left(\vec{r}\right)$, accordingly, the solutions of equations (Eqs. (29) and (30)) depend on the values of these parameters. In fact, we turn to the case of one-dimensional propagation of acoustic radiation along the OX axis passing through the point $\mathbf{r}_0$ $(\boldsymbol{X}_0, \boldsymbol{Y}_0\ , \boldsymbol{Z}_0)$. Let us choose $\Psi_0\left(\vec{r}, \omega\right)$ and $\vec{U}_0\left(\vec{r}, \omega\right)$ both in the form of plane waves and propagating along the X axis. We can put the moduli of these plane waves $\Psi_0(\omega)$ and $\vec{\mathrm{U}}_0(\omega)$ equal to the moduli of the acoustic pressure P_0 and the component $\vec{v}_x$ of the particle velocity vector on the OX axis. In this case, the component of the vector Green's function will be equal to the one-dimensional Green's function

$\mathrm{G}(\mathrm{x} - \mathrm{x}_1) = \frac{1}{2\mathrm{ik}_0} \exp\left[\mathrm{ik}_0|\mathrm{x} - \mathrm{x}_1|\right]$ we find the solution corresponding to the first approximation at the point X

$$\Psi(x, y_0, z_0, \omega) = P_0 \exp ik_0 x + \int_{-\infty}^{x} \mathrm{k}_{1\Psi}^2(x_1, y_0, z_0) \frac{P_0}{2ik_0} \exp(ik_0 x) dx_1 + \\ + \int_{x}^{\infty} \mathrm{k}_{1\Psi}^2(x_1, y_0, z_0) \frac{P_0}{2ik_0} \exp(-ik_0 x + 2ik_0 x_1) dx_1 = P_0 \exp ik_0 x + \Psi_1(x) + \Psi_2(x) \tag{31}$$

$$\vec{U}_x(x, y_0, z_0, \omega) = \vec{V}_x \exp ik_0 x + \int_{-\infty}^{x} k_{1U}^2(x, y_0, z_0) \frac{\vec{V}_x \exp(ik_0 x)}{2ik_0} dx_1 + \\ + \int_{x}^{\infty} k_{1U}^2(x_1, y_0, z_0) \frac{\vec{V}_x}{2ik_0} \exp(-ik_0 x + 2ik_0 x_1) dx_1 = \vec{V}_x \exp ik_0 x + \vec{U}_1(x) + \vec{U}_2(x) \tag{32}$$

Here $P_0 \exp ik_0 x$ и $\vec{\mathbf{U}}_0 \exp ik_0 x$ represent the primary radiation, the second terms are the radiation scattered forward in the region $-\infty \leq x_1 \leq x$, and the third terms are the radiation scattered back. Solutions (Eqs. (31) and (32)) and the relations (Eq. (21)) allow us, in the first approximation, to find an expression for the projection of the complex intensity vector on the on the OX axis

$$\vec{\mathbf{I}}_x(x, y_0, z_0, \omega) = \sqrt{\frac{Z_p(x, y_0, z_0) Z_v(x, y_0, z_0)}{Z_p^0 Z_v^0}} \begin{pmatrix} P_0 \vec{\mathbf{V}}_x + + P_0 \vec{\mathbf{U}}_1^*(x) + P_0 \vec{\mathbf{U}}_2^*(x) + \Psi_1(\mathbf{x}) \vec{\mathbf{V}}_x + \\ + \Psi_1(\mathbf{x}) \vec{\mathbf{U}}_1^*(x) + \Psi_1(\mathbf{x}) \vec{\mathbf{U}}_2^*(x) + \Psi_2(\mathbf{x}) \vec{\mathbf{V}}_x + \\ + \Psi_2(\mathbf{x}) \vec{\mathbf{U}}_1^*(x) + \Psi_2(\mathbf{x}) \vec{\mathbf{U}}_2^*(x) \end{pmatrix} \tag{33}$$

In this expression, the first term describes the complex intensity vector of the primary radiation, the fifth term corresponds to the forward propagating secondary radiation, and the ninth term corresponds to the backscattered radiation. The other terms describe the mutual energy of the primary and scattered radiation. If the field is measured arriving at a point x only from the region, $x_1 \leq x$ then the dependence of the projection of the complex intensity vector on the OX axis takes the following form:

$$\vec{\mathbf{I}}_x(x, y_0, z_0 \omega) \exp i\Phi(x, y_0, z_0) = \frac{C_0(x_0, y_0, z_0)}{C(x, y_0, z_0)} P_0 \vec{\mathbf{V}}_x \begin{bmatrix} 1 + \frac{i}{2k_0}(\alpha_v(x_0, y_0, z_0) - \alpha_p(x, y_0, z_0)) + \\ + \frac{1}{4k_0^2} \alpha_v(x, y_0, z_0) \alpha_p(x, y_0, z_0) \end{bmatrix}. \tag{34}$$

In this expression $\alpha_p(x) = \int_{-\infty}^{x} k_{1\Psi}^2(x_1)dx_1$ and $\alpha_v(x) = \int_{-\infty}^{x} k_{1U}^2(x_1)dx_1$. The modulus $\mathbf{I_0}(\mathbf{x}, y_0, z_0, \boldsymbol{\omega})$ and phase $\boldsymbol{\Phi}(\mathbf{x}, y_0, z_0, \boldsymbol{\omega})$ of the complex of acoustic intensity vector are respectively equal:

$$\mathrm{I}_0(x,\omega) = \frac{C_0 P_0 |\mathbf{V}_x|}{4k_0^2 C(x,y_0,z_0)} \sqrt{\left(4k_0^2 + \alpha_p(x,y_0,z_0)^2\right)\left(4k_0^2 + \alpha_v(x,y_0,z_0)^2\right)}, \Phi(x,y_0,z_0,\omega)$$
$$= arctg \frac{2k_0\left[\alpha_v(x,y_0,z_0) - \alpha_p(x,y_0,z_0)\right]}{4k_0^2 + \alpha_v(x,y_0,z_0)\alpha_p(x,y_0,z_0)}. \tag{35}$$

The field intensity vector is

$$\mathrm{Re}\,\frac{1}{2}\vec{\mathbf{I}}_x(x,y_0,z_0,\omega) = \frac{1}{2}\frac{C_0}{C(x,y_0,z_0)} P_0 \vec{\mathbf{V}}_x \left[1 + \frac{1}{4k_0^2}\alpha_v(x,y_0,z_0)\alpha_p(x,y_0,z_0)\right], \tag{36}$$

and for the average field energy density we have the following expression:

$$\varepsilon(x,y_0,z_0\omega) = \frac{P(x,y_0,z_0\omega)P^*(x,y_0,z_0\omega)}{\rho(x,y_0,z_0)C^2(x,y_0,z_0)} =$$
$$= \frac{P_0^2}{\rho_0 C^2(x,y_0,z_0)}\left[1 + \frac{1}{4k_0^2}\alpha_p(x,y_0,z_0)\alpha_p(x,y_0,z_0)\right]. \tag{37}$$

From expressions (Eqs. (22) and (23)) it is seen that the inhomogeneous nature of the speed of sound will have a more significant effect on the particle velocity vector than on the acoustic pressure. This makes it possible in principle to create methods for separately measuring the contribution to the acoustic field in an inhomogeneous medium of the density of the medium and the speed of sound in it.

In conclusion, we note that the proposed method makes it possible to analytically and numerically solve the problems of mathematical modeling of a shallow sea, remote sensing of natural media, problems of acoustics of a shallow sea, modeling acoustic and optical metamaterials, etc. Note that for applied problems of acoustics, both fields of the particle velocity vector and the intensity vector in any inhomogeneous medium have a vortex character. Therefore, the algorithms for solving applied problems of ocean and especially shallow sea acoustics, problems of modeling the propagation of acoustic energy in composite media and metamaterials should take into account the vortex component of the vector acoustic field intensity and curvature of the streamlines of the acoustic field.

Acknowledgements

The research is supported by the grant from the Russian Science Foundation (project # 19-12-00323).

Author details

Vladimir P. Dzyuba* and Roman Romashko
Institute of Automation and Control Processes of FEB Russian Academy of Sciences, Vladivostok, Russia

*Address all correspondence to: vdzyuba@iacp.dvo.ru

References

[1] Caloz Ch., Itoh T. Electromagnetic metamaterials: transmission line theory and microwave applications (the engineering approach). A John Wiley & Sons, Inc., 2006. 352 p.

[2] Prather, D. W., S.Shi. Formulation and application of the finite-difference time-domain method for the analysis of axially symmetric diffractive. J. Opt. Soc. Am. A.1999. Vol.16, Iss. 5. P. 1131-1142. Doi:10.1364/JOSAA.16.001131

[3] Taflove A. and Hagness S. C. Computational Electrodynamics: The Finite-Difference Time-Domain Method. Publisher: Artech House; 3 edition ,2005. 1038 pages.

[4] C.T. Tai, Diadic Green, Function in Electromagnetic Teory, 2nd ed. Piscataway, NJ, USA IEEEPress,1994.

[5] George W. Hanson Dyadic Green's Function for a Multilayered Planar Medium—A Dyadic Eigenfunction Approach. IEEE Transactions on Antennas and Propagation 2005, 52(12), P.3350 – 3356. DOI:10.1109/TAP.2004.836409

[6] Dzyuba, V.P., Zapolsky, A.M. The G, reen function for the elastic half-space with a curvilinear boundary. Doklady Earth Sciences V. 372, May 2000, Pages 709-711.

[7] V.A. Gordienko Vector-phase methods in acoustics. Moscow: Fizmatlit; 2007. 480 p. ISBN978-5-92210864

[8] F. J. Fahy Sound intensity. 2nd ed. London: Elsevier; 1989.

[9] Dzyuba V.P. Scalar-vector methods of theoretical acoustics. Vladivostok: Dalnauka; 2006. 192 p. ISBN 5-8044-0559-4.

[10] Akulichev V.A., Dzyuba V.P., Gladkov P. V. On Acoustic Tomography Scheme for Reconstruction of Hydrophysical Parameters for Marine Environment. Theoretical and Computational Acoustics. Ed. Er- Chang Shang and Qihi Li. Published by World Scientific; 2002.P.107 -114.

[11] Baster K.J. , Lanchle L.C., Mc. ConnelJ.A. Development of a Velocity Gradient Underwater Acoustic Intensity Sensor. JASA .1999.Vol. 106. P. 3178-3188. doi:10.1121/1.428172.

[12] Mann J.A., Tichy J., Romano A.J. Instantaneous and time-average energy transfer in acoustic fields. JASA. 1987. Vol. 82. №1. P. 17-30. DOI:10.1121/1.395562.

[13] K. J. Taylor Absolute measurement of acoustic particle velocity. The Journal of the Acoustical Society of America. 1976 Vol. 59, 691 DOI:10.1121/1.380896

[14] Nelson, P.A. and Elliott, S.J. (1991) Active control of sound: Academic Press; 1993. 250pp. ISBN: 0125154259.

[15] Medwin H.. Sounds in the sea: from ocean acoustics to acoustical oceanography. Cambridge, UK: Cambridge Univ. Press. 2005. 643 pp.

[16] David R. Dall'Ostoa and Peter H. Dahl Properties of the acoustic intensity vector field in a shallow water waveguide. The Journal of the Acoustical Society of America 2012. 131, 2023; DOI: 10.1121/1.3682063

[17] Woon Siong Gan New Acoustics Based on Metamaterials . Springer Nature Singapore Pte Ltd. 2018. P. 314. DOI:10.1007/978-981-10-6376-3.

[18] Morse, P.N., Ingard, K.U. Theoretical Acoustics. New York: McGraw-Hill; 1987. 949P.

[19] Jan-Niklas Hau1, and Björn Müller Acoustic wave propagation in a temporal evolving shear-layer for low-Mach number perturbations. Physics of Fluids 2018; 30, 016105. DOI:10.1063/1.4999044.

[20] L. Friedland, G. Marcus, J. S. Wurtele, and P. Michel3, Excitation and control of large amplitude standing ion acoustic waves. Phys. Plasmas 2019; 26, 092109 DOI: 10.1063/1.5122300

[21] Zhu, X., Li, K., Zhang, P. *et al.* Implementation of dispersion-free slow acoustic wave propagation and phase engineering with helical-structured metamaterials. Nat Commun. 2016; 11731. DOI:10.1038/ncomms11731

[22] N. Anders Petersson & Björn Sjögreen High Order Accurate Finite Difference Modeling of Seismo-Acoustic Wave Propagation in a Moving Atmosphere and a Heterogeneous Earth Model Coupled Across a Realistic Topography.Journal of Scientific Computing 2018; volume 74, pages 290–323. DOI:10.1016/j.jcp.2020.109386

[23] S.MishrabCh.SchwabaJ. ŠukyscMulti- level Monte Carlo finite volume methods for uncertainty quantification of acoustic wave propagation in random heterogeneous layered medium. Journal of Computational Physics. Volume 312, 1 May 2016, Pages 192-217. DOI:10.1016/j.jcp.2016.02.014.

[24] Elias Perras1, and Chuanzeng Zhang1 Analysis of acoustic wave propagation in composite laminates via aspectral element method// PAMM • Proc. Appl. Math. Mech. 2019;19: e201900282. DOI:10.1002/pamm.201900282

[25] Yijun Mao, Jiancheng Cai, Yuanyuan Gu and Datong Qi Direct Evaluation of Acoustic-Intensity Vector Field Around an Impedance Scattering Body AIAA JOURNA. 2015; Vol. 53, No. 5, May. DOI:10.2514/1.J053431.

[26] Granino Arthur Korn, Theresa M. Korn Mathematical handbook for scientists and engineers, McGraw-Hill, 1968. P. 1130

Chapter 12

Chaotic Systems with Hyperbolic Sine Nonlinearity

Jizhao Liu, Yide Ma, Jing Lian and Xinguo Zhang

Abstract

In recent years, exploring and investigating chaotic systems with hyperbolic sine nonlinearity has gained the interest of many researchers. With two back-to-back diodes to approximate the hyperbolic sine nonlinearity, these chaotic systems can achieve simplicity of the electrical circuit without any multiplier or sub-circuits. In this chapter, the genesis of chaotic systems with hyperbolic sine nonlinearity is introduced, followed by the general method of generating nth-order (n > 3) chaotic systems. Then some derived chaotic systems/torus-chaotic system with hyperbolic sine nonlinearity is discussed. Finally, the applications such as random number generator algorithm, spread spectrum communication and image encryption schemes are introduced. The contribution of this chapter is that it systematically summarizes the design methods, the dynamic behavior and typical engineering applications of chaotic systems with hyperbolic sine nonlinearity, which may widen the current knowledge of chaos theory and engineering applications based on chaotic systems.

Keywords: chaotic systems, torus chaos, hyperbolic sine nonlinearity, spread spectrum communication, image encryption

1. Introduction

Since Lorenz discovered chaos in a third-order ordinary differential equations, a new field of science has been launched [1]. The fact that simple equations can exhibit incredible complex behavior continues enthrall engineers to apply chaotic systems to cryptosystem, secure communication, spread spectrum communication, etc. [2].

There is no doubt that nonlinear term is very important to design chaotic systems, which has peculiar complex properties such as ergodicity, highly initial value sensitivity, non-periodicity and long-term unpredictability. According to the literature, the nonlinearities can be piecewise nonlinear function [3], trigonometric function [4], absolute value function [5], or power function [6]. With different nonlinearities, the chaotic system can have various strange attractors as single-scroll [7], double-scroll [8], multi-scroll [9], etc. The majority of such chaotic systems are known for many years, and some chaotic systems with hidden attractors are derived from them [10–12].

In recent years, chaotic systems with hyperbolic sine nonlinearities have gained the interest of many researchers. With two back-to-back diodes to approximate the hyperbolic sine nonlinearity, these chaotic systems can achieve simplicity of the electrical circuit without any multiplier or sub-circuits. Compared to single-scroll chaotic systems, the chaotic system with hyperbolic sine nonlinearity has richer dynamic behavior because it is symmetrical and can exhibit symmetry breaking, and offers the possibility that attractors will split or merge as some bifurcation parameter is changed [13].

In this chapter, we will systematically summarize the design method, the dynamic behavior and typical engineering applications of chaotic systems with hyperbolic sine nonlinearity. The genesis and general method of generating nth-order ($n > 3$) chaotic systems with hyperbolic sine nonlinearity are introduced in Section II. Some derived chaotic systems/torus-chaotic system with hyperbolic sine nonlinearity is discussed in Section III. The application such as random number generator algorithm, spread spectrum communication and image encryption schemes are introduced in Section IV. Conclusions are finally drawn in Section V.

2. General chaotic systems with hyperbolic sine nonlinearity

2.1 The genesis of chaotic systems with hyperbolic sine nonlinearity

In 2011, Sprott and Munmuangsaen proposed an exponential chaotic system [14], which happens to be an example of the simplest chaotic system [15]. In the same year, Sprott used common resistors, capacitors, operational amplifiers, and a diode to successfully implement this system in a circuit [16]. Few years later, the simplest hyperbolic sine chaotic system is proposed [17]. Compared to the exponential chaotic system, the hyperbolic sine chaotic system changed the nonlinearity from exponential function (asymmetric function) to hyperbolic sine function (symmetric function), which can exhibit symmetry breaking, and offers the possibility that attractors will split or merge as some bifurcation parameter is changed [18].

The simplest chaotic system with a hyperbolic sine is described as follows:

$$\dddot{x} + c\ddot{x} + x + \rho * \sinh(\varphi\dot{x}) = 0 \tag{1}$$

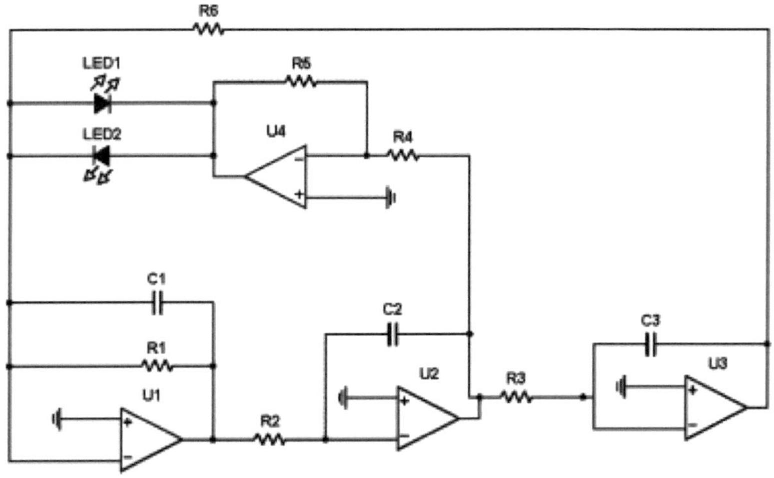

Figure 1.
The corresponding circuit schematic diagram of Eq. (1).

Where c is considered as the bifurcation parameter, $\sinh(\varphi\dot{x}) = \frac{e^{\varphi\dot{x}} - e^{-\varphi\dot{x}}}{2}$, $\rho = 1.2 * 10^{-6}$ and $\varphi = \frac{1}{0.026}$, which have been chosen to facilitate circuit implementation using diodes. The corresponding circuit schematic diagram of Eq. (1) is shown as **Figure 1**.

When $c = 0.75$, the Eq. (1) can exhibit chaotic behavior, which is shown as **Figure 2**.

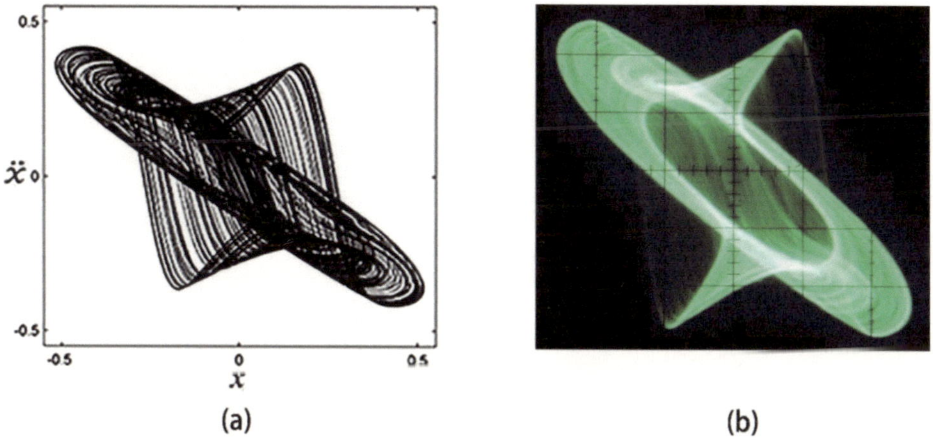

Figure 2.
Numerical and actual circuit state space plot in $x - \ddot{x}$ plane.

2.2 The general equations of generating chaotic systems with hyperbolic sine nonlinearity

It is obvious that Eq. (1) can be written in the form with jerk equations:

$$\begin{cases} \dot{x}_1 = x_2 \\ \dot{x}_2 = x_3 \\ \dot{x}_3 = -cx_3 - f(x_2) - x_1 \end{cases} \tag{2}$$

where $f(x_2) = \rho * \sinh(\varphi x_2)$. Therefore, the higher order chaotic systems with hyperbolic sine nonlinearity can be generated by adding jerk cabins, which is described by:

$$\begin{cases} \dot{x}_1 = x_2 - x_1 \\ \dot{x}_2 = x_3 - x_2 \\ \cdots \\ \dot{x}_{n-3} = x_{n-2} - x_{n-3} \\ \dot{x}_{n-2} = x_{n-1} \\ \dot{x}_{n-1} = x_n \\ \dot{x}_n = -cx_n - f(x_{n-1}) - nx_{n-2} - nx_{n-3} - \cdots - \dfrac{1}{2n}x_1 \end{cases} \tag{3}$$

where $\dot{x}_{k-1} = x_k - x_{k-1}$ is the jerk cabin. With Eq. (3), we can construct nth-order ($n > 3$) chaotic systems with hyperbolic sine nonlinearity.

When n = 4, the equations of fourth-order chaotic systems will be:

$$\begin{cases} \dot{x}_1 = x_2 - x_1 \\ \dot{x}_2 = x_3 \\ \dot{x}_3 = x_4 \\ \dot{x}_4 = -x_4 - f(x_3) - 5x_2 - 0.125x_1 \end{cases} \tag{4}$$

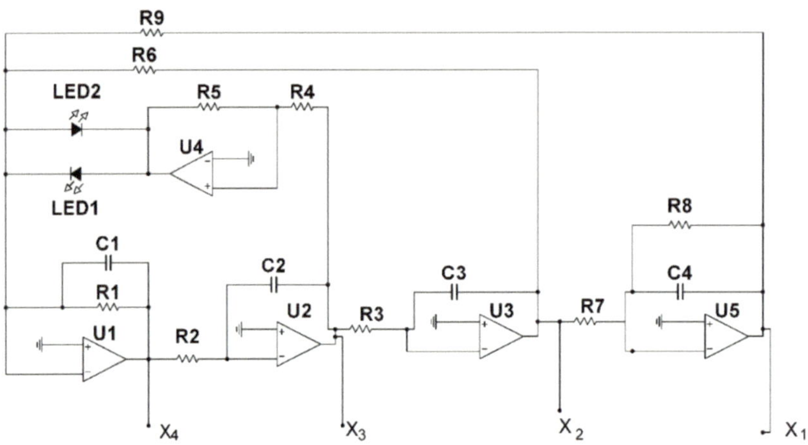

Figure 3.
The corresponding circuit schematic diagram of Eq. (4).

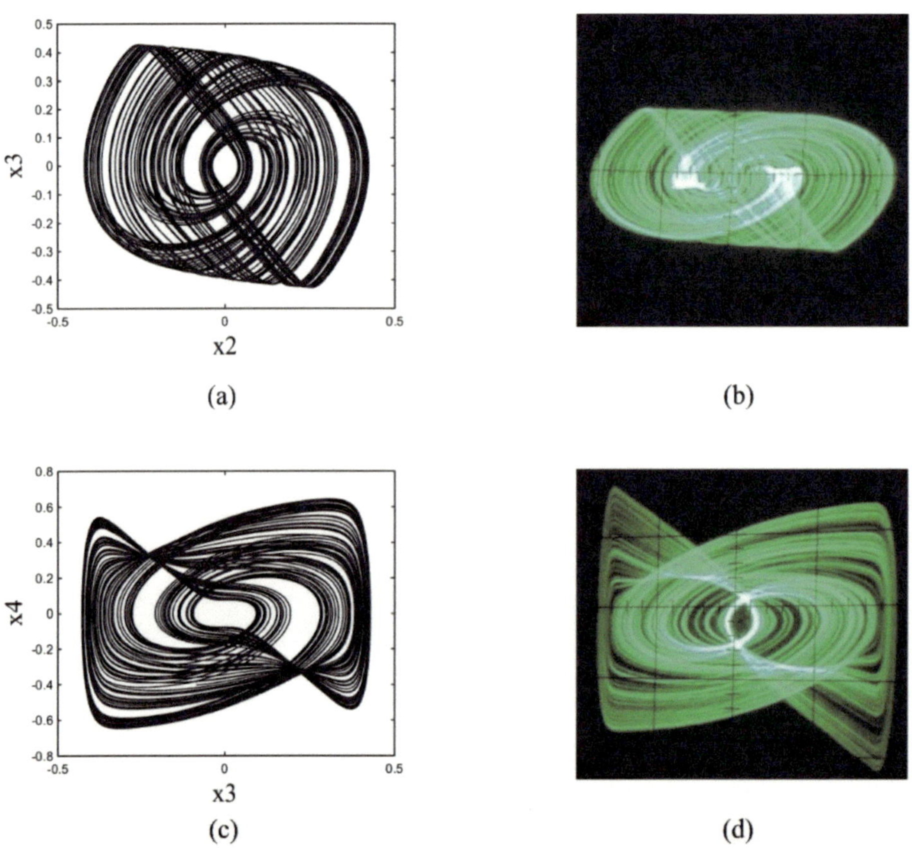

Figure 4.
Numerical and actual circuit state space plot in $x_2 - x_3$ plane and $x_3 - x_4$ plane.

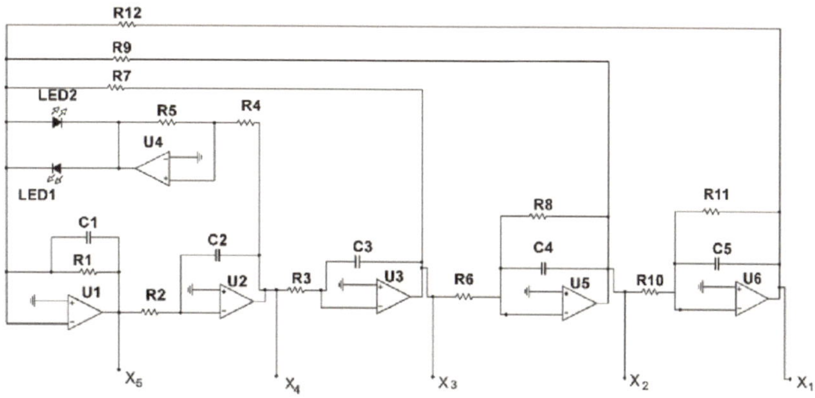

Figure 5.
The corresponding circuit schematic diagram of Eq. (5).

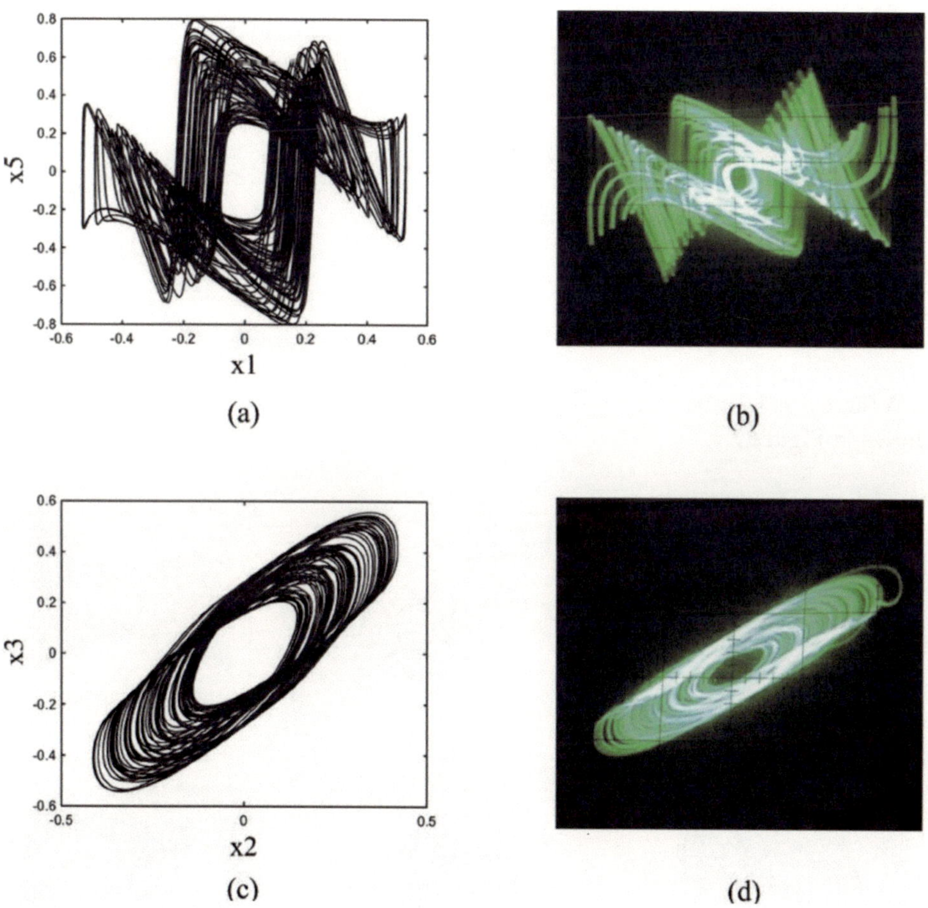

Figure 6.
Numerical and actual circuit state space plot in $x_1 - x_5$ plane and $x_2 - x_3$ plane.

The corresponding circuit schematic diagram of Eq. (4) is shown as **Figure 3**. Its numerical and actual circuit state space plot is shown as **Figure 4**.

When n = 5, the equations of fifth-order chaotic systems will be:

$$\begin{cases} \dot{x}_1 = x_2 - x_1 \\ \dot{x}_2 = x_3 - x_2 \\ \dot{x}_3 = x_4 \\ \dot{x}_4 = x_5 \\ \dot{x}_5 = -x_5 - f(x_4) - 5x_3 - 5x_2 - 0.1x_1 \end{cases} \tag{5}$$

The corresponding circuit schematic diagram of Eq. (5) is shown as **Figure 5**. Its numerical and actual circuit state space plot is shown as **Figure 6**.

3. Derived chaotic systems/torus-chaotic system with hyperbolic sine nonlinearity

3.1 Multi-nonlinearities hyperbolic sine chaotic system

One way to construct the derived chaotic systems is to add more nonlinear terms of the equations. For example, the new chaotic system can be constructed by Eq. (4), which is described as follows:

$$\begin{cases} \dot{x}_1 = x_2 - \rho sinh(\varphi x_1) \\ \dot{x}_2 = x_3 - 0.3x_2 - \rho sinh(\varphi x_2) \\ \dot{x}_3 = x_4 \\ \dot{x}_4 = -0.25x_4 - \rho sinh(\varphi x_3) - 0.5x_2 - 4x_1 \end{cases} \tag{6}$$

Where $\rho = 1.2 * 10^{-6}$, $\varphi = \frac{1}{0.026}$. These equations can exhibit chaotic behavior as shown in **Figure 7**.

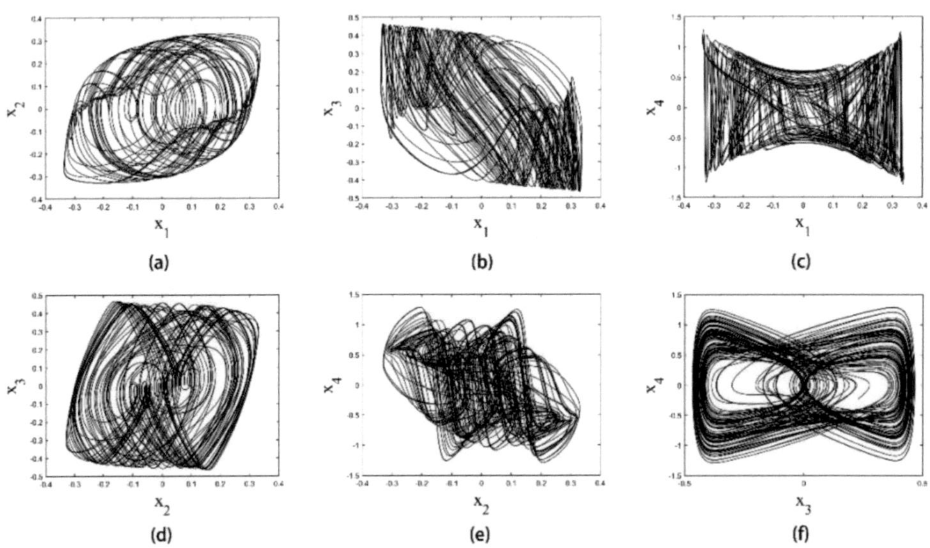

Figure 7.
Numerical phase space plot of Eq. (6).

3.2 Simple chaotic system with hyperbolic sine nonlinearity

The other way to construct the derived chaotic systems is to simplify the known chaotic systems. For example, if we remove the parameter ρ and φ, search the parameter space, we will have the following chaotic system:

$$\begin{cases} \dot{x}_1 = 6x_2 - x_1 \\ \dot{x}_2 = x_3 \\ \dot{x}_3 = x_4 \\ \dot{x}_4 = -x_4 - sinh\,(x_3) - x_1 \end{cases} \tag{7}$$

When initial conditions are set to be $(x_1, x_2, x_3, x_4) = (0.7, 0.9, 1.0, 1.3)$, or $(x_1, x_2, x_3, x_4) = (-0.7, -0.9, -1.0, -1.3)$, the system exhibits period behavior. When the initial conditions are set to be $(x_1, x_2, x_3, x_4) = (7, 9, 10, 13)$ and $(x_1, x_2, x_3, x_4) = (-7, -9, -10, -13)$, the system exhibits chaotic behavior. Therefore, this system has four coexistence attractors [19], as shown in **Figure 8**.

3.3 Torus-chaotic system with hyperbolic sine nonlinearity

By introducing a nonlinear feedback controller to system Eq. (5), the following system is obtained:

$$\begin{cases} \dot{x}_1 = x_2 - \rho\sinh\,(\varphi x_3) \\ \dot{x}_2 = x_3 - x_2 \\ \dot{x}_3 = x_4 \\ \dot{x}_4 = x_5 \\ \dot{x}_5 = -cx_5 - \rho\sinh\,(\varphi x_4) - 5x_3 - 5x_2 - 0.1x_1 \end{cases} \tag{8}$$

When c = 1, the Lyapunov exponents are $(\lambda_1, \lambda_2, \lambda_3, \lambda_4, \lambda_5) = (0.47, 0, 0, -1.10, -1.37)$, which suggests Eq. (8) is exhibiting torus-chaos behavior [20].

When c = 1.55 and the initial conditions are set to be $(x_1, x_2, x_3, x_4, x_5) = (-0.1, -0.1, -0.1, -0.1, -0.1)$ and $(x_1, x_2, x_3, x_4, x_5) = (0.1, 0.1, 0.1, 0.1, 0.1)$, the system has two coexisting attractors as shown in **Figure 9**.

Figure 10 shows the Lyapunov exponent spectrum, Kaplan–Yorke dimension spectrum and bifurcations of Eq. (8) as the coefficient c is varied over the range $c \in [0.3, 2]$. Those figures suggest there is an interesting route leading to chaos [21].

1. When $c \in [0.3, 0.4639]$, there exists a period-doubling behavior along with $\dot{x}_2$ and $\dot{x}_3$ subspace. However, the system shows torus behavior along with $\dot{x}_2$ and $\dot{x}_3$ subspace. It is like saddle point: the system is stable in one direction but unstable in the other direction.

2. When $c \in [0.4640, 0.5574]$, the system exhibits two-torus-chaos behavior except for some 2-torus windows. When the parameter passed c = 0.4639 to c = 0.4640, two-torus-chaos is born by replacing the 2-torus behavior. The Lyapunov exponents at these two critical values are $(\lambda_1, \lambda_2, \lambda_3, \lambda_4, \lambda_5) = (0, 0, -0.01, -0.57, -0.88)$ for c = 0.4639 and $(\lambda_1, \lambda_2, \lambda_3, \lambda_4, \lambda_5) = (0.02, 0, 0, -0.60, -0.88)$ for c = 0.4640. This may cause by the period-doubling route along with $\dot{x}_2$ and $\dot{x}_3$ subspace.

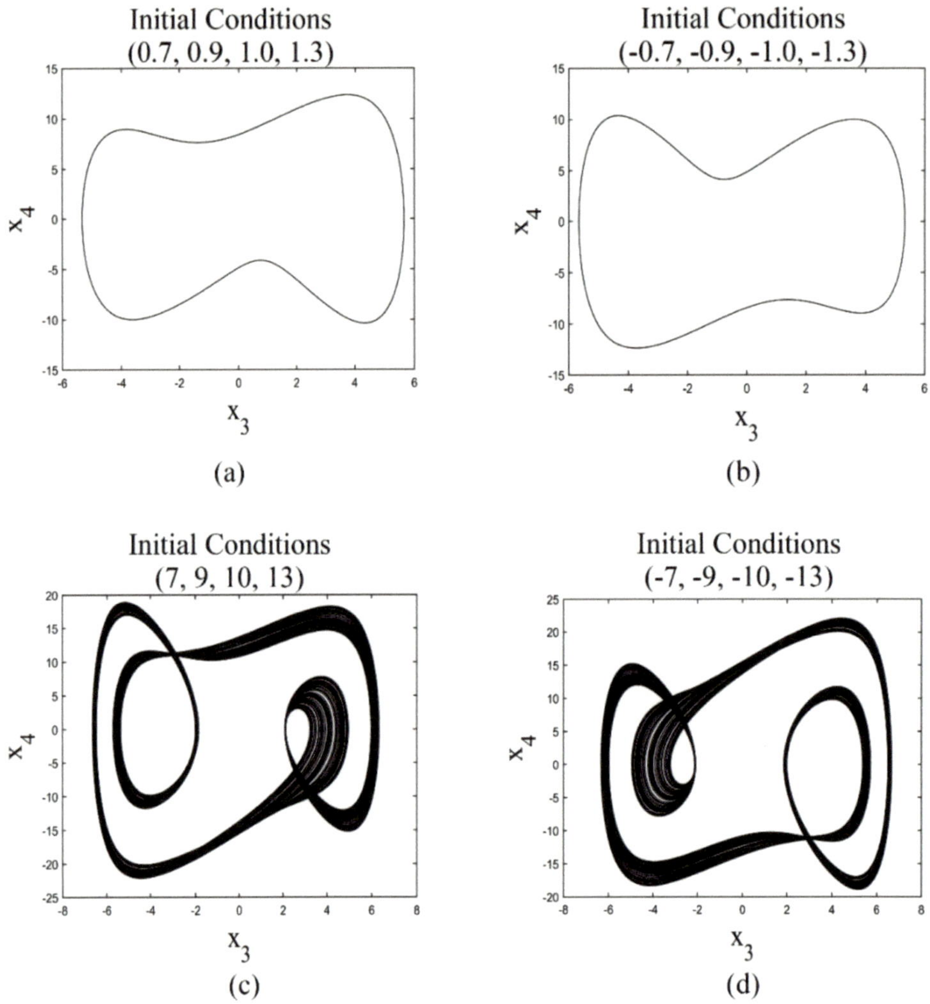

Figure 8.
Coexistence attractors of Eq. (7).

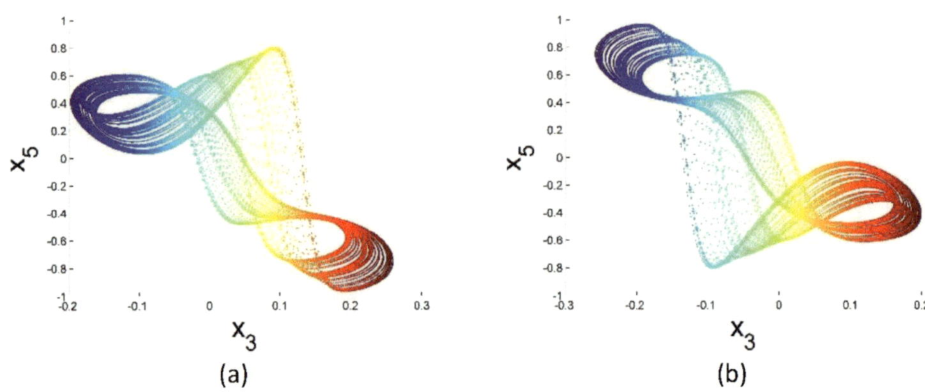

Figure 9.
Coexistence attractors of Eq. (8).

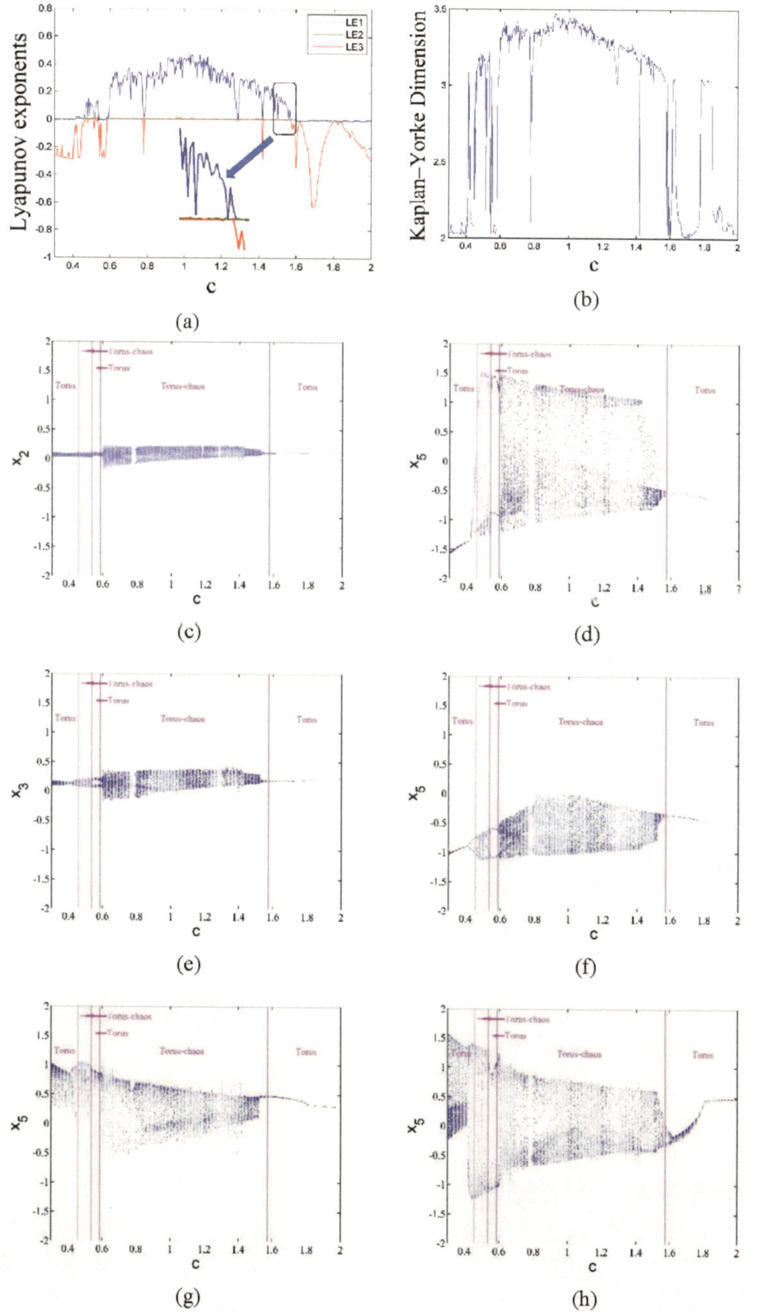

Figure 10.
LEs spectrum, Kaplan–Yorke dimension spectrum and bifurcations of Eq. (8) as the coefficient c is varied over the range $c \in [0.3, 2]$.

3. When $c \in [0.5575, 0.5901]$, the system exhibits 2-torus behavior.

4. When $c \in [0.5902, 1.5575]$, the system exhibits 2-torus-chaos behavior except for 2-torus windows. The route leading to chaos is same to point 3.

5. When $c \in [1.5575, 2]$ the system exhibits 2-torus behavior, except for some 3-torus windows like c = 1.6157.

4. Engineering applications with chaotic systems with hyperbolic sine nonlinearity

4.1 Random number generator

Sensitivity to initial conditions is one of the most important property of chaotic systems. Therefore, chaotic systems are very suitable for the cryptography purpose. But before that, it should be noticed that the probability density distributions (PDD) of chaotic systems are not uniform distribution. **Figure 11(a)** and **11(b)** are the waveform and PDD of x_4 of Eq. (4). It shows that PDD of the output sequences has physical characteristic. The cryptosystem with these sequences cannot resist side channel attack.

To remove physical characteristic, one can use the following de-correlation operation:

$$S_{out} = S_{in} * 10^6 - floor(S_{in} * 10^6) \tag{9}$$

In fact, Eq. (9) can be applied in all chaotic/torus-chaotic/hyperchaotic systems. The output sequences can pass fifteen random tests of NIST 800-22, as shown as in **Table 1**, which indicated the proposed method can provide high security Level. This proposed method can be used as a part of some cyber security systems such as the verification code, secure QR code and some secure communication protocols.

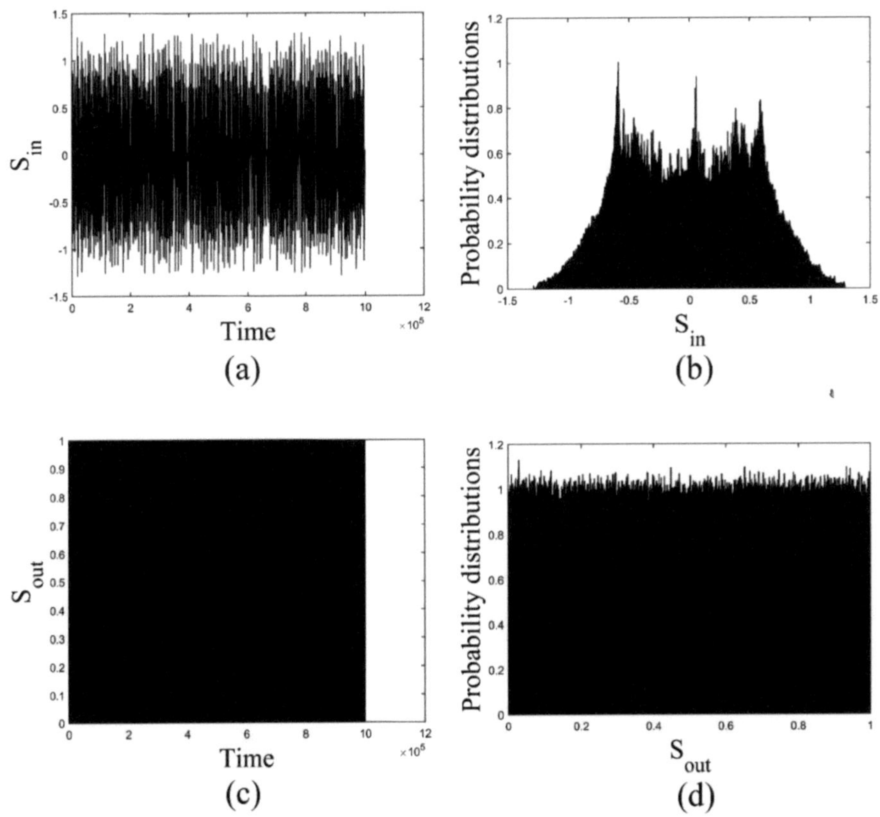

Figure 11.
Waveform and PDD before and after de-correlation operation of x_4 of Eq. (4): (a) is the waveform of x_4 before de-correlation operation; (b) is the PDD of x_4 before de-correlation operation; (c) is the waveform of x_4 after de-correlation operation; (b) is the PDD of x_4 after de-correlation operation.

Test	P-value	Result
Frequency	0.841481	Success
Block frequency	0.900704	Success
Runs	0.744455	Success
Longest run	0.172897	Success
Rank	0.368065	Success
FFT	0.762020	Success
Non-overlapping template	0.813121	Success
Overlapping template	0.532736	Success
Universal	0.856573	Success
Linear complexity	0.408679	Success
Serial	0.967366	Success
Approximate entropy	0.433157	Success
Cumulative sums	0.688582	Success
Random excursions	0.075229	Success
Random excursions variant	0.102049	Success

Table 1.
Pseudo-random properties of x_3 of Eq. (8) after de-correlation operation.

4.2 Image encryption

Image encryption is another widely used engineering application of chaotic system. In this section, we will use Eq. (7) for image encryption purpose.

A flowchart of the encryption scheme is shown in **Figure 12**.

The detailed encryption process includes the following steps.

Input: Plain image; Initial conditions for the chaotic system; Control parameters of the chaotic system.

Output: Ciphered image.

Step 1: Calculate the average pixel value of the plain image and generate the pseudorandom sequence.

Step 2: Transform the pseudorandom sequence and change pixel value of the image via XOR.

Step 3: Sort the pseudorandom sequence for permutation.

Step 4: Shift the pixel positions by column using the sorted elements.

Step 5: Shift the pixel positions by row using the sorted elements.

To provide a better understanding of this scheme, the pseudocode is provided in **Table 2**.

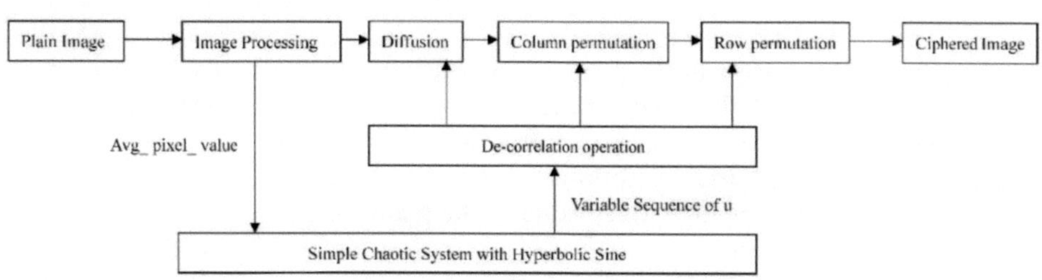

Figure 12.
A flowchart of the encryption scheme.

```
Input: Plain image Org_Img, Initial conditions for the chaotic system, Control parameter for the chaotic
system,
Output: Ciphered Image En_Img
---------------------------------------------------------------------------------------------------
[m,n] ← size(Org_Img);

Avg_pixel_value ← mean2(Org_Img)*10^(-5) % mean2 is a function that
returns the

                              % average value of a matrix

x(1) ← x(1) + Avg_pixel_value
y(1) ← y(1)
z(1) ← z(1)
u(1) ← u(1)
s(1) ← u(1)*10^4 – floor(u(1)*10^4)

For i=1:1:m*n             % Generate pseudorandom sequence that will
                                  % be used for diffusion and permutation
   [dx, dy, dz, du] ← Runge-Kutta (x(i), y(i), z(i), u(i))
   x(i+1) ← x(i) +dx
   y(i+1) ← y(i) +dy
   z(i+1) ← z(i) +dz
   u(i+1) ← u(i) +du
s(i+1) ← u(i+1)*10^4 – floor(u(i+1)*10^4)
End

Count=1         % Count flag
For i=1:m        % Diffusion Operation
  For j=i:n
   diff(Count) ← mod (s(Count)*10^14, 256) % transform s, which could be used for XOR
  En_Dif(i,j)=bitxor(Org_Img(i,j), diff (Count)); % Bitwise exclusive OR
    Count= Count+1;
  End
End

S_index ← Sort(s)
For i=1:n              % Column-wise permutation
 For j=1:m
 En_per_col (i,j) ← Sort (En_Dif, S_index)
 End
End
For i=1:m              % Row-wise permutation
 For j=1:n
 En_Img (i,j) ← Sort (En_per_col, S_index)
 End
End
```

Table 2.
Image encryption scheme.

The decryption process of the proposed algorithm is the reverse process of the encryption algorithm. A flowchart of the decryption process is shown in **Figure 13**.

The detailed decryption process includes the following steps.

Input: Plain image; Initial conditions for the chaotic system; Control parameter of the chaotic system; Average pixel value of the plain image

Output: Decrypted image

Step 1: Generate the pseudorandom sequence via the initial conditions and the average pixel values of the plain image

Step 2: Sort the pseudorandom sequence for row and column recovery.

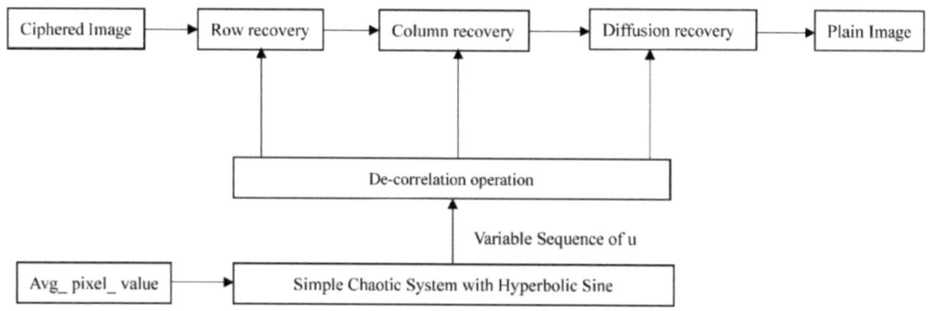

Figure 13.
A flowchart of the decryption scheme.

Input: Ciphered image En_Img, Initial conditions for the chaotic system, control parameter for the chaotic system, Avg_pixel_value of Org_Img
Output: Plain Image Org_Img

```
[m,n] ← size(En_Img);
x(1) ← x(1) + Avg_pixel_value
y(1)← y(1)
z(1) ← z(1)
u(1) ← u(1)
s(1) ← u(1)*10^4 – floor(u(1)*10^4)

For i=1:1:m*n              % Generate a pseudorandom sequence that will
% be used for decryption
[dx, dy, dz, du] ← Runge-Kutta (x(i), y(i), z(i), u(i))
x(i+1) ← x(i) +dx
y(i+1) ← y(i) +dy
z(i+1) ← z(i) +dz
u(i+1) ← u(i) +du
  s(i+1) ← u(i+1)*10^4 – floor(u(i+1)*10^4)
End

S_index ← Sort(s)
For i=1:m                  % Row-wise permutation recovery
 For j=1:n
 De_per_row (i,j) ← Sort (En_Img, S_index)
 End
End

For i=1:n                  % Column-wise permutation recovery
 For j=1:m
 De_per_col (i,j) ← Sort (De_per_row, S_index)
 End
End

Count=1                    % Count flag
For i=1:m                  % Diffusion recovery
  For j=i:n
  diff(Count) ← mod (s(Count)*10^14, 256) % transform s, which could be used for XOR
  Org_Img (i,j)=bitxor(De_per_col (i,j), diff (Count)); % Bitwise exclusive OR
    Count= Count+1;
  End
 End
```

Table 3.
Image decryption scheme.

Step 3: Shift the pixel positions by row

Step 4: Shift the pixel positions by column

Step 5: Transform the pseudorandom sequence and recover the pixel values of the image via XOR

To provide a better understanding of this scheme, the pseudo-code is provided in **Table 3**

The testing results of encryption and decryption are shown in **Figure 14**.

In this system, all the initial conditions and control parameters can be considered as secret keys. Because the basin of attraction of each initial condition is greater than 1, it could have more than $10^{15*4} = 10^{60}$ choices via a resolution of 10^{-15}, in terms of a numeric calculation. Moreover, if a range of control parameters are considered for the key space, the key space of this system would far exceed 10^{90}. Such a large key space provides sufficient security against brute-force attacks.

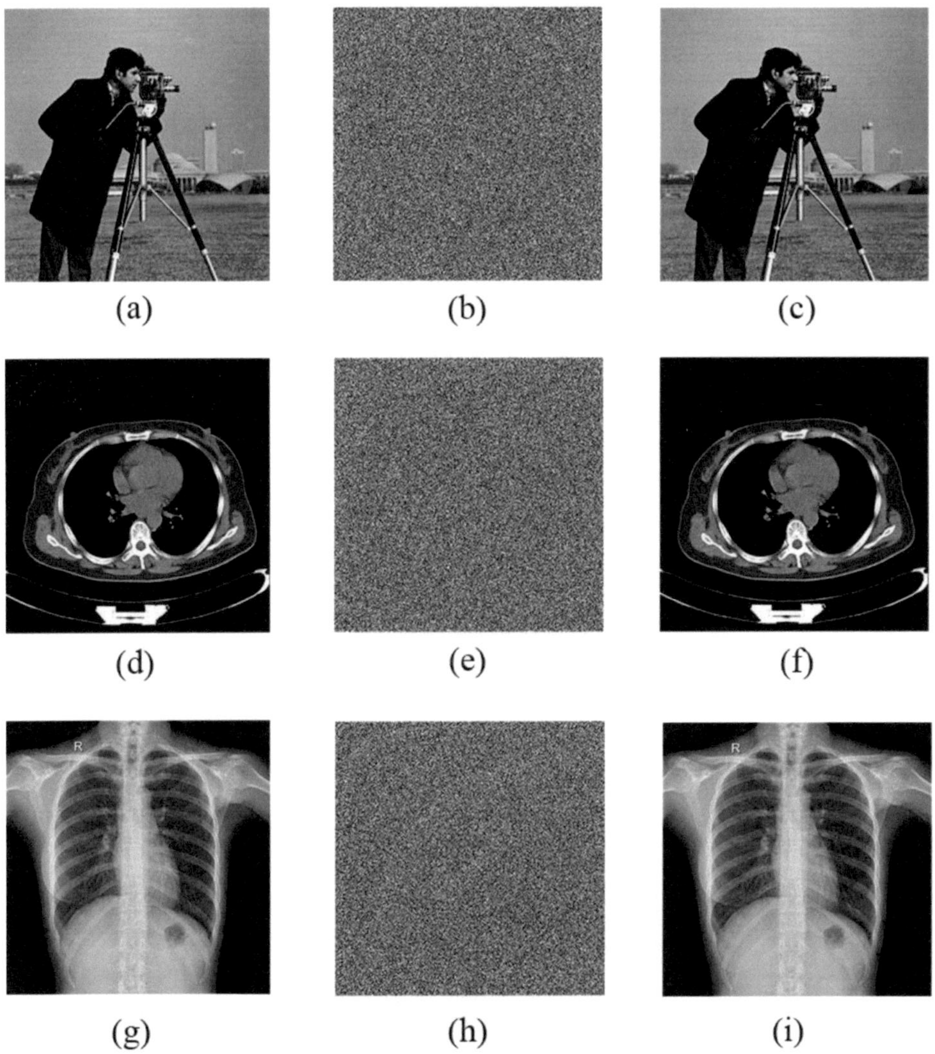

Figure 14.
The testing results of encryption and decryption: (a) is the plain image of cameraman; (b) is the encrypted image of cameraman; (c) is the decrypted image of cameraman; (d) is the plain image of breast CT image; (e) is the encrypted image of breast CT image; (f) is the decrypted image of breast CT image; (g) is the plain image of thorax CT image; (h) is the encrypted image of thorax CT image; (i) is the decrypted image of thorax CT image.

Correlation coefficients of adjacent pixels in the plain and encrypted image are shown in **Table 4**.

The NPCR and UACI score of CT image are 99.5804% and 33.3227%.

From the above security analysis, the proposed scheme can provide high security for cryptographic applications.

4.3. Spread spectrum communication

Chaotic systems can also use for spread spectrum communication propose. Different chaos shift keying (DCSK) technology employs nonperiodic and wideband chaotic signals as carriers so as to achieve the effect of spectrum spreading in the process of digital modulation. **Figure 15** shows the scheme of modulation for DCSK.

In this scheme, every bit has two time slots. The first time slot is used for transmission of a chaotic sequence for the reference signal. The second time slot is used for transmission of another chaotic sequence for the reference signal which has the same length as the first time slot. If the information bit is +1, then the information signal is exactly the same as the reference signal. If the information signal bit is −1, then the information signal is the negative of the reference signal. For bits b_k, the signal at time k is:

$$s_i = \begin{cases} x_i & 2k\beta < i \le (2k+1)\beta \\ b_k x_{i-\beta} & (2k+1)\beta < i \le 2(k+1)\beta \end{cases} \tag{10}$$

Where β is the number of sampling points. The spreading factor (SF) in the DCSK system is $SF = 2\beta$.

Figure name	Direction	Plain-image	Ciphered image
Cameraman Image	Horizontal	0.983146	0.001731
Cameraman Image	Vertical	0.990025	0.004141
Cameraman Image	Diagonal	0.973249	0.000324
Breast CT image	Horizontal	0.978292	0.002500
Breast CT image	Vertical	0.955481	0.006207
Breast CT image	Diagonal	0.940737	0.003071
Thorax CT image	Horizontal	0.994585	0.001267
Thorax CT image	Vertical	0.994761	0.001267
Thorax CT image	Diagonal	0.991973	0.001558

Table 4.
Correlation coefficients of adjacent pixels in the plain and encrypted image.

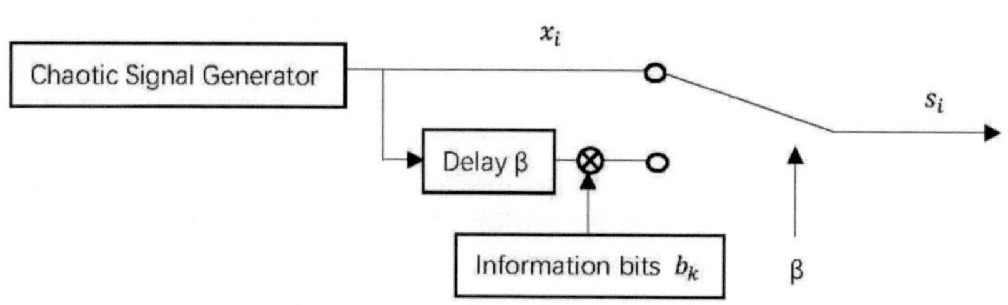

Figure 15.
Scheme of DCSK modulation.

For demodulation as shown in **Figure 16**, the receiver calculates the correlation between the received signal r_i and the signal $r_{i-\beta}$, which is r_i delayed by β . After a time k, the output of the correlator is:

$$Z_k = \sum_{2(k+1)\beta}^{i=(2k+1)\beta+1} r_i r_{i-\beta} \tag{11}$$

Thus, the information bit b_k can be restored by the sign of the decision variable:

$$\hat{b}_k = sgn\,[Z_k] \tag{12}$$

The obtained BER performance under additive white Gaussian noise (AWGN) channels for spreading factor $2\beta = 200$ is shown in **Figure 17**. From the comparison results, DCSK can have a lower BER when using this system as a carrier signal in the presence of noise.

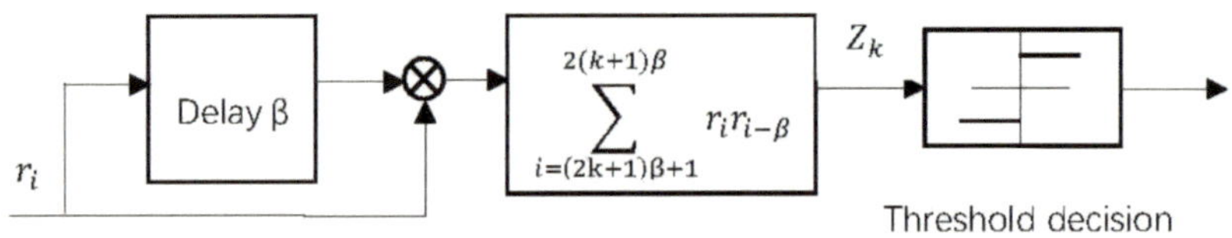

Figure 16.
Scheme of the DCSK demodulation.

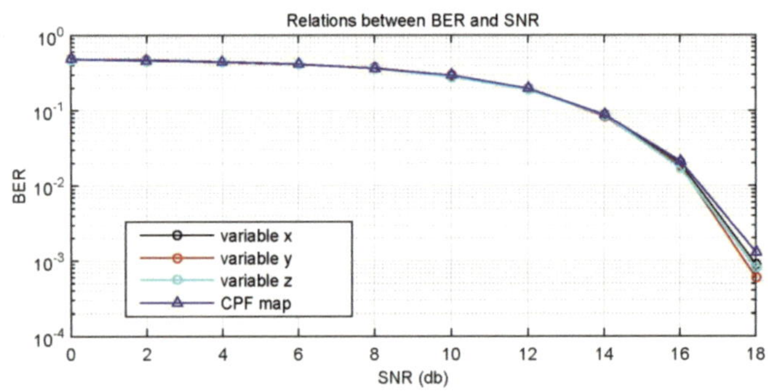

Figure 17.
Comparison of the bit error rate for a Chebyshev sequence and the hyperbolic sine system with DCSK.

5. Conclusions

In this chapter, we first described a third order chaotic system with hyperbolic sine nonlinearity, then we introduced the method to expend this chaotic system to high order chaotic systems. After that, we introduced the method to construct the derived chaotic torus-chaotic systems. Finally, we introduced some applications such as random number generator algorithm, spread spectrum communication and image encryption schemes. The contribution of this chapter is that it systematically summarizes the design method, the dynamic behavior and typical engineering application of chaotic systems with hyperbolic sine nonlinearity, which may widen the current knowledge of chaos theory and engineering applications based on chaotic systems.

Acknowledgements

Jizhao Liu has received research grants from Sun Yat-sen University.

This study was supported by the Fundamental Research Funds for the Central Universities. No. 19lgpy230.

Conflict of interest

The authors declare that they have no conflict of interest.

Notes/thanks/other declarations

The authors would like to thank professor Julien Clinton Sprott for helpful discussion.

Author details

Jizhao Liu[1,2*], Yide Ma[2], Jing Lian[3] and Xinguo Zhang[2]

1 School of Data and Computer Science, Sun Yat-sen University, Guangzhou, Guangdong, China

2 School of Information Science and Engineering, Lanzhou University, Lanzhou, Gansu, China

3 School of Electronics and Information Engineering, Lanzhou Jiaotong University, Lanzhou, Gansu, China

*Address all correspondence to: liujizhao@mail.sysu.edu.cn

References

[1] Sprott, Julien C. "Elegant chaos: algebraically simple chaotic flows." World Scientific, 2010.

[2] Zhang Xinguo and Li Yide, Ma. Shouliang. "Nonlinear Circuit-Based Analysis and Design". Higher Education Press, 2011.

[3] Li, Shujun, Guanrong Chen, and Xuanqin Mou. "On the dynamical degradation of digital piecewise linear chaotic maps." International journal of Bifurcation and Chaos 15.10 (2005): 3119-3151.

[4] Zhu, Hegui, et al. "Analyzing Devaney chaos of a sine–cosine compound function system." International Journal of Bifurcation and Chaos 28.14 (2018): 1850176.

[5] Sprott, J. C. "A new class of chaotic circuit." Physics Letters A 266.1 (2000): 19-23.

[6] Sprott, J. Clint. "Some simple chaotic flows." Physical review E 50.2 (1994): R647.

[7] Wang, Chunhua, Hu Xia, and Ling Zhou. "A memristive hyperchaotic multiscroll jerk system with controllable scroll numbers." International Journal of Bifurcation and Chaos 27.06 (2017): 1750091.

[8] Xiong, Li, et al. "Design and hardware implementation of a new chaotic secure communication technique." PloS one 11.8 (2016): e0158348.

[9] Yu, Simin, et al. "Design and implementation of n-scroll chaotic attractors from a general jerk circuit." IEEE Transactions on Circuits and Systems I: Regular Papers 52.7 (2005): 1459-1476.

[10] Li, Chunbiao, and Julien Clinton Sprott. "Coexisting hidden attractors in a 4-D simplified Lorenz system." International Journal of Bifurcation and Chaos 24.03 (2014): 1450034.

[11] Zaamoune, Faiza, et al. "Symmetries in Hidden Bifurcation Routes to Multiscroll Chaotic Attractors Generated by Saturated Function Series." Journal of Advanced Engineering and Computation 3.4 (2019): 511-522.

[12] Tlelo-Cuautle, Esteban, et al. "Dynamics, FPGA realization and application of a chaotic system with an infinite number of equilibrium points." Nonlinear Dynamics 89.2 (2017): 1129-1139.

[13] Liu, Jizhao, et al. "An approach for the generation of an nth-order chaotic system with hyperbolic sine." Entropy 20.4 (2018): 230.

[14] Munmuangsaen, Buncha, Banlue Srisuchinwong, and Julien Clinton Sprott. "Generalization of the simplest autonomous chaotic system." Physics Letters A 375.12 (2011): 1445-1450.

[15] Piper, Jessica R., and Julien Clinton Sprott. "Simple autonomous chaotic circuits." IEEE Transactions on Circuits and Systems II: Express Briefs 57.9 (2010): 730-734.

[16] Sprott, Julien Clinton. "A new chaotic jerk circuit." IEEE Transactions on Circuits and Systems II: Express Briefs 58.4 (2011): 240-243.

[17] Liu, Jizhao, et al. "Simplest chaotic system with a hyperbolic sine and its applications in DCSK scheme." IET Communications 12.7 (2018): 809-815.

[18] Liu, Jizhao, et al. "An approach for the generation of an nth-order chaotic system with hyperbolic sine." Entropy 20.4 (2018): 230.

[19] Sprott, Julien Clinton, Xiong Wang, and Guanrong Chen. "Coexistence of point, periodic and strange attractors." International Journal of Bifurcation and Chaos 23.05 (2013): 1350093.

[20] Kinsner, Witold. "Characterizing chaos through Lyapunov metrics." IEEE Transactions on Systems, Man, and Cybernetics, Part C (Applications and Reviews) 36.2 (2006): 141-151.

[21] Liu, Jizhao, et al. "A Torus-Chaotic System and Its Pseudorandom Properties." Complexity 2020 (2020).